Hemifacial Spasm

Kwan Park • Jae Sung Park

Editors

Hemifacial Spasm

A Comprehensive Guide

 Springer

Editors
Kwan Park
Department of Neurosurgery
Konkuk University Medical Center
Seoul
Korea (Republic of)

Jae Sung Park
Department of Neurosurgery
Konkuk University School of Medicine
Chungju Hospital
Chungju
Korea (Republic of)

ISBN 978-981-15-5419-3   ISBN 978-981-15-5417-9   (eBook)
https://doi.org/10.1007/978-981-15-5417-9

This Springer imprint is published by the registered company Springer Nature Singapore Pte Ltd. The registered company address is: 152 Beach Road, #21-01/04 Gateway East, Singapore 189721, Singapore

Contents

Overview of Hemifacial Spasm 1
Jae Sung Park

Natural History of Hemifacial Spasm 7
Jeong-A Lee and Kwan Park

Novel Classification Systems for Hemifacial Spasm 13
Jae Sung Park and Kwan Park

Pathogenesis of Hemifacial Spasm 21
Min Ho Lee and Jae Sung Park

Clinical Symptoms and Differential Diagnosis
of Hemifacial Spasm 27
Jong Hyeon Ahn and Jin Whan Cho

The Electrophysiological Study for Hemifacial Spasm 33
Byung-Euk Joo

Magnetic Resonance Imaging Evaluation of Hemifacial Spasm 43
Hyung-Jin Kim and Minjung Seong

Surgical Principles of Hemifacial Spasm:
How We Do Microvascular Decompression 57
Seunghoon Lee and Kwan Park

Technical Difficulties of Microvascular Decompression
Surgery for Hemifacial Spasm 67
Kwan Park and Seunghoon Lee

Various Applications of Microvascular Decompression
Other than for Hemifacial Spasm 75
Min Ho Lee and Jae Sung Park

Intraoperative Neurophysiological Monitoring
in Microvascular Decompression for Hemifacial Spasm 83
Sang-Ku Park

Anesthetic Management of MVD 111
Jeong Jin Lee

Botulinum Toxin Injection in Hemifacial Spasm 119
Jinyoung Youn, Wooyoung Jang, and Jong Kyu Park

**Medical Treatment of Hemifacial Spasm
and Other Involuntary Facial Movement Disorders** 127
Wooyoung Jang and Jinyoung Youn

Possible Complications of Microvascular Decompression 135
Doo-Sik Kong

**Prognosis of Symptoms After Microvascular Decompression
for Hemifacial Spasm** 141
Jeong-A Lee and Kwan Park

**Redo Surgery for Failed Microvascular Decompression
for Hemifacial Spasm** 151
Seunghoon Lee and Kwan Park

Overview of Hemifacial Spasm

Jae Sung Park

Definition of Hemifacial Spasm

The term hemifacial spasm (HFS) is self-explanatory: contractions on one side of the face. More specifically, the clinical term HFS refers to involuntary facial contractions that are unilateral, irregular, and tonic or clonic. The twitches usually start with the periorbital muscles, and then they can spread to perinasal, perioral, zygomaticus, and platysma muscles [1]. The diagnosis of HFS is mainly based on clinical history and physical examination, although adjunctive use of electromyographic (EMG) and radiological evaluation methods is commonly acknowledged. Conditions that need to be differentiated from HFS include blepharospasm, facial myokymia, and post-facial palsy synkinesis, etc.

History

Esmail Jorjani (1042–1137), a Persian physician, described syndromes that were probably consistent with trigeminal neuralgia, HFS, and Bell's palsy in his book *Treasure of the Khawarazm Shah* [2]. Also, he implicated an artery–nerve conflict as an etiology of trigeminal neuralgia. In the more modern period, the prototype of HFS was described by Schultze in 1875, when a verte-bral artery aneurysm was found to compress the seventh nerve [3]. One of the first descriptions on HFS with a picture of the patient was provided by Edouard Brissaud in 1893 [4]. In 1905, Babinski described a phenomenon called "other Babinski sign" that referred to a paradoxical synkinesis in HFS patients, and this is typically observed in HFS, but not in blepharospasm patients [5, 6]. A modern-day concept of vascular compression syndrome that included trigeminal neuralgia, HFS, and glossopharyngeal neuralgia was introduced by McKenzie in 1936. Based on its pathophysiological background, vascular decompression for HFS was first introduced by Gardner in 1962, following which, a more modern technique with a minimal approach, i.e., microvascular decompression (MVD) via retrosigmoid craniotomy, was first performed by Bremond in 1974 [7, 8]. The current concept of pathophysiology and the surgical treatment of HFS was established and popularized by Jannetta, and it started with his article in 1975, titled as "Neurovascular cross-compression in patients with hyperactive dysfunction symptoms of the eighth cranial nerve" [9].

Epidemiology

According to an epidemiological study based on Norwegian population, the prevalence of HFS was about 9.8 per 100,000 persons [10]. Another study from the USA reported the prevalence rate

J. S. Park (✉)
Department of Neurosurgery, Konkuk University School of Medicine, Chungju Hospital, Chungju, Korea (Republic of)

© Springer Nature Singapore Pte Ltd. 2020
K. Park, J. S. Park (eds.), *Hemifacial Spasm*, https://doi.org/10.1007/978-981-15-5417-9_1

of HFS as 7.4 per 100,000 men and 14.5 per 100,000 women [11]. Data from our own institute revealed the male-to-female ratio being 1:2.28 and the average age of 52.2 years [12]. Concerning the ethnic distribution, HFS has been reported to be more prevalent in Asian population than others [13–15]. Concomitant psychological issues such as anxiety or depression are noticeable and they are thought to influence the prognosis as well [16]. Except for a few familial cases, HFS does not occur in a hereditary manner, and it predominantly occurs to adults [17–19].

Etiology and Pathophysiology

Vascular compression on the root entry zone (REZ) of the facial nerve is acknowledged to be responsible for primary HFS, whereas any impairment of the facial nerve due to a preexisting condition can constitute a secondary HFS: facial palsy, cerebellopontine angle (CPA) tumors, Chiari I malformations, demyelinating diseases, infections, etc. [20]. Primary HFS is 3–4 times more prevalent than secondary HFS [20, 21]. When a vascular curvature causes the compression on the REZ, anterior inferior cerebellar artery (AICA) is most commonly involved one, followed by posterior inferior cerebellar artery (PICA) and the vertebral artery (VA). A single artery could be the sole cause of the neurovascular compression, but it was rather infrequent (4.7%) according to our previous report [22]. In consideration of other additional factors, a total of six compressive patterns in HFS were proposed: loop, arachnoid, perforator, branch, sandwich, and tandem types [22].

Microscopic disruption of myelin in the REZ or its proximal vicinity where an offending vessel compresses has been acknowledged as the pathophysiology of HFS [23]. Regarding a more detailed mechanism of HFS, there are two major hypotheses: central (hyperexcitability of the facial motor nucleus) vs. peripheral (ephaptic transmission between the facial nerve bundles) hypothesis. Increasing number of microanatomical and neurophysiological research is dedicated to elucidate the precise pathway of HFS; but one hypothesis cannot explain all the phenomena without the other.

Diagnosis

Clinical evaluation including history and physical examination is the key to the diagnosis of HFS. The definition of HFS itself is the most important clue; involuntary facial contractions that are unilateral, irregular, and tonic or clonic. In addition to a close observation of patients' face, a physical maneuver called "other Babinski sign" may be handy. This maneuver, also known as Babinski-2 sign, refers to a synchronized activity of the frontalis or orbicularis oculi muscle that is induced by a self-lifting of one's eyebrow while it is closed. This is reported to yield 100% of specificity and 86% of sensitivity for diagnosis of HFS [24]. EMG, magnetic resonance image (MRI), or computed tomography (CT) also can be used to confirm the diagnosis. Time of flight of MR angiography may display the anatomical relationship between the REZ and an offending vessel. More recent studies using 3D MRI volumetric analysis suggested that CSF space in the posterior fossa of HFS patients was smaller than that of the control group [25]. Also, an analysis using color-duplex ultrasound demonstrated that the mean flow velocity of AICA and PICA on the HFS side was greater than that on the contralateral side [26]. EMG in HFS would show spontaneous and high-frequency synchronized firing, and this finding may be helpful to differentiate HFS from other movement disorders, such as myokymia, blepharospasm, post-facial palsy synkinesia, tic disorders, myokymia, partial motor seizures, craniocervical dystonia (Meige syndrome), tardive dyskinesias (TD) and neuromyotonia, as well as phychogenic HFS.

Treatment

Nonsurgical Treatment

No pharmaceutical medicine has succeeded to provide long-term benefits for HFS. Anticonvulsants or GABAergic medicines may improve symptoms partially and temporarily, but the effectiveness of these is not comparable to botulinum neurotoxin (BTX) injection, not to mention to MVD. BTX injection is the most preferred nonsurgical treatment for HFS, yielding up to 85% of symptomatic relief. Among seven serotypes of BTX, serotypes A and B are currently commercialized. Following injections, symptomatic improvement occurs in 1–3 days and it usually reaches its peak effect in 5 days [27]. The duration of clinical benefit varies from centers to centers by 3–6 months [28, 29]. Repeated injections of BTX is unavoidable, and tolerance can naturally develop in some subjects, although a 10-year multicenter study reported that the average of duration of improvement did not change from the first year of injection to the 10th year of treatment with the similar dose of BTX [30]. Also, they stated that the BTX-induced adverse responses decreased throughout the 10-year course. Local complications of BTX injection include ptosis, blurred vision, and diplopia, but they are rarely permanent [31]. Incidence of any adverse effect is estimated from 20 to 53% (the mode being around 30–40%), and ptosis is universally the most frequent one [28, 29, 32]. Despite its relatively high success rate of symptomatic improvement, one cannot ignore the fact that BTX injection fundamentally requires repeated sessions, which lead to emotionally and financially non-negligible burden on the patients.

Surgical Treatment

MVD is the only curative treatment option for HFS with high success rate and with low incidence of recurrence and complications. According to a systemic review on 22 studies with 5700 patients who underwent MVD, the complete resolution was achieved in 91.1% (95% CI: 90.3–91.8%) of patients [33]. Recurrence occurred in 2.4% (95% CI: 1.9–2.9%) of patients and postoperative complications included transient complications included facial palsy (9.5% [95% CI:8.8–10.3%]), hearing deficit (3.2% [95% CI: 2.7–3.7%]), and cerebrospinal fluid leak (1.4%

[95% CI: 1.1–1.7%]). Permanent complications included hearing deficit in 2.3% (95% CI: 1.9–2.7%) and facial palsy in 0.9% (95% CI: 0.7–1.2%) of patients. The risk of stroke was 1 in 1800 and risk of death was 1 in 5500 [33].

The basic technique of MVD is well described in the literature, but the detailed maneuver varies depending on surgeons. Once a lateral retrosigmoid suboccipital craniectomy or craniotomy is performed under a general anesthesia, the dura is incised to reveal the cerebellar cortex. With or without traction of the flocculus, the REZ of the facial nerve is observed. Upon the identification of the compressing vessels, or the offending arteries, they are separated from the seventh nerve, which then can be perpetuated by insertion of Teflon pieces. A few more additional techniques, including transposition of the vessels, snare technique, vascular sling, etc., have been proposed [34–36].

Intraoperative EMG monitoring can be beneficial for improvement of surgical outcomes. Lateral spread response (LSR) is one of the most popularly applied neurophysiologic tests for HFS, since Moller and Jannetta advocated that disappearance of LSR would indicate properly performed decompression [37]. However, persistence of LSR did not necessarily indicate a poor outcome, which precludes LSR from being a reliable predictor for long-term prognosis of HFS after MVD [38]. Also, to properly monitor the integrity of the eighth nerve (CN VIII) during MVD, intraoperative brain stem auditory evoked potential (BAEP) can be employed, which has been accepted by numerous institutions in decreasing the risk of hearing impairment during MVD.

Clinical course following MVD is not identical. According to our own report, 737 (92.8%) of 807 patients who had undergone MVD for HFS became absolutely or nearly spasm-free by the 2-year postoperative follow-up. However, not everyone started to be asymptomatic immediately after the surgery; 140 (19.0%) of 737 patients still experienced residual spasms more than a month, and some of them lasted more than a year [12]. No universally acknowledged explanation on this disparate clinical course is available so far; therefore, more electrophysiologic microanatomic researches are needed to elucidate it in the future.

References

1. Jankovic J, Brin MF. Botulinum toxin: historical perspective and potential new indications. Muscle Nerve. 1997;20:129–45.
2. Shoja MM, Tubbs RS, Khalili M, Khodadoost K, Loukas M, Cohen-Gadol AA. Esmail Jorjani (1042–1137) and his descriptions of trigeminal neuralgia, hemifacial spasm, and Bell's palsy. Neurosurgery. 2010;67:431–4.
3. de Abreu Junior L, Kuniyoshi CH, Wolosker AB, et al. Vascular loops in the anterior inferior cerebellar artery, as identified by magnetic resonance imaging, and their relationship with otologic symptoms. Radiol Bras. 2016;49:300–4.
4. Colosimo C, Berardelli A. An early image of hemifacial spasm: Edouard Brissaud contribution. Mov Disord. 2010;25:531–3.
5. Devoize J. "The other" Babinski's sign: paradoxical raising of the eyebrow in hemifacial spasm. J Neurol Neurosurg Psychiatry. 2001;70:516.
6. Stamey W, Jankovic J. The other Babinski sign in hemifacial spasm. Neurology. 2007;69:402–4.
7. Rand RW. Gardner's neurovascular decompression for hemifacial spasm. Arch Neurol. 1982;39:510–1.
8. Bremond G, Garcin M, Magnan J, Bonnaud G. L'abord a minima de l'espace pontocerebelleux. Cah ORL. 1974;19:443–60.
9. Jannetta P. Neurovascular cross-compression in patients with hyperactive dysfunction symptoms of the eighth cranial nerve. Surge Forum. 1975;26:467–8.
10. Nilsen B, Le K-D, Dietrichs E. Prevalence of hemifacial spasm in Oslo, Norway. Neurology. 2004;63:1532–3.
11. Auger RG, Whisnant JP. Hemifacial spasm in Rochester and Olmsted county, Minnesota, 1960 to 1984. Arch Neurol. 1990;47:1233–4.
12. Park JS, Lee S, Park S-K, Lee J-A, Park K. Facial motor evoked potential with paired transcranial magnetic stimulation: prognostic value following microvascular decompression for hemifacial spasm. J Neurosurg. 2018;1:1–8.
13. Wu Y, Davidson AL, Pan T, Jankovic J. Asian overrepresentation among patients with hemifacial spasm compared to patients with cranial–cervical dystonia. J Neurol Sci. 2010;298:61–3.
14. Jankovic J. Peripherally induced movement disorders. Neurol Clin. 2009;27:821–832, vii.
15. Tan E-K, Chan L. Clinico-radiologic correlation in unilateral and bilateral hemifacial spasm. J Neurol Sci. 2004;222:59–64.
16. Jin Y, Zhao C, Su S, Zhang X, Qiu Y, Jiang J. Residual hemifacial spasm after microvascular decompression: prognostic factors with emphasis on preoperative psychological state. Neurosurg Rev. 2015;38:567–72.
17. Wilkins RH. Hemifacial spasm: a review. Surg Neurol. 1991;36:251–77.
18. Carter JB, Patrinely JR, Jankovic J, McCrary JA, Boniuk M. Familial hemifacial spasm. Arch Ophthalmol. 1990;108:249–50.
19. Jho HD, Jannetta PJ. Hemifacial spasm in young people treated with microvascular decompression of the facial nerve. Neurosurgery. 1987;20:767–70.
20. Colosimo C, Bologna M, Lamberti S, et al. A comparative study of primary and secondary hemifacial spasm. Arch Neurol. 2006;63:441–4.
21. Batla A, Goyal C, Shukla G, Goyal V, Srivastava A, Behari M. Hemifacial spasm: clinical characteristics of 321 Indian patients. J Neurol. 2012;259:1561–5.
22. Park J, Kong D-S, Lee J-A, Park K. Hemifacial spasm: neurovascular compressive patterns and surgical significance. Acta Neurochir (Wien). 2008;150:235–41.
23. Campos-Benitez M, Kaufmann AM. Neurovascular compression findings in hemifacial spasm. Case Rep Neurol Med. 2008;109:416–20.
24. Pawlowski M, Gess B, Evers S. The Babinski-2 sign in hemifacial spasm. Mov Disord. 2013;28:1298–300.
25. Chan L-L, Ng K-M, Fook-Chong S, Lo Y-L, Tan E-K. Three-dimensional MR volumetric analysis of the posterior fossa CSF space in hemifacial spasm. Neurology. 2009;73:1054–7.
26. Perren F, Magistris MR. Is hemifacial spasm accompanied by hemodynamic changes detectable by ultrasound? Acta Neurochir (Wien). 2014;156:1557–60.
27. Dutton JJ, Buckley EG. Long-term results and complications of botulinum A toxin in the treatment of blepharospasm. Ophthalmology. 1988;95:1529–34.
28. Dutton JJ, Fowler AM. Botulinum toxin in ophthalmology. Surv Ophthalmol. 2007;52:13–31.
29. Ababneh OH, Cetinkaya A, Kulwin DR. Long-term efficacy and safety of botulinum toxin A injections to treat blepharospasm and hemifacial spasm. Clin Exp Ophthalmol. 2014;42:254–61.
30. Defazio G, Abbruzzese G, Girlanda P, et al. Botulinum toxin A treatment for primary hemifacial spasm: a 10-year multicenter study. Arch Neurol. 2002;59:418–20.
31. Kong D-S, Park K. Hemifacial spasm: a neurosurgical perspective. J Korean Neurosurg Soc. 2007;42:355.
32. Czyz CN, Burns JA, Petrie TP, Watkins JR, Cahill KV, Foster JA. Long-term botulinum toxin treatment of benign essential blepharospasm, hemifacial spasm, and Meige syndrome. Am J Ophthalmol. 2013;156:173–177.e172.
33. Miller LE, Miller VM. Safety and effectiveness of microvascular decompression for treatment of hemifacial spasm: a systematic review. Br J Neurosurg. 2012;26:438–44.
34. Kurokawa Y, Maeda Y, Toyooka T, Inaba K-I. Microvascular decompression for hemifacial spasm caused by the vertebral artery: a simple and effective transposition method using surgical glue. Surg Neurol. 2004;61:398–403.

35. Masuoka J, Matsushima T, Kawashima M, Nakahara Y, Funaki T, Mineta T. Stitched sling retraction technique for microvascular decompression: procedures and techniques based on an anatomical viewpoint. Neurosurg Rev. 2011;34:373–80.

36. Lee SH, Park JS, Ahn YH. Bioglue-coated teflon sling technique in microvascular decompression for hemifacial spasm involving the vertebral artery. J Korean Neurosurg Soc. 2016;59:505.

37. Møller AR, Jannetta PJ. Physiological abnormalities in hemifacial spasm studied during microvascular decompression operations. Exp Neurol. 1986;93:584–600.

38. Von Eckardstein K, Harper C, Castner M, Link M. The significance of intraoperative electromyographic "lateral spread" in predicting outcome of microvascular decompression for hemifacial spasm. J Neurol Surg B Skull Base. 2014;75:198–203.

Natural History of Hemifacial Spasm

Jeong-A Lee and Kwan Park

Hemifacial spasm (HFS) is characterized by unilateral, paroxysmal, and involuntary movements of muscles distributed by the ipsilateral facial nerve. Involuntary contractions usually start from the orbicularis oculi muscle and gradually spread to other muscles associated with facial expressions [1, 2]. At present, since HFS rarely improves spontaneously, it has been agreed that most patients need to be treated [3]. However, the natural history of HFS is not well documented yet.

The overall clinical course of HFS is shown in Fig. 1. Previous studies have focused on the clinical course after microvascular decompression (MVD) (I) and botulinum toxin treatment (II). This chapter illustrates the natural history of untreated HFS (III). To illustrate this, we introduce our two previous papers that investigated the clinical course at points IV and V in Fig. 1.

J.-A. Lee (✉)
Departments of Neurosurgery, Samsung Medical Center, Sungkyunkwan University School of Medicine, Seoul, Republic of Korea
e-mail: naja.lee@samsung.com

K. Park
Department of Neurosurgery, Konkuk University Medical Center, Seoul, Korea (Republic of)
e-mail: kwanpark@skku.edu

Natural History of Hemifacial Spasm Until Visit to Hospital [4]

This study was to set an objective parameter for determining the severity of HFS. We investigated the relationship between the severity of spasms and other factors, including the duration of symptoms. A total of 121 HFS patients who visited an outpatient clinic in our hospital between April and August 2010 were enrolled. The following criteria were included: (a) a clinical diagnosis of primary HFS and (b) no evidence of cognitive impairment. Patients with other movement disorders such as myokymia or blepharospasm or chronic debilitating or life-threatening diseases such as malignancy were excluded. Moreover, two patients who treated with botulinum toxin and one patient who lost to follow-up were excluded. The patients were classified into four groups depending on the severity of spasms (Table 1) [5].

Finally, a total of 118 patients were included in the study. There were 90 women (76.3%) and 28 men (23.7%), with a mean age of 51 years ranging from 22 to 79 years. Preoperative evaluation using the SMC grading system for HFS was divided into 25 patients with grade I, 48 patients with grade II, 33 patients with grade III, and 12 patients with grade IV. Overall, the median duration of symptoms was 48 months, with interquartile ranges of 24–90 months. On the basis of the SMC grade, the mean duration of symptoms was 18 months (range

© Springer Nature Singapore Pte Ltd. 2020
K. Park, J. S. Park (eds.), *Hemifacial Spasm*, https://doi.org/10.1007/978-981-15-5417-9_2

Fig. 1 Research focus on clinical course of HFS. I, clinical course after microvascular decompression (MVD); II, clinical course after botulinum toxin treatment; III, natural history of hemifacial spasm (HFS), IV (until visit to hospital) + V (untreated)

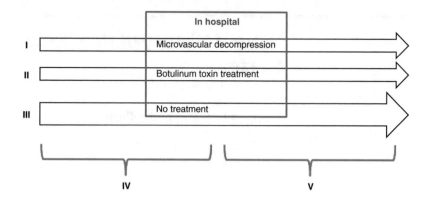

Table 1 SMC grading system for HFS

Grade	Characteristics
I	Localized spasm around the periocular area
II	Involuntary movements spreading to other parts of the ipsilateral face and affecting other muscle groups, including the orbicularis oris, zygomaticus, frontalis, or platysma muscle
III	Frequent tonic spasms affecting vision
IV	Disfiguring asymmetry with continuous contractions of the orbicularis oculi muscle affecting the opening of the eye

Table 2 Spasm severity and symptom duration ($N = 118$)

Spasm severity (n)	Symptom duration, month, mean (range)
Grade I (25)	18 (2–80)
Grade II (48)	39 (2–180)
Grade III (33)	84 (7.5–240)
Grade IV (12)	171 (24–396)

2–80 months) in grade I, 39 months (range 2–180 months) in grade II, 84 months (range 7.5–240 months) in grade III, and 171 months (range 24–396 months) in grade IV patients (Table 2). We observed that the higher the SMC grade, the longer the duration of symptoms that last ($p < 0.05$). This result indicates that the longer the duration of symptoms, the more severe the spasms.

Natural History of Untreated Hemifacial Spasm [6]

This study was to characterize the natural history of untreated HFS over a 5-year period. All 2155 patients initially visited the outpatient clinic of our hospital between 2001 and 2010, and were

diagnosed with HFS after a neurological evaluation according to published criteria. Of these patients, 205 patients were selected who met the following criteria: (a) primary HFS diagnosed by one experienced neurosurgeon (K.P.), (b) identification of vascular compression of the facial nerve on magnetic resonance imaging (MRI), and (c) no botulinum toxin or surgical treatment since the initial diagnosis. Other movement disorders such as myokymia or blepharospasm or secondary HFS were excluded. Follow-up was done in 113 of the 205 patients, but the other 92 were not in contact. Nine of these 113 patients were excluded; 6 died and 3 suffered from other diseases such as malignancies and dementia. This is summarized in Fig. 2.

The course of symptoms was divided into four categories: worsened in frequency, duration and intensity, stationary, improved partially, and in remission (little or no spasm). Patients who no longer followed were contacted by phone. These outcomes were determined not by direct medical examination, but by reminding the patients of changes in symptoms since the onset.

Finally, a total of 104 patients were included in the study. There were 62 women (59.6%) and 42 men (60.4%). The mean age of the patients was 62 years (range 34–86 years) at the initial diagnosis of HFS and 50 years (range 22–76 years) at the onset of HFS. The average duration of symptoms was 10.1 years, with a range of 0.2–42.0 years. Changes in the condition were tracked for 5–42 years (mean 12 years) from the onset of symptoms. In 11 out of 104 patients (10.6%), their symptoms worsened from 6 to 42 years (average 16 years). Forty patients

Fig. 2 Study enrollment
and follow-up data of
untreated HFS patients
(*n*, %). *HFS* hemifacial
spasm, *MRI* magnetic
resonance imaging,
MVD microvascular
decompression

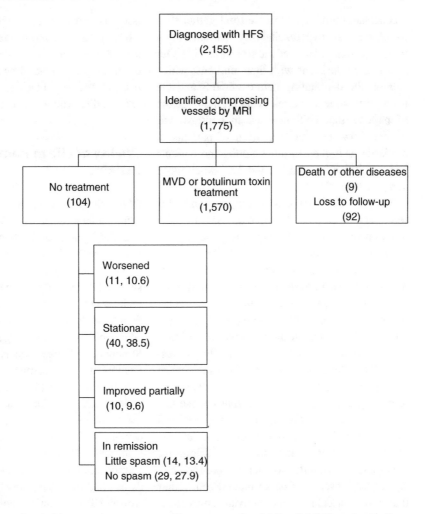

(38.5%) were stationary for 6–23 years (average 12 years). On the other hand, 10 patients (9.6%) improved partially over 7–18 years (average 11 years). Between 2 months and 23 years (mean 6.4 years) after onset, 43 patients (41.3%) were alleviated and no additional treatment was needed for 5 months to 13 years (mean 5.7 years) (Fig. 2). Despite no improvement in symptoms, these patients were conservatively followed due to other health conditions, involved costs, their anxiety and concerns related to risks and limitations, and their adaptation to the symptoms. Thirty-seven patients continued to receive other treatments such as treatments of Traditional Chinese Medicine including acupuncture or herbal remedies, medication, and physiotherapy. Thirty-eight patients did not respond properly and were not treated anymore and 29 patients did not receive any treatment.

However, only 4.8% of 2155 patients diagnosed with HFS were included in this study. Excluded patients may later have surgical intervention due to worsening of symptoms. Given this, the remission rate, or 41% of the population, is expected to decrease.

The follow-up studies after botulinum toxin treatment were conducted by several researchers [3, 7–9]. In the study by Conte et al., spread of spasms to the other facial muscles of the same side of the face existed in 93.4% of HFS patients, and longer latency of spread (the time interval between the onset of muscle spasm and the subsequent involvement of other part of the face) was associated with longer duration of disease and younger age at onset. They concluded that the development of spread likely represents the natural course of HFS. They also assumed that spread in HFS can occur through an ephaptic mechanism on the facial nerve trunk rather than facial nucleus hyperexcitability, as botulinum toxin injection did not reduce spread rate [7]. Our results are consistent with their findings, which support the debilitating nature of HFS and the need for treatment in a relatively high proportion of patients that HFS persists. In contrast, our results show that the proportion of patients in remission is higher. In other studies, no patients had remission or only 2.3–4.2% experienced symptomatic relief [3, 8, 9].

On the other hand, disease progression is different across patients. In our data, based on the duration and severity of symptoms, patients were classified into rapidly and slowly progressive groups and the factors related to the differences between the two groups were investigated. As a result, the younger age at the time of surgery, the older age at the onset of symptoms, and the absence of indentation on the facial nerve intraoperatively were associated with rapidly progressive HFS. The reason for the young patients in the rapidly progressive group is that if the symptoms progress faster, the patients may choose surgery earlier. Rapid progression of late-onset HFS may be related to aging or the cortical influences of the facial motor nucleus. In addition, we hypothesized that mild vascular compression and consequent absence of nerve indentation, rather than severe vascular compression and consequent nerve indentation, lead to rapid HFS progression. These results may be useful for understanding and informing patients on the differences in disease progression in HFS [10].

In summing up the above results, HFS is often associated with the gradual deterioration of the spasm pattern and the reduction of the spasm-free intervals. Nevertheless, there were no previous studies on why HFS patients have different courses. Even though it is unclear why this happens, our findings suggest that the degree and progression rate of symptoms due to demyelination of the facial nerve [10, 11] or hyperexcitability in the facial motor nucleus [11–13] or the denervation of the affected muscles [14] are diverse, according to individual characteristics. If this inference is correct, further research is needed to clarify the contributing factors. Another

issue is whether it is spasm-free interval or whether the symptoms actually improve. Our data showed that 35 of 43 patients (81.4%) in remission were spasm-free for more than 3 years, but this was not clarified. Future trials should explore this evaluation standard.

Quality of Life in Hemifacial Spasm Patients

Although HFS symptoms naturally improve in some patients, patients with HFS have problems with their quality of life (QoL) during their symptoms. First, HFS patients showed a high frequency of social anxiety disorder compared to healthy subjects, especially in young and depressed patients [15]. And HFS patients had higher scores than focal dystonia or healthy-control subjects for "contamination" and "aggressiveness" on the Structured Clinical Interview for Obsessive-Compulsive Spectrum Self-Report [16]. Additionally, sexual function was affected in HFS patients than in healthy controls [17]. Furthermore, the more severe the spasms, the worse are the depression, headache, and QoL [4, 18–20]. Conversely, HFS patients with anxiety reported a significant improvement in symptoms after proper care. Therefore, because stress and anxiety can worsen HFS, the authors concluded that diagnosis and early management of anxiety symptoms can improve QoL in HFS patients [21].

On the other hand, treatments of HFS improved QoL in HFS patients. In many studies of botulinum toxin treatment for HFS, QoL has changed significantly before and after the treatment [22–26]. Also about MVD, many studies have demonstrated that the post-surgery improvements of the HFS symptoms were associated with decreased social anxiety and improved QoL [27–29]. In some cases, the QoL improvement was prolonged due to delayed improvement of symptoms after surgery [30–32]. According to the above findings, patients with HFS benefited from botulinum toxin treatment or MVD in symptoms associated with long-term QoL.

Thus, even if the symptoms improve, appropriate treatment will be necessary for well-selected patients because such situations do not often occur and QoL is impaired while the symptoms are present.

References

1. Barker FG 2nd, Jannetta PJ, Bissonette DJ, Shields PT, Larkins MV, Jho HD. Microvascular decompression for hemifacial spasm. J Neurosurg. 1995;82:201–10.
2. Wang A, Jankovic J. Hemifacial spasm: clinical findings and treatment. Muscle Nerve. 1998;21:1740–7.
3. Mauriello JA Jr, Leone T, Dhillon S, Pakeman B, Mostafavi R, Yepez MC. Treatment choices of 119 patients with hemifacial spasm over 11 years. Clin Neurol Neurosurg. 1996;98:213–6.
4. Lee JA, Jo KW, Kong DS, Park K. Using the new clinical grading scale for quantification of the severity of Hemifacial spasm: correlations with a quality of life scale. Stereotact Funct Neurosurg. 2012;90:16–9.
5. Hyun SJ, Kong DS, Park K. Microvascular decompression for treating hemifacial spasm: lessons learned from a prospective study of 1,174 operations. Neurosurg Rev. 2010;33:325–34. discussion 334
6. Lee JA, Kim KH, Park K. Natural history of untreated Hemifacial spasm: a study of 104 consecutive patients over 5 years. Stereotact Funct Neurosurg. 2017;95:21–5.
7. Conte A, Falla M, Diana MC, Bologna M, Suppa A, Fabbrini A, Colosimo C, Berardelli A, Fabbrini G. Spread of muscle spasms in Hemifacial spasm. Mov Disord Clin Pract. 2014;2(1):53–5.
8. Defazio G, Abbruzzese G, Girlanda P, Vacca L, Curra A, De Salvia R, Marchese R, Raineri R, Roselli F, Livrea P, Berardelli A. Botulinum toxin a treatment for primary hemifacial spasm: a 10-year multicenter study. Arch Neurol. 2002;59:418–20.
9. Mauriello JA, Aljian J. Natural history of treatment of facial dyskinesias with botulinum toxin: a study of 50 consecutive patients over seven years. Br J Ophthalmol. 1991;75:737–9.
10. Sanders DB. Ephaptic transmission in hemifacial spasm: a single-fiber EMG study. Muscle Nerve. 1989;12:690–4.
11. Valls-Solé J. Electrodiagnostic studies of the facial nerve in peripheral facial palsy and hemifacial spasm. Muscle Nerve. 2007;36:14–20.
12. Kuroki A, Møller AR. Facial nerve demyelination and vascular compression are both needed to induce facial hyperactivity: a study in rats. Acta Neurochir. 1994;126:149–57.
13. Møller AR, Jannetta PJ. On the origin of synkinesis in hemifacial spasm: results of intracranial recordings. J Neurosurg. 1984;61:569–76.
14. Tunç T, Cavdar L, Karadag YS, Okuyucu E, Coskun O, Inan LE. Differences in improvement between patients with idiopathic versus neurovascular hemifacial spasm after botulinum toxin treatment. J Clin Neurosci. 2008;15:253–6.
15. Ozel-Kizil ET, Akbostanci MC, Ozguven HD, Atbasoglu EC. Secondary social anxiety in hyperkinesias. Mov Disord. 2008;23(5):641–5.
16. Mula M, Strigaro G, Marotta AE, Ruggerone S, Tribolo A, Monaco R, Cantello F. Obsessive-compulsive-spectrum symptoms in patients with focal dystonia, hemifacial spasm, and healthy subjects. J Neuropsychiatry Clin Neurosci. 2012;24(1):81–6.
17. Perozzo P, Salatino A, Cerrato P, Ricci R. Sexual Well-being in patients with blepharospasm, spasmodic torticollis, and hemifacial spasm: a pilot study. Front Psychol. 2016;7:1492.
18. Tan EK, Lum SY, Fook-Chong S, Chan LL, Gabriel C, Lim L. Behind the facial twitch: depressive symptoms in hemifacial spasm. Parkinsonism Relat Disord. 2005;11(4):241–5.
19. Peeraully T, Tan SF, Fook-Chong SM, Prakash KM, Tan EK. Headache in hemifacial spasm patients. Acta Neurol Scand. 2013;127(5):e24–7.
20. Cheng J, Lei D, Hui X, Zhang H. Improvement of quality of life in patients with Hemifacial spasm after microvascular decompression: a prospective study. World Neurosurg. 2017;107:549–53.
21. Tan EK, Fook-Chong S, Lum SY. Case-control study of anxiety symptoms in hemifacial spasm. Mov Disord. 2006;21(12):2145–9.
22. Tan EK, Fook-Chong S, Lum SY, Lim E. Botulinum toxin improves quality of life in hemifacial spasm: validation of a questionnaire (HFS-30). J Neurol Sci. 2004;219(1–2):151–5.
23. Setthawatcharawanich S, Sathirapanya P, Limapichat K, Phabphal K. Factors associated with quality of life in hemifacial spasm and blepharospasm during long-term treatment with botulinum toxin. Qual Life Res. 2011;20(9):1519–23. https://doi.org/10.1007/s11136-011-9890-y.
24. Kongsengdao S, Kritalukkul SJ. Quality of life in hemifacial spasm patient after treatment with botulinum toxin A; a 24-week, double-blind, randomized, cross-over comparison of Dysport and Neuronox study. Med Assoc Thai. 2012;95(Suppl 3):S48–54.
25. Weiss D, Sturm J, Hieber L, Börtlein A, Mayr I, Appy M, Kühnler B, Buchthal J, Dippon C, Arnold G, Wächter T. Health-related quality of life outcomes from botulinum toxin treatment in hemifacial spasm. Ther Adv Neurol Disord. 2017;10(4):211–6. https://doi.org/10.1177/1756285616682676. Epub 2017 Feb 1
26. Yuksel B, Genc F, Yaman A, Goksu EO, Ak PD, Gomceli YB. Evaluation of stigmatization in hemifacial spasm and quality of life before and after botulinum toxin treatment. Acta Neurol Belg. 2019;119(1):55–60. https://doi.org/10.1007/s13760-018-1018-5. Epub 2018 Sep 3
27. Kim YG, Jung NY, Kim M, Chang WS, Jung HH, Chang JW. Benefits of microvascular decompression on social anxiety disorder and health-related

quality of life in patients with hemifacial spasm. Acta Neurochir. 2016;158(7):1397–404.

28. Montava M, Rossi V, CurtoFais CL, Mancini J, Lavieille JP. Long-term surgical results in microvascular decompression for hemifacial spasm: efficacy, morbidity and quality of life. Acta Otorhinolaryngol Ital. 2016;36(3):220–7.

29. Lawrence JD, Frederickson AM, Chang YF, Weiss PM, Gerszten PC, Sekula RF Jr. An investigation into quality of life improvement in patients undergoing microvascular decompression for hemifacial spasm. J Neurosurg. 2017;10:1–9.

30. Ray DK, Bahgat D, McCartney S, Burchiel KJ. Surgical outcome and improvement in quality of life after microvascular decompression for hemifacial spasms: a case series assessment using a validated disease-specific scale. Stereotact Funct Neurosurg. 2010;88(6):383–9.

31. Shibahashi K, Morita A, Kimura T. Surgical results of microvascular decompression procedures and patient's postoperative quality of life: review of 139 cases. Neurol Med Chir. 2013;53(6):360–4.

32. Lee JA, Kong DS, Lee S, Park SK, Park K. Clinical outcome after microvascular decompression according to the progression rates of hemifacial spasm. World Neurosurg. 2020;134:e985-e990. https://doi.org/10.1016/j.wneu.2019.11.052. Epub 2019 Nov 14.

Novel Classification Systems for Hemifacial Spasm

Jae Sung Park and Kwan Park

The accumulated experience in performing microvascular decompression (MVD) for hemifacial spasm (HFS) at our institute for more than two decades has allowed us to broaden the scope of the concept in treating HFS patients. During the journey, through more than 5000 cases of MVD for HFS, the results of our observation were quantified and categorized, which are to be introduced in this chapter. We believe that they are intuitive, clinically pertinent, and pragmatic. Also, if used by more and more researchers around the globe, our classification systems can be a foundation for deeper theoretical knowledge and understanding of HFS.

Compression Patterns

Vascular compression on the root entry zone (REZ) of the seventh nerve (CN VII) is the most predominantly accepted etiology of HFS. The term neurovascular compression seems straightforward, but the actual intraoperative findings during MVDs show much more variety of com-

Table 1 Number of compression vessels involved in each compression pattern

	VA ($p < 0.001$)	AICA ($p = 0.005$)	PICA ($p = 0.003$)
Loop type	2	1	8
Arachnoid type	3	40	25
Perforator type	0	49	9
Branch type	0	14	4
Sandwich type	0	26	13
Tandem type	32	43	25
Miscellaneous type	0	2	0
Total	37	175	84

pressing patterns. In terms of the involving arteries, anterior inferior cerebellar artery (AICA) was the most prevalent (51.7%), followed by posterior inferior cerebellar artery (PICA, 21.6%) [1]. The vertebral artery (VA) was rarely responsible for the compression by itself (1.7%), but in combination with AICA, PICA, or both, it could account for 14% of the compressions (Table 1). Compressive patterns that we discuss in this section depict the ways in which a vessel or vessels compressed the REZ of the CN VII, purely based on the microsurgical findings.

Loop Type (Fig. 1a)

The loop type refers to a type of compression on the REZ caused by a single artery without any other concomitant factors; the excessive curvature of a vessel that directly contacts the REZ

J. S. Park (✉)
Department of Neurosurgery, Konkuk University School of Medicine, Chungju Hospital, Chungju, Korea (Republic of)

K. Park
Department of Neurosurgery, Konkuk University Medical Center, Seoul, Korea (Republic of)
e-mail: kwanpark@skku.edu

© Springer Nature Singapore Pte Ltd. 2020
K. Park, J. S. Park (eds.), *Hemifacial Spasm*, https://doi.org/10.1007/978-981-15-5417-9_3

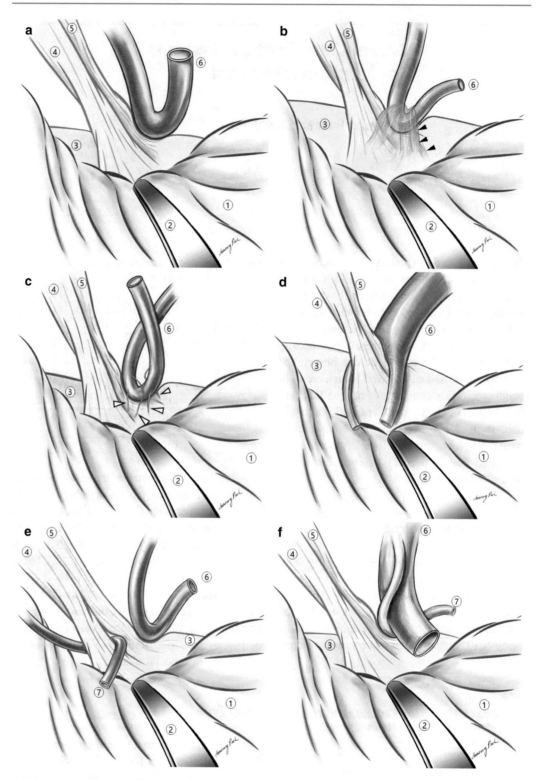

Fig. 1 Various compression patterns of hemifacial spasm: intraoperative view for the right side lesion. (**a**) loop type, (**b**) arachnoid type, (**c**) perforator type, (**d**) branch type, (**e**) sandwich type, and (**f**) tandem type. (1) Cerebellum, (2) retractor, (3) brain stem (pons), (4) CNVIII, the ves- tibulocochlear nerve, (5) CN VII, the facial nerve, (6) compressing vessel #1, (7) compressing vessel #2; arrow- heads, arachnoid membrane; hollow arrowheads, perfora- tors to the brain stem (©Jae Sung Park 2020. All rights reserved)

conveys the throbbing pulse to it. This is indeed the simplest type, and yet it is far from the most common one. According to our previous report in 2008, this specific type accounted for only 11 (4.7%) of 236 cases [1]. What interested us was that the involvement of PICA among the loop type compressions (8 of 11, 72.7%) was disproportionately greater than that of the whole cases (51 of 236, 21.6%). Given that there was no additional structure to cause the compression other than the vessel curvature itself, the postoperative outcome for this specific type was excellent; the success rate was 100% with no noticeable complications.

Arachnoid Type (Fig. 1b)

The arachnoid type is defined by a compression where the compressing vessel is tethered tightly to the REZ or brainstem by thickened arachnoid trabeculae, and this was the most common type of all (28.0%). Unlike the loop type, the mobilization of the vessel is hardly achievable without sufficient release of the tethering between the vessel and the REZ. Cautious and meticulous dissection of the thickened arachnoid trabeculae is the key to the successful decompression, and once it is done, the compressing vessel is usually freed from the REZ. More often than not, the insertion of Teflon sponge was not even necessary merely for the purpose of decompression; yet, it was performed for the sake of prevention of recurrence.

Perforator Type (Fig. 1c)

This type is analogous to the arachnoid type in that the tethering of the vessel is attributable to the perforating arteries from the main compressing vessel, instead of the thickened arachnoid trabeculae in the arachnoid type. However, instead of surgical dissection of arachnoid trabeculae in the arachnoid type, much more cautious approach to the REZ is crucial in the perforator type, since any disruption of perforators must be avoided at all costs. In our experience, though, this type was

not necessarily related to worse outcome compared to others. To be aware that the neurovascular compression might involve perforators as the tethering mechanism has helped us tremendously, and we believe that many other young neurosurgeons may benefit from it. Perforator type was the second most popular type, following the arachnoid one.

Branch Type (Fig. 1d)

When the REZ is encircled by two branches of an offending artery, this is called a branch type. Although the range of immobilization may be limited compared to other types, no statistically significant difference was noted in terms of employed surgical techniques, postoperative outcomes, and complications.

Sandwich Type (Fig. 1e)

This type depicts a compression where the REZ is caught between two offending arteries: a dorsal and a ventral one. A ventrally compressing artery is easy to miss, unless a thorough inspection around the REZ is carried out. Even after a successful decompression is achieved for the dorsally positioned artery, the patient may end up suffering the unresolved symptom, if the ventral artery is still causing the compression. Existence of this type with the frequency of 11.9% would be the rationale for the necessity of after-decompression examination around the REZ. One of 236 subjects who underwent MVD developed a permanent hearing loss; and she happened to harbor the sandwich type. No statistical analysis was available for this one case, but one could assume that the decompression for a ventrally residing artery may pose technical challenges.

Tandem Type (Fig. 1f)

This type is defined by a compression where a vessel compresses another vessel, and in turn, their combined force affects the REZ. Tandem type accounted for 22.0% of whole cases, and this had a strong predilection for the VA, and vice versa. The VA was involved in 61.5% of the tandem type and the tandem type accounted for 86.5% of all the cases where the VA was identified as an offending artery. When the VA is suspected to be an offending artery, one must consider the possibility of a tandem type, and an exhaustive exploration around the REZ must be carried out in search of another offending artery.

Pattern of Hearing Loss

Despite a high success rate of MVD for HFS (more than 90%), it is not without postoperative complications, and hearing loss is one of the most serious ones [2–5]. Thus, it is imperative to monitor the functional integrity of the eighth nerve (CN VIII) during the surgery. Intraoperative brain stem auditory evoked potential (BAEP) is one of the most commonly performed monitoring methods, and it is acknowledged to reduce the risk of hearing impairment during a surgery involving the CN VIII [6, 7]. In this section, a new grading system of hearing loss is to be introduced, and its association with intraoperative BAEP is also to be presented.

Hearing Evaluations

Our previous reports on hearing loss following MVD employed pure tone audiometry (PTA) along with speech discrimination score (SDS) [4, 8]. All patients went through both tests prior to and after the surgery; the postoperative one was performed within 10 postoperative days. The average of PTA thresholds on the frequencies of 0.5, 1, 2, and 3 kHz was calculated for each subject, and their postoperative results were compared with the preoperative ones. For those who experienced a change of PTA thresholds greater than 15 dB or 20% or more decrease in SDS, another session of PTA and SDS was performed. The great majority of all patients (1098 of 1144, 96.0%) showed no or negligible change in PTA thresholds, and they were categorized as group 1.

Those who developed total deafness fell into group 4 (10, 0.9%), and the remaining 36 belonged to group 2 or 3. Group 2 consisted of subjects whose postoperative hearing had deteriorated by more than 15 dB in PTA threshold, and the decrease in their SDS was proportionate to the degree of PTA. Patients whose SDS was disproportionately worsened compared to the changes of PTA comprised group 3. In accordance with Speech Recognition Interpretation (SPRINT) chart, when SDS score fell in the shaded area, it was considered as "disproportionate decrease" compared to PTA thresholds [8–10] (Fig. 2). Table 2 shows the definition of each group. The pattern of hearing loss of group 2 was considered as "cochlear type," whereas that of group 3 as "retrocochlear type" [4, 8]. In terms of recovery from the hearing loss, the cochlear type was associated with better prognosis than the retrocochlear one.

Intraoperative BAEP in Relation to Postoperative Hearing Loss

Throughout the surgery (from the skin incision to the dural closure), BAEP was monitored for all patients. A decrease in the amplitude in peak V by more than 50% or delayed latency of peak V longer than 1.0 ms was considered as a significant change of BAEP. Whenever either one or both took place, the surgeon was alarmed; the cerebellar retractors were thereupon released or repositioned until the change of BAEP recovered. 280 of 1144 patients experienced decreased amplitude of peak V by 50% or more at some point during the surgery, and upon release of the cerebellar retractor, 268 (95.7%) of 280 recovered their amplitude of peak V above 50% of the reference grand average (Table 3). Whether or not the amplitude of peak V was restored to above 50% of the reference was statistically associated with the four groups of hearing loss based on PTA and SDS. Table 3 shows the relationship of intraoperative BAEP with the categorization of hearing loss.

On the usefulness of intraoperative BAEP, Sindou proposed that an increase in latency of

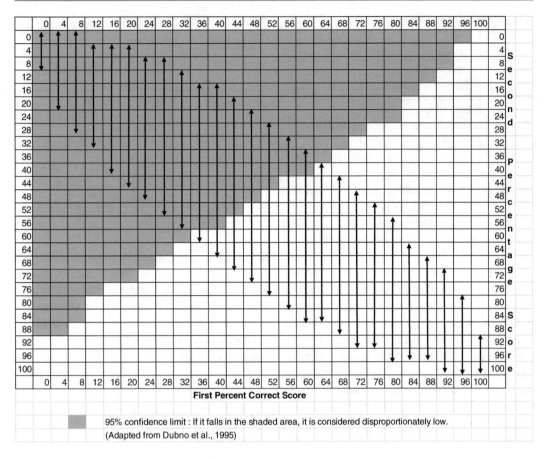

Fig. 2 Speech Recognition Interpretation (SPRINNT) chart

Table 2 Patterns of hearing loss

	Definition
Group 1	None or negligible change
Group 2	Deterioration in PTA >15 dB and proportionate decrease in SDS
Group 3	Deterioration in PTA >15 dB and disproportionate decrease in SDS
Group 4	Total deafness

Table 3 Relationship between intraoperative BAEP change and clinical hearing loss

	No change in BAEP	>50% decrease in peak V amplitude (recovery to >50% vs. no recovery)	
Group 1	841	257	(257 vs. 0)
Group 2	18	8	(7 vs. 1)
Group 3	2	8	(3 vs. 5)
Group 4	3	7	(1 vs. 6)

peak V as well as a decrease in the amplitude of peak I could be a warning sign for excessively stretched cochlear nerve, whereas Polo et al. emphasized the delay in peak V as a watching, warning, or critical sign, depending on the length of the delay [7, 11]. Our previous report suggested that a decrease in the amplitude of peak V was statistically consistent with the clinical severity of hearing loss [4].

Extent of Compression

As previously mentioned, the intraoperative findings greatly vary. One of the diverse features of the compression is the indentation of the REZ. Based on surgical findings of 293 MVDs,

Table 4 Grades of indentation on the REZ

	Definition	Number of patients (%)
Grade 1	No recognizable indentation	64 (21.6)
Grade 2	Indentation without discoloration	114 (39.2)
Grade 3	Indentation with discoloration	115 (39.2)

Table 5 Definition of LSR grades

LSR grades	Definition
0	Nonexistent LSR
1	Disappearance of LSR before decompression
1a	Upon dural opening
1b	Between dural opening and decompression
2	Disappearance of LSR after decompression
2a	Upon decompression
2b	Delayed disappearance after decompression
3	Persistent LSR throughout MVD

we proposed a new grading system for the extent of the compression (Table 4) [12].

The grades of indentation were associated with the postoperative outcomes. Our initial hypothesis was that a more severe indentation would yield a poorer outcome; however, the opposite result was drawn. Concerning this rather odd result, a couple of assumptions were suggested. A more severe, that is, more distinct indentation may have enabled the surgeon recognize the optimal site for decompression. Another explanation would be that group 1 (with no or negligible indentation) may have included those with secondary HFS or HFS mimicking conditions such as myokymia, post-facial palsy synkinesis, etc.

Lateral Spread Response

Lateral spread response (LSR) is one of the most frequently researched neurophysiological methods for HFS; and yet there are still many controversies if it actually reflects the long-term outcomes. LSR, a phenomenon observed during the intraoperative facial electromyogram, refers to abnormal responses in a facial muscle group when an irrelevant branch of the facial nerve is stimulated. Given that LSR, in general, disappears during the MVD procedure, numerous researchers postulated that the disappearance of LSR could indicate a successful decompression; this is true in most cases, but not all. There are non-negligible numbers of exceptions: no LSR before a decompression, persistent LSR after a successful decompression, disappearance of LSR followed by an unsatisfactory outcome, etc. No universally accepted consensus is available concerning how to analyze the intraoperative LSR results; factors that can affect the conclusion include the use of neuromuscular blockade (NMB), the dose of NMB, minimal amplitude to constitute LSR, and precise timing of the LSR disappearance. These factors may vary from institute to institute.

The grading system we used to analyze LSR is presented in Table 5. If this grading system is to be prevalently used among institutes, a less controversial conclusion on LSR may be drawn in the near future.

Clinical Grade: SMC Grading System

HFS is mainly diagnosed by clinical observation. Thus, the severity of the disease is also determined by patients' verbatim and physicians' examination, not by any objective measures including laboratory data, electrophysiological test results, or radiological evaluations. Although changes in symptoms of HFS can vary from person to person, there is a typical pattern of progress. Based on this pattern of progress, we suggested a new grading system for symptoms of HFS, named "SMC grading system" in 2005 [13] (Table 6). This was created in accordance with patients' symptoms and their chronological changes, and our previous report demonstrated that this grading system was consistent with HFS-7 questionnaire, also known as short self-rating quality of life scale ($p < 0.001$) [13]. More and more scholars are adopting this scale to their

Table 6 Description of SMC grading system

Grade	Detailed description
I	Localized spasm around the periocular area
II	Involuntary movement spreads to other parts of the ipsilateral face and affects other muscle groups: the orbicularis oris, zygomaticus, frontalis or platysma muscle
III	Interference with vision because of frequent tonic spasms
IV	Disfiguring asymmetry: continuous contraction of the orbicularis oculi muscles affects opening of the eye

researches, which we wish to be a foundation for a greater understanding of HFS in the future.

References

1. Park J, Kong D-S, Lee J-A, Park K. Hemifacial spasm: neurovascular compressive patterns and surgical significance. Acta Neurochir. 2008;150(3):235–41.
2. Chung S, Chang J, Kim S, Chang J, Park Y, Kim D. Microvascular decompression of the facial nerve for the treatment of hemifacial spasm: preoperative magnetic resonance imaging related to clinical outcomes. Acta Neurochir (Wien). 2000;142(8):901–7.
3. Samii M, Günther T, Iaconetta G, Muehling M, Vorkapic P, Samii A. Microvascular decompression to treat hemifacial spasm: long-term results for a consecutive series of 143 patients. Neurosurgery. 2002;50(4):712–9.
4. Jo K-W, Kim J-W, Kong D-S, Hong S-H, Park KJ. The patterns and risk factors of hearing loss following microvascular decompression for hemifacial spasm. Acta Neurochir (Wien). 2011;153(5):1023–30.
5. Sindou M, Fobe J, Ciriano D, Fischer C. Intraoperative brainstem auditory evoked potential in the microvascular decompression of the 5th and 7th cranial nerves. Rev Laryngol Otol Rhinol (Bord). 1990;111(5):427–31.
6. Ojemann RG, Levine RA, Montgomery WM, et al. Use of intraoperative auditory evoked potentials to preserve hearing in unilateral acoustic neuroma removal. J Neurosurg. 1984;61(5):938–48.
7. Sindou M, Ciriano D, Fischer C. Lessons from brainstem auditory evoked potential monitoring during microvascular decompression for trigeminal neuralgia and hemifacial spasm. Intraoperative Neurophysiologic Monitoring in Neurosurgery. Cham: Springer; 1991. p. 293–300.
8. Park K, Hong S, Hong S, Cho Y, Chung W, Ryu NG. Patterns of hearing loss after microvascular decompression for hemifacial spasm. J Neurol Neurosurg Psychiatry. 2009;80(10):1165–7.
9. Roeser RJ, Valente M, Hosford-Dunn H. Audiology diagnosis. 2nd ed. New York: Thieme; 2007. p. 302–3.
10. Dubno JR, Lee F-S, Klein AJ, Matthews LJ, Lam CF. Confidence limits for maximum word-recognition scores. J Speech Hear Res. 1995;38(2):490–502.
11. Polo G, Fischer C, Sindou MP, Marneffe VJN. Brainstem auditory evoked potential monitoring during microvascular decompression for hemifacial spasm: intraoperative brainstem auditory evoked potential changes and warning values to prevent hearing loss—prospective study in a consecutive series of 84 patients. Neurosurgery. 2004;54(1):97–106.
12. Kim HR, Rhee D-J, Kong D-S, Park KJ. Prognostic factors of hemifacial spasm after microvascular decompression. J Korean Neurosurg Soc. 2009;45(6):336.
13. Lee JA, Jo KW, Kong D-S, Park K. Using the new clinical grading scale for quantification of the severity of hemifacial spasm: correlations with a quality of life scale. Stereotact Funct Neurosurg. 2012;90(1):16–9.

Pathogenesis of Hemifacial Spasm

Min Ho Lee and Jae Sung Park

Hemifacial spasm (HFS), a hyperactive motor dysfunction of the facial nerve, is believed to be caused by vascular compression at the root exit zone (REZ) of the facial nerve, which offered the basis for microvascular decompression (MVD) to become the standard treatment for HFS [1–5]. However, questions have yet to be fully answered about the microscopic pathophysiology of HFS, beyond the compression on the REZ: How come the compression on the REZ causes a hyperactivity instead of hypoactivity of the facial nerve? What can explain disparate clinical courses following a successful MVD? Are there additional contributing factors to the symptoms other than the compression on the REZ?

Anatomy of Facial Nerve

The facial nerve (seventh pair of cranial nerves) is a mixed nerve with efferent (motor) and afferent (general and special sensory) nerve fibers. It emerges from pons of the brainstem and reaches to the internal auditory canal. The segment

M. H. Lee (✉)
Department of Neurosurgery, Uijeongbu St. Mary's Hospital, The Catholic University of Korea, Seoul, Republic of Korea
e-mail: minho919.lee@catholic.ac.kr

J. S. Park
Department of Neurosurgery, Konkuk University School of Medicine, Chungju Hospital, Chungju, Korea (Republic of)

between the brainstem and the internal auditory canal, that is, intracranial or cisternal segment, is estimated to be 17.93 ± 2.29 mm of length. The proximal end of the cisternal segment where it adheres to the pons is called the REZ. About 0.96 mm (range 2.86–1.9 mm) proximal to the REZ is located the "transition zone." Transition zone, that is, Obersteiner-Redlich zone, refers to the area where the central myelin is gradually replaced by peripheral Schwann cell–derived myelin. The transition zone, not surrounded by perineurium or epineurium, is very susceptible to mechanical disruption when it is compressed by a vessel or vessels [6–10].

Hypotheses for the Pathogenesis of Hemifacial Spasm: Peripheral vs. Central

There are two main hypotheses for the detailed pathogenesis of HFS: peripheral (ephaptic transmission) vs. central (antidromic conduction to the facial motor nucleus) hypothesis (Fig. 1). In 1962, Gardner [1] postulated that the local irritation of the facial nerve caused by vascular compression may lead to ephaptic impulse transmission. The term "ephaptic transmission" depicts a situation where an impulse in a group of nerve fibers initiates additional impulse in the adjacent fibers [1, 11]. On the contrary, the central hypothesis refers to hyperexcitability of the facial motor nucleus (FMN) [12, 13]. In 1987, Møller and Jannetta

© Springer Nature Singapore Pte Ltd. 2020
K. Park, J. S. Park (eds.), *Hemifacial Spasm*, https://doi.org/10.1007/978-981-15-5417-9_4

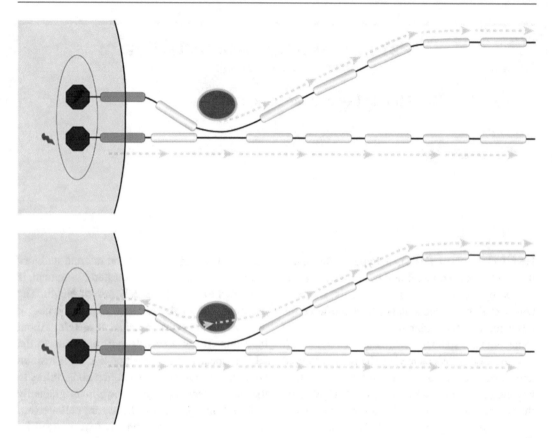

Fig. 1 Hypotheses for the pathogenesis of hemifacial spasm. (**a**) Above: peripheral hypothesis with ephaptic impulse transmission. Adjacent nerve fiber compressed by vessel may transmit the nerve conduction. (**b**) Below: cen-tral hypothesis with antidromic nerve conduction to the nucleus. The stimulated signal transmits backward to the FMN, and then it travels forward

reported that there was a characteristic electrophys-iological wave that was confined to HFS patients: abnormal muscle response (AMR) or lateral spread response (LSR) [14]. LSR consists of two electro-physiological waves: the initial one followed by a smaller, delayed one. To explain the phenomenon of LSR, the authors advocated that the backfiring of the facial nerve might have antidromically acti-vated the FMN, which, in turn, emitted a delayed signal to the facial nerve [15, 16]. The hyperexcit-ability of the FMN as a result of the antidromic backfiring constitutes the central hypothesis of HFS. Although the central hypothesis appears to be opposite to the peripheral one, these two are not mutually exclusive; Moller and Jannetta stated that the phenomenon of LSR might be attributable to ephaptic transmission alone or in combination with the hyperexcitability of the FMN [17–20].

Sympathetic Hypothesis

Generally, once the offending artery was removed away from the facial nerve, the LSR wave dimin-ishes immediately and the symptom of spasm ceased postoperatively in most of the cases [2, 21]. This may not be explained by the peripheral or central hypotheses as above, for neither the histological changes at the compression sites nor the hyper-excitability of facial motor neurons was able to repair right after decompression. Of course, if anatomical change or damage is not severe, we can expect a quick improvement, but it is difficult to explain everything with this. Dou et al. proposed that the offending artery may play a more important role than the effect of merely mechanical compression [22]. Anatomically, arteries are covered with adventitia that contains sympathetic nerve endings as well as vasa vasorum. Normally, the sympathetic endings release neurotransmitters. These neurotransmitters released from the sympathetic nervous endings in the adventitia may spill over from the artery wall and spread to the demyelinated nerve fibers in close contact. And it may induce an ectopia action potential in those demyelinated facial nerve fibers expanding to the neuromuscular conjunction and trigger an attack of HFS. This was revealed using the animal model by Zhou et al. [23]. They called it sympathetic hypothesis. But this still needs more evidence to prove it.

Other Factors That May Be Related with Pathogenesis of Hemifacial Spasm

Vertebrobasilar dolichoectasia may cause vascular crowding in the limited space of the posterior fossa, increasing the risk of vascular compression of the facial nerve. The direction of lateral deviation of the vertebral artery has been reported to be significantly related to the symptomatic side of HFS [24, 25]. Pathogenesis of dolichoectasia is thought to be associated with rarity of the elastic tissue in the tunica media with fragmentation of the internal elastic lamina, and it became elongated and arteriosclerotic. The main causes of such a process are vascular risk factors, like old age, male sex, and hypertension [26, 27]. Several studies have shown a positive correlation between hypertension and HFS [28–34]. Atherosclerosis was reported to be relevant in some studies [35], but it was not found in other studies [36].

In addition to compression of the vessel, another contributing factor leading to HFS is adhesion and traction to vessel of adjacent arachnoid membranes. Some studies have stated that complete separation and dissection of arachnoid membranes can result in sufficient MVD for HFS [5, 37, 38]. Moreover, thickened arachnoid membranes have sometimes been observed during surgery for HFS. Among the classifications suggested by Park et al. [5], it may be regarded as arachnoid type, which thick arachnoid trabeculae between the vessel and the brainstem cause the vessel to be tethered tightly to the nerve.

Previous studies have reported that these abnormal thick arachnoid membranes are caused by inflammation (infection, hemorrhage, etc.) [39–41]. There is no known link between inflammation and HFS yet. Recently, Chen et al. [42] compared inflammatory factors (IL-γ, IL-2 receptor, IL-6, IL-8, IL-10, TNF-α) between patients with HFS, and other control groups (patients with lumbar disc herniation, and healthy control subjects). And they suggested that IL-6 is involved in pathogenesis of HFS.

Meanwhile, a study has been published that HFS is caused by not only simple mechanical compression but also exacerbation by sympathetic activity of autonomic nervous system [43]. They performed simultaneous recordings of the electrocardiography and facial electromyography in patients with HFS and revealed that a transient increase in the heart rate occurred a few seconds before the onset of spasm.

Although these studies still show preliminary results, it is a very interesting approach to the cause of neurovascular conflict.

Secondary Hemifacial Spasm

The incidence of cerebellopontine angle (CPA) tumor-induced HFS ranges from 0.3 to 2.5% of all patients with HFS [44, 45]. HFS can also be induced by not only CPA tumors (meningioma, or epidermoid cyst), even stem glioma, aneurysm, vascular malformation, or infection [46–50]. Some authors suggested that displaced vessels by tumor triggered the HFS [51–53]. And others suggested that tumor may compress the nerve or nucleus directly [37, 54]. It is too primitive for us to conclude here. It is too early to conclude the actual pathogenesis of secondary HFS, yet. But, we should consider various possibilities for these causes of HFS and review the images meticulously. These studies are also expected to be one of the keys to elucidate the pathogenesis of HFS.

Conclusion

As mentioned above, the pathophysiology of HFS is not yet clear. But, the fact is that the compression of the facial nerve by an offending vessel is definitely the origin of the increased excitability of the facial nerve that leads to HFS.

References

1. Gardner WJ. Concerning the mechanism of trigeminal neuralgia and hemifacial spasm. J Neurosurg. 1962;19:947–58.
2. Lee MH, Jee TK, Lee JA, Park K. Postoperative complications of microvascular decompression for hemifacial spasm: lessons from experience of 2040 cases. Neurosurg Rev. 2016;39:151–8. discussion 158
3. Jho HD, Jannetta PJ. Hemifacial spasm in young people treated with microvascular decompression of the facial nerve. Neurosurgery. 1987;20:767–70.
4. Sindou M, Mercier P. Microvascular decompression for hemifacial spasm: outcome on spasm and complications. A Rev Neurochir. 2018;64:106–16.
5. Park JS, Kong DS, Lee JA, Park K. Hemifacial spasm: neurovascular compressive patterns and surgical significance. Acta Neurochir. 2008;150:235–41. discussion 241
6. Mercier P, Brassier G, Fournier H-D, Delion M, Papon X, Lasjaunias P. Anatomie morphologique des nerfs crâniens dans leur portion cisternale (du III au XII). Neurochirurgie. 2009;55:78–86.
7. Haller S, Etienne L, Kovari E, Varoquaux AD, Urbach H, Becker M. Imaging of neurovascular compression syndromes: trigeminal neuralgia, Hemifacial spasm, vestibular Paroxysmia, and glossopharyngeal neuralgia. AJNR Am J Neuroradiol. 2016;37:1384–92.
8. Guclu B, Sindou M, Meyronet D, Streichenberger N, Simon E, Mertens P. Cranial nerve vascular compression syndromes of the trigeminal, facial and Vagoglossopharyngeal nerves: comparative anatomical study of the central myelin portion and transitional zone; correlations with incidences of corresponding hyperactive dysfunctional syndromes. Acta Neurochir. 2011;153:2365–75.
9. Tomii M, Onoue H, Yasue M, Tokudome S, Abe T. Microscopic measurement of the facial nerve root exit zone from central glial myelin to peripheral Schwann cell myelin. J Neurosurg. 2003;99:121–4.
10. Skinner HA. The origin of acoustic nerve tumours. Br J Surg. 1929;16:440–63.
11. Gardner WJ. Cross talk--the paradoxical transmission of a nerve impulse. Arch Neurol. 1966;14:149–56.
12. Ferguson JH. Hemifacial spasm and the facial nucleus. Ann Neurol. 1978;4:97–103.
13. Esteban A, Molina-Negro P. Primary hemifacial spasm: a neurophysiological study. J Neurol Neurosurg Psychiatry. 1986;49:58–63.
14. Moller AR, Jannetta PJ. Monitoring facial EMG responses during microvascular decompression operations for hemifacial spasm. J Neurosurg. 1987;66:681–5.
15. Kuroki A, Moller AR. Facial nerve demyelination and vascular compression are both needed to induce facial hyperactivity: a study in rats. Acta Neurochir. 1994;126:149–57.
16. Moller AR, Jannetta PJ. Physiological abnormalities in hemifacial spasm studied during microvascular decompression operations. Exp Neurol. 1986;93:584–600.
17. Moller AR, Jannetta PJ. On the origin of synkinesis in hemifacial spasm: results of intracranial recordings. J Neurosurg. 1984;61:569–76.
18. Moller AR. Hemifacial spasm: ephaptic transmission or hyperexcitability of the facial motor nucleus? Exp Neurol. 1987;98:110–9.
19. Moller AR. Interaction between the blink reflex and the abnormal muscle response in patients with hemifacial spasm: results of intraoperative recordings. J Neurol Sci. 1991;101:114–23.
20. Roth G, Magistris MR, Pinelli P, Rilliet B. Cryptogenic hemifacial spasm. A neurophysiological study. Electromyogr Clin Neurophysiol. 1990;30:361–70.
21. Lee MH, Lee JA, Park K. Different roles of microvascular decompression in Hemifacial spasm and trigeminal neuralgia. J Neurol Surg B Skull Base. 2019;80:511–7.
22. Dou NN, Zhong J, Zhou QM, et al. The mechanism of hemifacial spasm: a new understanding of the offending artery. Neurol Res. 2015;37:184–8.
23. Zhou QM, Zhong J, Jiao W, et al. The role of autonomic nervous system in the pathophysiology of hemifacial spasm. Neurol Res. 2012;34:643–8.
24. Guan H-X, Zhu J, Zhong J. Correlation between idiopathic hemifacial spasm and the MRI characteristics of the vertebral artery. J Clin Neurosci. 2011;18:528–30.
25. Kim KJ, Kim JM, Bae YJ, et al. The association between vertebrobasilar dolichoectasia and hemifacial spasm. Parkinsonism Relat Disord. 2016;32:54–9.
26. Pico F, Labreuche J, Amarenco P. Pathophysiology, presentation, prognosis, and management of intracranial arterial dolichoectasia. Lancet Neurol. 2015;14:833–45.
27. Pico F, Labreuche J, Touboul P-J, Amarenco P. Intracranial arterial dolichoectasia and its relation with atherosclerosis and stroke subtype. Neurology. 2003;61:1736–42.
28. Defazio G, Berardelli A, Abbruzzese G, et al. Primary hemifacial spasm and arterial hypertension: a multicenter case–control study. Neurology. 2000;54:1198–200.
29. Tan E, Chan L, Lum S, et al. Is hypertension associated with hemifacial spasm? Neurology. 2003;60:343–4.
30. Nakamura T, Osawa M, Uchiyama S, Iwata M. Arterial hypertension in patients with left primary hemifacial spasm is associated with neurovascular compression of the left rostral ventrolateral medulla. Eur Neurol. 2007;57:150–5.

31. Sandell T, Holmen J, Eide PK. Hypertension in patients with cranial nerve vascular compression syndromes and comparison with a population-based cohort. J Neurosurg. 2013;119:1302–8.

32. Rudzińska M, Wójcik-Pędziwiatr M, Malec M, Grabska N, Hartel M, Szczudlik A. Small volume of the posterior cranial fossa and arterial hypertension are risk factors of hemifacial spasm. Neurol Neurochir Pol. 2014;48:383–6.

33. Leong J-L, Li H-H, Chan L-L, Tan E-K. Revisiting the link between hypertension and hemifacial spasm. Sci Rep. 2016;6:21082.

34. Abbruzzese G, Berardelli A, Defazio G. Hemifacial spasm. In: Handbook of clinical neurology. Amsterdam: Elsevier; 2011. p. 675–80.

35. Kurokawa Y, Maeda Y, Toyooka T, Inaba K. Microvascular decompression for hemifacial spasm caused by the vertebral artery: a simple and effective transposition method using surgical glue. Surg Neurol. 2004;61:398–403.

36. Ohta M, Kobayashi M, Terano N, Wakiya K, Suzuki K, Fujimaki T. Does arteriosclerosis contribute to hemifacial spasm? Acta Neurochir. 2016;158:181–7.

37. Higashi S, Yamashita J, Yamamoto Y, Izumi K. Hemifacial spasm associated with a cerebellopontine angle arachnoid cyst in a young adult. Surg Neurol. 1992;37:289–92.

38. Kobata H, Kondo A, Kinuta Y, Iwasaki K, Nishioka T, Hasegawa K. Hemifacial spasm in childhood and adolescence. Neurosurgery. 1995;36:710–4.

39. Torres-Corzo J, Vinas-Rios JM, Viana Rojas JA, et al. Endoscopic transventricular exploration with biopsy of the basal cisterns and the role of endoscopic third ventriculostomy in patients suffering with basal cistern meningitis and consecutive hydrocephalus. Neurol Res. 2016;38:593–9.

40. Dolan RA. Spinal adhesive arachnoiditis. Surg Neurol. 1993;39:479–84.

41. Dragunow M, Feng S, Rustenhoven J, Curtis M, Faull R. Studying human brain inflammation in leptomeningeal and choroid plexus explant cultures. Neurochem Res. 2016;41:579–88.

42. Chen M, Yang M, Zhou WP, Li ST. Preliminary study on the relationship between inflammation and Hemifacial spasm. World Neurosurg. 2019;125:e214–20.

43. Hamasaki T, Morioka M, Fujiwara K, et al. Is hemifacial spasm affected by changes in the heart rate? A study using heart rate variability analysis. Clin Neurophysiol. 2018;129:2205–14.

44. Han H, Chen G, Zuo H. Microsurgical treatment for 55 patients with hemifacial spasm due to cerebellopontine angle tumors. Neurosurg Rev. 2010;33:335–40.

45. Sprik C, Wirtschafter JD. Hemifacial spasm due to intracranial tumor: an international survey of botulinum toxin investigators. Ophthalmology. 1988;95:1042–5.

46. Cancelli I, Cecotti L, Valentinis L, Bergonzi P, Gigli G. Hemifacial spasm due to a tentorial paramedian meningioma: a case report. Neurol Sci. 2005;26:46–9.

47. Kobata H, Kondo A, Iwasaki K. Cerebellopontine angle epidermoids presenting with cranial nerve hyperactive dysfunction: pathogenesis and long-term surgical results in 30 patients. Neurosurgery. 2002;50:276–86.

48. Nagata S, Matsushima T, Fujii K, Fukui M, Kuromatsu C. Hemifacial spasm due to tumor, aneurysm, or arteriovenous malformation. Surg Neurol. 1992;38:204–9.

49. Sandberg DI, Souweidane MM. Hemifacial spasm caused by a pilocytic astrocytoma of the fourth ventricle. Pediatr Neurol. 1999;21:754–6.

50. Han MS, Jung S, Kim IY, Moon KS, Jung TY, Jang WY. Hemifacial spasm caused by small epidermoid tumor, misinterpreted as delayed secondary Hemifacial spasm caused by vestibular schwannoma treated with gamma knife surgery. World Neurosurg. 2019;130:410–4.

51. Zhang X, Wang X-H, Zhao H, et al. Surgical treatment of secondary hemifacial spasm: long-term follow-up. World Neurosurg. 2019;125:e10–5.

52. Choi S-K, Rhee B-A, Lim YJ. Hemifacial spasm caused by epidermoid tumor at cerebello pontine angle. J Korean Neurosurg Soc. 2009;45:196.

53. Harada A, Takeuchi S, Inenaga C, et al. Hemifacial spasm associated with an ependymal cyst in the cerebellopontine angle: case report. J Neurosurg. 2002;97:482–5.

54. Iwai Y, Yamanaka K, Nakajima H. Hemifacial spasm due to cerebellopontine angle meningiomas. Neurol Med Chir. 2001;41:87–9.

Clinical Symptoms and Differential Diagnosis of Hemifacial Spasm

Jong Hyeon Ahn and Jin Whan Cho

Introduction

Hemifacial spasm (HFS) is characterized by brief, repetitive, and involuntary tonic–clonic contraction of unilateral facial expression muscles. The diagnosis of HFS should be made based on the clinical history and typical symptoms of the disease. At the onset of the disease, HFS is usually insidious; the spasms can be brief and localized in the lower eyelid muscle (orbicularis oculi) before progressing to the upper and lower face [1–3]. In advanced cases, the spasms become more tonic and involve whole unilateral facial expression muscles, such as the frontalis, corrugator, procerus, zygomaticus major, zygomaticus minor, nasalis, levator labii superioris, levator labii superioris alaeque nasi, orbicularis oris, mentalis, depressor labii inferioris, levator anguli

oris, risorius, and platysma. HFS is not a life-threatening disease but may induce social embarrassment and withdrawal. Finally, it can decrease the patient's quality of life [3–5]. Sometimes, establishing a diagnosis of HFS is challenging, especially when the symptoms are subtle or before the development of definite symptoms of HFS. Misdiagnosis of HFS as another hyperkinetic facial movement condition can occur and HFS should be differentiated from other hyperkinetic facial movement disorders to ensure the delivery of precise treatment. In these cases, following up on clinical symptoms, performing closed investigations of specific signs of HFS, and taking home videos can help with the precise diagnosis of HFS. In this chapter, we reviewed the clinical features and the differential diagnoses of HFS to ensure a precise diagnosis.

J. H. Ahn
Department of Neurology, Samsung Medical Center, Seoul, Republic of Korea

Neuroscience Center, Samsung Medical Center, Seoul, Republic of Korea
e-mail: jonghyeon.ahn@samsung.com

J. W. Cho (✉)
Department of Neurology, Samsung Medical Center, Seoul, Republic of Korea

Neuroscience Center, Samsung Medical Center, Seoul, Republic of Korea

Department of Neurology, Sungkyunkwan University School of Medicine, Seoul, Republic of Korea
e-mail: jinwhan.cho@samsung.com

Clinical Symptoms of Hemifacial Spasm

The facial nerve consists of the temporal, zygomatic, buccal, marginal mandibular, and cervical branches. HFS can involve any of these five branches, although the lateral and lower part of orbicularis oculi muscle, which is innervated by the zygomatic branch [6], is most frequently affected (75–90%) in the beginning. Therefore, visible symptoms of HFS often initiate with twitching of the lower eyelid. In contrast, only 5–11.7% of patients first show the condition in

© Springer Nature Singapore Pte Ltd. 2020
K. Park, J. S. Park (eds.), *Hemifacial Spasm*, https://doi.org/10.1007/978-981-15-5417-9_5

the lower face, while just a few patients reported contractions involving the upper and lower face simultaneously [1, 3–5]. Generally, symptoms beginning in the periocular area spread to the upper and lower facial areas.

With the progression of the disease, the temporal branch, which innervates to the upper part of orbicularis oculi, frontalis, procerus, and corrugator muscles, becomes involved. The contraction of the orbicularis oculi and frontalis muscles lead to repetitive eye-closing movements and elevation of the eyebrows, while the involvement of the procerus and corrugator muscles facilitate spasm in the middle of the forehead.

Also with the progression of HFS, the midface and lower facial muscles also can be affected and the contraction of theses muscles triggers various facial symptoms. Muscles placed at midface are innervated by the zygomatic and buccal branches. The zygomaticus major and minor muscles innervated by the zygomatic and buccal branches support upward and lateral movements of the mouth angle. Separately, the levator labii superioris aleque nasi, levator labii superioris, and levator anguli oris muscles are also innervated by the zygomatic and buccal branches and trigger elevation of the upper lip and spasm in the nasal area. Finally, the risorius muscle, innervated by the buccal branch of the facial nerve, pulls the corner of the mouth laterally.

Lower facial symptoms can also present when HFS involves the marginal mandibular or buccal branches of the facial nerve. Both of these branches innervate to the orbicularis oris and contraction of the orbicularis oris leads to perioral spasm and twitching. Moreover, they also innervate to the depressor labii inferioris, depressor anguli oris, and mentalis muscles, triggering involuntary chin movement. In some patients, the platysma muscle, a neck muscle innervated by the cervical branch, is also involved and presents spasm [7] (Fig. 1).

Usually, HFS involves only one side of the face, with left-sided HFS being slightly more common. However, bilateral HFS is rarely reported (less than 3% of cases) and other hyperkinetic movement disorders such as blepharospasm should be differentiated in this context.

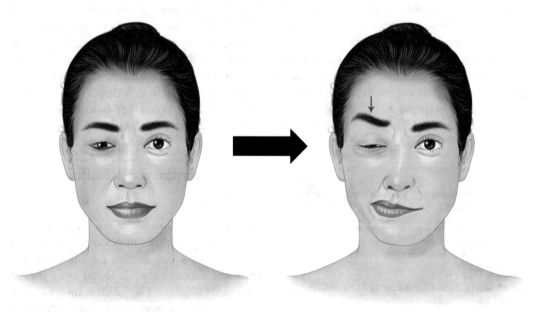

Fig. 1 Symptoms of hemifacial spasm. Visible symptoms of hemifacial spasm often initiate with twitching or spasm of the periocular area without the upper and lower face involvement (Left). With the progression of the disease, the upper, midface, and lower face are also affected (Right). The paradoxical involuntary upward replacement of the ipsilateral eyebrow during involuntary eye closing, known as "the other Babinski sign" or "Babinski sign-2," is observed in HFS patients with the presentation correlated with the severity of symptoms. As such, it is less common in the early stage (Left) of the disease and typically appears as the disease progresses (Right)

About 2% of primary HFS patients report a positive family history of HFS, but genetic factors associated with HFS have not yet been elucidated. The age of onset of familial HFS is variable, and, occasionally, this condition can occur early in life. Vascular decompression is also effective in familial HFS, with study results suggesting that vascular compression is involved in generating HFS even in familial cases [8].

The symptoms of HFS are fluctuated by various factors. Anxiety and stress are the most common aggravating or trigger factors in HFS. Patients have also reported a worsening of symptoms with sleep deprivation, fatigue, reading, light exposure, wind, chewing, reading, or talking. Most patients reported their symptoms were alleviated by relaxation, alcohol intake, touching the area, or facial massage. HFS remains persistent during sleep. One study using polysomnography reported that all the included patients ($n = 12$) showed persistent HFS during sleep, albeit of a decreased severity [9].

Patients with HFS may complain of various symptoms other than hyperkinetic movement including social embarrassment, depression, interference with vision, dysarthria, and sialorrhea [3]. Along these lines, the presence of a "clicking ear sound" that was simultaneous and synchronous with the involuntary facial muscle contractions was reported by 22.7% of patients with HFS. It is believed that co-contraction of the stapedius muscle leads to rarefaction and compression in the middle ear [10].

In rare cases, trigeminal neuralgia is preceded by or accompanies HFS and is called tic convulsif, which was coined by Cushing in 1920. This rare condition is also induced by vascular compression of the nerve root, with some cases induced by cerebellopontine angle tumor [11].

Provocation Maneuver and Home Video

The diagnosis of HFS is often hampered and contingent by whether patients show the subtle involuntary movement in question at the clinic. Furthermore, the severity of involuntary contraction is variable depending on the situation or the emotional state of the patient. One way to overcome this drawback is by using provocation maneuvers and taking home videos.

Provocation maneuvers such as prolonged voluntary contraction, repetitive forceful contraction of eyes, and repetitive smiling can help observers to visualize involuntary muscle contractions. Forceful, prolonged contraction induces involuntary contraction of the affected muscles and makes it easy to detect the typical features of HFS. Physicians can observe synkinesis of the lower facial muscles during eye closing. Having patients chew or whistle is a simple and practical maneuver inducing synkinesis in the upper facial muscles.

Given the existence of paroxysmal and fluctuating features of HFS, taking home videos is also a helpful tool to facilitate precise initial diagnosis as well as monitor disease progression and the treatment response. It is important to take videos so that physicians can check changes in the treatment response, alterations in symptoms according to various situations, and the progression of the disease over time.

Specific Signs and Red-Flag Signs of HFS

The Other Babinski Sign

The paradoxical involuntary upward replacement of the ipsilateral eyebrow during involuntary eye closing, known as "the other Babinski sign" or "Babinski sign-2," is a specific sign of HFS, with 85–90% of primary HFS patients presenting with such and with the presentation correlated with the severity of symptoms. As such, it is less common in the early stage of the disease and typically appears as the disease progresses [12–15]. Surgical treatment of HFS should be carefully decided if the patient does not show the other Babinski sign. The other Babinski sign may be observed not only during involuntary periocular movement but also in the resting state; therefore, the eyebrow of the affected side is relatively raised more than the unaffected side in the resting state.

Synchronous Contraction of the Upper and Lower Facial Muscles

In HFS, a highly synchronous involuntary contraction in all involved muscles can be observed. This synchronous contraction is a feature of HFS that distinguishes it from other hyperkinetic movements. Patients with postparalysis facial synkinesis (PPFS) can show synchronous contraction, but what is different from the synkinesis of HFS is that the synkinesis in patients with PPFS is seen during voluntary movement of facial muscles, in contrast with that in HFS induced by involuntary contraction [16, 17].

Red-Flag Signs

Despite HFS having characteristic symptoms, sometimes it is difficult to distinguish the condition from other hyperkinetic facial movement disorders, particularly in the early stage of the disease. As such, there are several "red-flag signs" of involuntary facial movements that suggest another hyperkinetic condition other than HFS is at play. The physician should consider the possibility of other hyperactive facial movement disorders if the patient is of a young age at the time of onset, has bilateral involvement, had a past history of peripheral facial palsy, the condition initiated with the lower facial muscles, there are accompanying hyperkinetic movements in body parts other than the face, Charcot's sign is visible, and there is involvement of the masseter muscle or involuntary movement of the tongue. However, the presence of identified red-flag signs does not always indicate a diagnosis than HFS because HFS can also appear paired with unusual, atypical symptoms.

Differential Diagnosis of Hemifacial Spasm

Blepharospasm

Blepharospasm is an involuntary, spasmodic, synchronized contraction of the bilateral orbicu-laris oculi muscles. Occasionally, involuntary contraction spreads to the lower face as part of a condition known as Meige syndrome. Few patients with blepharospasm show asymmetric symptoms and a small number of patients with HFS present with bilateral symptoms. Therefore, sometimes it can be difficult to distinguish the two conditions. In general, patients with blepharospasm show bilateral symptoms from the beginning of symptom onset, while those with HFS experience a delay between the involvement of one side and the other side. In addition, the contraction of bilateral orbicularis oculi muscles is asynchronous in HFS and is distinct from blepharospasm, which is characterized by synchronized contractions of the bilateral periorbital muscles. The other clinical sign of note is lowering of the eyebrow upon closure of the eyes (Charcot's sign), suggesting blepharospasm, whereas there is a raising of eyebrows with contraction in HFS (the other Babinski sign) [18].

Postparalysis Facial Synkinesis

Postparalysis facial synkinesis (PPFS) aberrant regeneration following peripheral facial palsy triggers synkinetic facial movement and can occur in 9.1% of patients with peripheral facial palsy within 6 months of disease onset [19]. This involuntary co-contraction of the upper and lower facial-expression muscles is often triggered by voluntary, automatic, or emotional movement and can mimic HFS. HFS and PPFS have similar appearances as they involve the same muscle territories of the unilateral face. The most apparent difference between HFS and PPFS is that the synkinesis in PPFS occurs primarily following voluntary contraction of the facial muscles, whereas the synkinesis in HFS is accompanied by involuntary facial movements. Careful history-taking of facial palsy and examination, especially of the frontalis and orbicularis oculi muscles (the other Babinski sign), is important in differential diagnosis. An electrophysiological study can help to differentiate PPFS from HFS.

Facial Motor Tics

Facial tics also can mimic HFS. Facial tics are brief, stereotyped movements that are complex, multifocal, and nonrhythmic, while motor tics can involve limbs and combine with vocalization (vocal tic) and other features of Tourette's syndrome. The mean onset age of tic disorders is 6 years and 93% of patients are symptomatic by the age of 10 years. However, adult-onset tic disorders have also been reported; therefore, they should be carefully investigated in young patients with hyperkinetic facial movement [20]. Abnormal facial movement of tics can affect unilateral or bilateral facial muscles and an urge to perform the movement usually precedes the actual abnormal movement.

Facial Myokymia

Myokymia is a small, undulating, rippling movement of muscles that appears as tiny snakes wriggling just beneath the skin. The majority of cases of facial myokymia involve common benign symptoms in the general population, yet some cases show symptoms of neurologic disease such as multiple sclerosis, Guillain–Barré syndrome, pontine glioma, or episodic ataxia 1 [21]. Needle electromyography is a useful test by which to differentiate myokymia from HFS, where myokymia appears as brief bursts of doublets, triplets, or multiplets of repetitively firing motor-unit potentials without lateral spread responses.

Oromandibular Dystonia

Oromandibular dystonia (OMD) is characterized as sustained, repetitive, stereotypic involuntary contractions of the lower face, tongue, jaw, mouth, and pharynx and can interfere with speaking, chewing, or swallowing. The majority of patients with HFS present with the condition initiating in the upper face, while, in those with OMD, the lower facial muscles are mainly affected and muscles not innervated by facial nerves (e.g., masseter muscle, tongue) may also become involved. In patients with a history of exposure to neuroleptic drugs or dopaminergic antagonists, tardive dyskinesia should be considered. Abnormal movements in tardive dyskinesia are usually irregular, bilateral, and asynchronous, unlike in HFS.

Facial Myoclonus

Facial myoclonus associated with focal motor seizure or cortical myoclonus can mimic HFS. Facial myoclonus is induced by various pathologies including Möebius syndrome, Rasmussen encephalitis, structural lesion of the frontotemporal lobe, olivopontocerebellar atrophy, or vertebrobasilar dolichoectasia. Facial myoclonus can share clinical features with HFS, having brief contractions of the perioral area. Therefore, if patients have uncommon features of HFS, electroencephalography and brain imaging study should be considered for differentiating facial myoclonus.

Hemimasticatory Spasm

Hemimasticatory spasm (HMS) is a rare movement disorder characterized by unilateral, involuntary, paroxysmal contractions of the jaw-closing muscles. Spasms in HMS are painful and triggered by activities like chewing, talking, clenching of teeth, or voluntary tapping of the involved muscles. HMS involves the masseter or temporalis muscles or both, innervated by the trigeminal nerve, unlike in HFS. Electromyography recordings demonstrate spontaneous irregular bursts of high-frequency motor-unit potentials and, in some patients, a loss of inhibition in the form of an absent silent period during spasms in HMS [22].

Myorhythmia

Myorhythmia is an involuntary, hyperkinetic, rhythmic, slow (1–4 Hz) movement, usually affecting the facial muscles but which may also affect the limb muscles. Isolated facial myorhythmia can occur in various contexts, including in patients with thalamic infarcts or Whipple's disease, patients receiving treatment with interferon alpha-2a, or in correlation with phenytoin intoxication [23]. Myorhythmia induced by Whipple's disease usually appears as oculomasticatory myorhythmia. Isolated facial myorhythmia is more rhythmic, slower, and continuous than HFS.

Functional Facial Spasm

Functional facial spasm is one of the most common presentations of functional movement disorders. Functional facial spasm is usually nonpatterned, varies in intensity, and is distractible. Patients with functional facial spasm demonstrate different clinical features from those of HFS, such as a younger age of onset (30s), more common bilateral involvement, no other Babinski sign, isolated lower facial involvement, downward deviation of the mouth's angle, and the disappearance of symptoms during sleep [24].

References

1. Conte A, Falla M, Diana MC, Bologna M, Suppa A, Fabbrini A, et al. Spread of muscle spasms in Hemifacial spasm. Mov Disord Clin Pract. 2015;2(1):53–5.
2. Digre K, Corbett JJ. Hemifacial spasm: differential diagnosis, mechanism, and treatment. Adv Neurol. 1988;49:151–76.
3. Wang A, Jankovic J. Hemifacial spasm: clinical findings and treatment. Muscle Nerve. 1998;21(12):1740–7.
4. Au WL, Tan LC, Tan AK. Hemifacial spasm in Singapore: clinical characteristics and patients' perceptions. Ann Acad Med Singap. 2004;33(3):324–8.
5. Wang L, Hu X, Dong H, Wang W, Huang Y, Jin L, et al. Clinical features and treatment status of hemifacial spasm in China. Chin Med J. 2014;127(5):845–9.
6. Tong J, Patel BC. Anatomy, head and neck, eye orbicularis oculi muscle. Treasure Island, FL: StatPearls Publishing LLC; 2019.
7. Marur T, Tuna Y, Demirci S. Facial anatomy. Clin Dermatol. 2014;32(1):14–23.
8. Miwa H, Mizuno Y, Kondo T. Familial hemifacial spasm: report of cases and review of literature. J Neurol Sci. 2002;193(2):97–102.
9. Incirli SU, Yilmaz R, Akbostanci MC. Hemifacial spasm in sleep—a polysomnographic study. J Clin Neurosci. 2019;64:160–2.
10. Batla A, Goyal C, Shukla G, Goyal V, Srivastava A, Behari M. Hemifacial spasm: clinical characteristics of 321 Indian patients. J Neurol. 2012;259(8):1561–5.
11. Revuelta-Gutiérrez R, Velasco-Torres HS, Hidalgo LOV, Martínez-Anda JJ. Painful tic convulsif: case series and literature review. Cir Cir (English Edition). 2016;84(6):493–8.
12. Devoize JL. "The other" Babinski's sign: paradoxical raising of the eyebrow in hemifacial spasm. J Neurol Neurosurg Psychiatry. 2001;70(4):516.
13. Pawlowski M, Gess B, Evers S. The Babinski-2 sign in hemifacial spasm. Mov Disord. 2013;28(9):1298–300.
14. Stamey W, Jankovic J. The other babinski sign in hemifacial spasm. Neurology. 2007;69(4):402–4.
15. Varanda S, Rocha S, Rodrigues M, Machado A, Carneiro G. Role of the "other Babinski sign" in hyperkinetic facial disorders. J Neurol Sci. 2017;378:36–7.
16. Eekhof JL, Aramideh M, Speelman JD, Devriese PP, Ongerboer De Visser BW. Blink reflexes and lateral spreading in patients with synkinesia after Bell's palsy and in hemifacial spasm. Eur Neurol. 2000;43(3):141–6.
17. Valls-Sole J. Electrodiagnostic studies of the facial nerve in peripheral facial palsy and hemifacial spasm. Muscle Nerve. 2007;36(1):14–20.
18. Tan EK, Chan LL, Koh KK. Coexistent blepharospasm and hemifacial spasm: overlapping pathophysiologic mechanism? J Neurol Neurosurg Psychiatry. 2004;75(3):494–6.
19. Yamamoto E, Nishimura H, Hirono Y. Occurrence of sequelae in Bell's palsy. Acta Oto-Laryngol. 1987;104(Suppl 446):93–6.
20. Knight T, Steeves T, Day L, Lowerison M, Jette N, Pringsheim T. Prevalence of tic disorders: a systematic review and meta-analysis. Pediatr Neurol. 2012;47(2):77–90.
21. Gutmann L, Gutmann L. Myokymia and neuromyotonia 2004. J Neurol. 2004;251(2):138–42.
22. Radhakrishnan DM, Goyal V, Shukla G, Singh MB, Ramam M. Hemi masticatory spasm: series of 7 cases and review of literature. Mov Disord Clin Pract. 2019;6(4):316–9.
23. Baizabal-Carvallo JF, Cardoso F, Jankovic J. Myorhythmia: phenomenology, etiology, and treatment. Mov Disord. 2015;30(2):171–9.
24. Baizabal-Carvallo JF, Jankovic J. Distinguishing features of psychogenic (functional) versus organic hemifacial spasm. J Neurol. 2017;264(2):359–63.

The Electrophysiological Study for Hemifacial Spasm

Byung-Euk Joo

Introduction

Hemifacial spasm (HFS) is an involuntary and irregular spasm of the facial muscles innervated by the facial nerve that typically progresses in severity an extent over time. The etiology of HFS has been attributed to vascular compression of the facial nerve at the root exit zone (REZ) [1]. This led to the development of surgical treatment called microvascular decompression (MVD) for HFS, and MVD has become established as the most effective treatment for HFS now [2]. Although much progress has been made on the cause and treatment for HFS, there has been a debate about the pathogenesis of HFS despite numerous electrophysiological studies on HFS until now. There are two hypotheses for the underlying mechanism (Fig. 1): (1) as the peripheral nerve mechanism, the compression of the facial nerve by a blood vessel causes an injury of the myelin sheath, facilitating ectopic excitation and ephaptic transmission between individual nerve fiber [3–7]; and (2) as the central mechanism, the hyperexcitability of the facial motor nucleus (FMN), triggered by antidromically propagated discharges, induces a spasm [8–14].

Many researchers have studied to elucidate the pathogenesis of HFS in clinical settings using electrophysiological studies involving the lateral spread response (LSR), blink reflex test, facial F-wave, and transcranial electrical stimulation (TES). Due to the efforts of many researchers, much progress has been made in elucidating the pathogenesis of HFS over the last 40 years. In this article, the previous main researches using each electrophysiological study for HFS will be discussed together.

Electrophysiological Study

Lateral Spread Response (LSR)

Lateral spread response (LSR) is an abnormal electromyographic findings in patients with HFS [3]. The LSR is the response of the muscles innervated by the other facial nerve branches by stimulating of one branch of the facial nerve. So, LSR is the most representative electrophysiological findings of HFS and thus has diagnostic value for HFS. Also, the disappearance of LSR usually occur immediately after identifying the offending vessels and performing sufficient decompression during MVD for HFS [15]. Therefore, LSR has been used not only as the diagnostic tool for HFS but also as an indicator of successful MVD. Until now, many electrophysiological studies using LSR have been conducted to elucidate the pathogenesis of HFS and to ensure sufficient MVD for HFS with the development of intraoperative monitoring. Surprisingly, however, there is still

B.-E. Joo (✉)
Department of Neurology, Soonchunhyang University Seoul Hospital, Soonchunhyang University College of Medicine, Seoul, Republic of Korea

© Springer Nature Singapore Pte Ltd. 2020
K. Park, J. S. Park (eds.), *Hemifacial Spasm*, https://doi.org/10.1007/978-981-15-5417-9_6

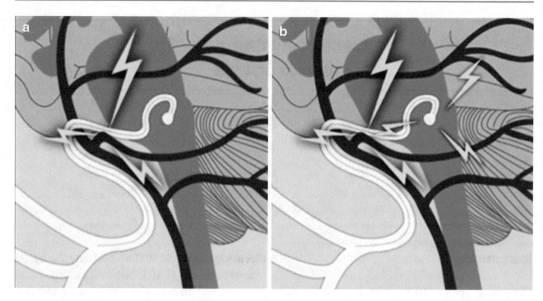

Fig. 1 The pathomegenesis for hemifacial spasm. (**a**) Peripheral ectopic excitation with ephaptic impulse transmission, (**b**) Hyperexcitability of the facial motor nucleus

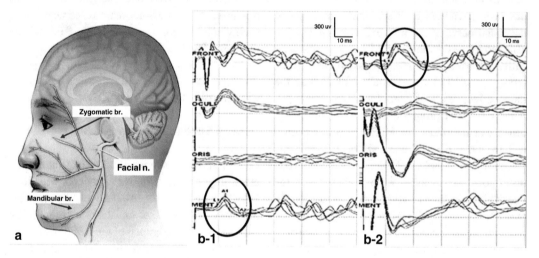

Fig. 2 The lateral spread response (LSR). (**a**) Facial nerve branch. (**b**). (1) LSR from mentalis muscle with stimulation of zygomatic branch of facial nerve. (2) LSR from frontalis muscle with stimulation of mandibular branch of facial nerve

much debate about the origin of the LSR with pathogenesis of HFS.

Methodology

Two types of LSR can be recorded by stimulating the upper and the lower branches of the facial nerve on the symptomatic side of HFS. By stimulating the zygomatic branches of the facial nerve on the symptomatic side, LSR can be recorded from the mentalis muscle. In addition, LSR can be also obtained from the orbicularis oculi muscle or the frontalis muscle by stimulating mandibular branch (Fig. 2). Constant current stimuli are applied for 0.1–0.2 ms with a bar electrode.

Resting motor threshold (rMT) is initially defined as the minimum intensity that could induce the amplitude of LSR of >10 μV in at least five successive trials. After defining the rMT, the LSR can be obtained by stimulating rectangular shock with a suprathreshold strength.

Results and Interpretation

Nielson reported that there was LSR on the symptomatic side of all 62 HFS patients, although there was no LSR on the asymptomatic side of those patients as in the healthy controls [3]. In this study, LSR on mentalis muscle was observed in all 62 patients, and that on the orbicularis oculi muscle in 60 of the 62 patients. The latency of LSR was an average of 9.3 ± 0.13 ms and 9.0 ± 0.13 ms for the orbicularis oculi and the mentalis muscle. Also, the amplitude of LSR was always much smaller about 20–30% than that of the maximal orthodromic response after stimulating the facial nerve. After the previous study, Nielson and Jannetta evaluated LSR for 59 patients with HFS before and after MVD [6]. LSR disappeared in 23% and changed from bidirectional to unidirectional in 45% patients within 1 week after MVD surgery. Within 2–8 months after MVD, LSR was observed in 27%, and unidirectional in 17%. Through these findings, Nielsen insisted that the peripheral mechanism

including ephaptic transmission is the main pathogenesis for HFS though the delayed disappearance of LSR after MVD could not exclude the hyperexcitability of FMN as pathogenesis for HFS [4]. To define the origin of LSR in HFS, Møller and Jannetta analyzed the latency of LSR from orbicularis oculi muscle under anesthesia during MVD [10]. After obtaining the latency of the LSR (11.03 ± 0.66 ms) by stimulating the mandibular branch, they simultaneously measured the latency of the response from the facial nerve near the REZ (3.87 ± 0.36 ms). They also measured the latency of the response from the orbicularis oculi muscle by stimulation the facial nerve near the REZ (4.65 ± 0.25 ms). They showed that the latency of the LSR from the orbicularis oculi muscle by stimulating the mandibular branch was larger than the sum of the conduction time from the points of stimulation of the mandibular branch to the REZ of facial nerve and from REZ of the facial nerve to the orbicularis oculi muscle (8.52 ± 0.38 ms) (Fig. 3). Through this difference of the latency, they insisted that the LSR from orbicularis oculi muscle was not a direct result of ephaptic conduction at the site of the lesion, and hyperexcitability of FMN was involved in the synthesis of the LSR. To identify the origin of the LSR, there was the study using double stimulation instead of a

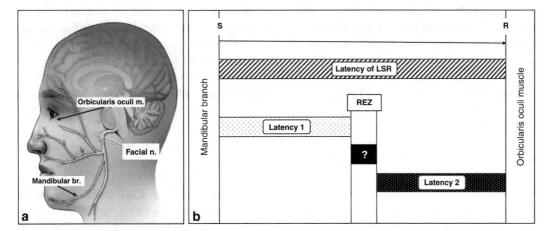

Fig. 3 The results of Møller and Jannetta's study about the origin of lateral spread response (LSR). (**a**) Schematic diagram of LSR method used in this study. The LSR was recorded from orbicularis oculi muscle by stimulating mandibular branch of facial nerve. (**b**) The difference between the actual measured value ant the calculated value

of the latency of the LSR. The latency of the LSR from the orbicularis oculi muscle by stimulating the mandibular branch was larger than the sum of the conduction time from the points of stimulation of the mandibular branch to the root exit zone (REZ) of facial nerve and from REZ of the facial nerve to the orbicularis oculi muscle single stimulus [16]. Yamashita et al. conducted double stimulation at interstimulus intervals (ISIs) ranging from 0.5 to 0.7 ms to evoke the LSR in patients with HFS. By this double stimulation, a total of 15 LSR consisting of two responses (R1 and R2) were obtained. R1 showed a constant latency and amplitude regardless of the ISIs, whereas R2 presented after a fixed refractory period without facilitation or depression in a recovery curve of latency and amplitude. As R2 showed no suppression, they suggested that LSR did not arise from FMN. To elucidate the origin of LSR, there were also studies using the central suppressive effect of anesthetics. Wilkinson et al. defined the changes in amplitude and latency of LSR according to the changes in the concentration of desflurane during MVD in 22 HFS patients [17]. According to their research, the LSR amplitude under 1 MAC desflurane with TIVA was significantly decreased than under TIVA. On the other hand, there was no change on the latency of LSR and on EEG according to the concentration of desflurane.

Blink Reflex Test

The blink reflex is the electrical correlate of the clinically evoked corneal reflex. The blink reflex is a true reflex with a sensory afferent limb, intervening synapses, and a motor efferent. The afferent limb of the blink reflex is mediated by sensory fibers of the supraorbital branch of the ophthalmic division of the trigeminal nerve (cranial nerve V1) and the efferent limb by motor fibers of the facial nerve (cranial nerve VII). Just as the corneal reflex, ipsilateral stimulation of the supraorbital branch of the trigeminal nerve elicits a facial nerve (eye blink) response bilaterally. Stimulation of the ipsilateral supraorbital nerve results in an afferent volley along the trigeminal nerve to both the main sensory nucleus of CN V (mid pons) and the nucleus of the spinal tract of CN V (lower pons and medulla) in the brainstem. Through a series of interneurons in the pons and lateral medulla, the nerve impulse next reaches the ipsilateral and contralateral facial nuclei, from which the efferent signal travels along the

facial nerve bilaterally. The blink reflex has two components, an early R1 and late R2 response. The R1 response is only present on the side of stimulation, while the R2 response typically is present bilaterally. The R1 response is thought to represent the disynaptic reflex pathway between the main sensory nucleus of V in the mid pons and the ipsilateral facial nucleus in the lower pontine tegmentum. The R2 responses are mediated by a multisynaptic pathway between the nucleus of the spinal tract of V in the ipsilateral pons and medulla [18]. As mentioned above, because the blink reflex pathway is well known, and the pathway includes the entire facial nerve including FMN, many studies performed blink reflex study to clarify the pathogenesis of HFS.

Methodology

The blink reflex test is basically conducted in the method suggested by Kimura [19]. The cathode of the stimulating electrode is placed at the supraorbital foramen and the anode was placed immediately above (on the forehead), using a bar stimulator. The recording electrodes are placed over the orbicular muscle of both eyes (the mid-lower eyelid and the temple). To avoid stimulation of the nerve during spasm, stimulation was applied when the muscles were electrically at rest. Constant current stimuli of 0.1–0.2 ms in duration were delivered. After defining the threshold that could cause the minimal constant response, a suprathreshold stimulation is applied. To ensure the reproducibility and accurate response, at least five stimuli are applied to each side and then averaged. The latencies of the ipsilateral R1 and R2 responses were defined as the shortest time to the onset of the response. The maximum amplitude and duration of each of R1 and R2 responses were measured.

Results and Interpretation

Nielsen conducted the blink reflex study in 62 patients with HFS [5]. In this study, the latency and amplitude of the R1 response on the symptomatic side were increased as compared with the asymptomatic side and controls ($p < 0.001$). The latency of R1 response on the symptomatic side was increased by 2.1 ms than that of asymptomatic side, which was interpreted by focal demyelination over the lesion. Also, all patients showed a synkinetic response in the mental muscle on the symptomatic side, and after-activity and late-activity was observed after the reflex response. Based on these findings, ephaptic/ectopic excitation due to compression and demyelination of the facial nerve was proposed as the primary pathogenesis for HFS [4, 5]. However, Esteban et al. presented other results from the previous study using the blink reflex study. They also measured the values of the blink reflex study in the 53 patients with HFS, and then compared with healthy controls [8]. In this study, the latency of R1 response was not different between groups, and the latency of R2 response was shortened on symptomatic side in HFS patients unlike the results of Nielsen's study. Also, the duration of R2 response was greater on the HFS side when compared with those of asymptomatic side and healthy controls. So, they insisted that the hyperexcitability of FMN was the main pathogenesis for HFS. In performing the blink reflex study for HFS patients, Eekhof et al. showed findings different from the previous studies. In this study, the latency and amplitude of R1 and R2 responses from orbicularis oculi muscle present no significant difference between the HFS patients and healthy controls. However, both R1 and R2 response from the orbicularis oris muscle occurred significantly more often on the symptomatic side in HFS patients, and showed higher amplitude significantly compared to healthy controls [20]. Valls-Sole et al. studied blink reflex response in patients with HFS by applying double stimulation as well as single stimulation [21]. By applying single stimulation, the area of R1 and R2 responses was greater on the symptomatic side in patients with HFS as compared with the asymptomatic side and normal controls. Also, with double stimulation, the inhibitory effect of the conditioning stimuli upon the test stimuli R2 response, which was always observed in healthy controls, was significantly less pronounced at short ISIs in HFS. They reported that this enhanced recovery curve of R2 response was attributed to enhanced excitability of FMN in HFS. Møller and Jannetta conducted the blink

reflex study under anesthesia using inhalational anesthetics (isoflurane and nitrous oxide) during MVD [22]. In this study, the R1 response on asymptomatic side was not evoked under anesthesia; however, the R1 response could be observed on the symptomatic side of HFS patients under anesthesia. Also, this R1 response of the symptomatic side was abolished after MVD. Through these findings under anesthesia, they insisted the hyperexcitability of FMN as the main mechanism for HFS.

F-Wave

F-wave is an antidromic pulse that propagates to an alpha motorneuron in the anterior horn cell of the spinal cord and then returns orthodromically down the same axon. So, the F-wave circuitry, both afferent and efferent, is pure motor. There is no synapse, so it is not a true reflex. In the extremities, F-waves have been considered as index of the excitability of anterior horn cell, and have been used as good reflection of lower motoneuron excitability. In the facial muscles, F-waves are also measurable, and those can be used for evaluation of the excitability of the facial motor nucleus. Therefore, there were many studies using facial F-waves to define the pathogenesis of HFS.

Methodology

Though facial F-waves could be obtained from orbicular oculi or the mentalis muscle by stimulating zygomatic branch or mandibular branch of the facial nerve, obtaining them from the mentalis muscle as long as possible by stimulating at the distal marginal mandibular branch is usually recommended to prevent an overlap between M-waves and F-waves. As the amplitude of facial F-wave is relatively small, it is necessary to perform the examination after the spasm has completely disappeared. After obtaining a flat baseline, stimulation was applied at the border or the mandible that was 10 cm from the stylomastoid foramen. Stimulation was performed with a bar electrode and was repeated 10–20 times using a 0.2 ms square wave at the frequency of 1 Hz. The stimulation intensity was set to supramaximum. When a facial F-wave had a distinct peak and amplitude above 30 μV, it was regarded as F-wave. The parameters that can be analyzed using facial F-waves are as follows [9]: F/M amplitude ratio (the percentage of the peak to peak amplitude of the F-waves to the M-waves), total duration (from the initial deflection from the baseline to the final return of the F-wave), F-wave frequency (the percentage of 10–20 stimuli that produced F-waves with a distinct peak and amplitude above 30 μV), minimum latency (from the onset of the stimulus artifact to the first deflection of the F-wave from the baseline), and F chronodispersion (the difference between minimal and maximal latencies of the F-wave in a series of 10–20 waves).

Result and Interpretation

Ishikawa et al. obtained facial F-wave from the mentalis muscle by stimulating distal mandibular branch in 20 patients with HFS before MVD, and 10 HFS patients after MVD and 10 healthy controls [9]. In their study, F-wave duration F/M amplitude and frequency of F-wave on the symptomatic side of HFS patients were significantly increased when compared with asymptomatic side of HFS patients and health controls before MVD. On the other hand, there was no difference in minimum latency and chronodispersion between groups. They also showed that the enhancement of the facial F-wave eventually decreased at the same time as disappearance with LSR after MVD surgery. In another study, they compared facial F-waves from mentalis muscle and LSR from orbicularis oculi muscle by stimulating the marginal mandibular branch to investigate the origin of LSR in 10 HFS patients [23]. In this study, the LSR showed an afterdischarge after a constant response, and the afterdischarge of LSR with the facial F-wave duration tended to increase on symptomatic side of patients. Also, a lineal correlation between the facial F-wave duration and the afterdischarge duration was observed. ($r^2 = 0.961$, $p < 0.0001$). So, they insisted that facial F-waves and the LSR would have the same origin. Hai et al. measured LSR and facial F-waves like the previous study after creating an HFS animal model in 10 rabbits [24]. This study also presented that linear correlations between the amplitude ratio of LSR/M-waves and F-waves/M-waves and between the duration of LSR and F-waves. They reported that the peripheral mechanism including ephaptic transmission could not alone explain the increase of facial F-wave duration in HFS as the transmission time of the ephapses between nerve fiber is below 100–200 μs [25]. In another study, Ishikawa et al. conducted facial F-wave study with blink reflexes and LSR before and after MVD in 20 patients with HFS [26]. In this study, the facial F-wave and blink reflex on symptomatic side showed increased values than those of the asymptomatic side before MVD, and facial F-waves and LSR were still recorded in some patients within 1 month after the HFS had disappeared completely. Through these findings, they suggested that hyperexcitability of FMN would be the main cause of HFS.

Transcranial Facial Motor Evoked Potential (TcFMEP)

Transcranial facial motor evoked potentials (TcMEPs) are one of the most powerful tools in the intraoperative monitoring to monitor motor function, particularly for spine surgery. TcMEP are obtained by stimulating the motor pathways rostral to the site surgery. Activation of the motor pathways can be measured by recording waveforms as the impulse descends along the corticobulbar tract and corticospinal tract.

Like the blink reflex, the TcFMEP study can be a tool for examining the complete efferent pathway of the facial nerve.

Methodology

TcFMEPs from the facial muscles are elicited by using transcranial anodal electrical stimulation. Electrodes are placed on the scalp over C3 and C4 according to the international 10–20 system bilaterally. Stimulation electrodes are typically subdermal needle electrodes or cork-screw electrodes. Stimuli are applied as single shocks with a pulse width of 150–200 μs and a voltage range of 90–305 V. For the recording, subdermal needles are placed in pairs in orbicularis oculi, orbicularis oris, and mentalis muscles. Though most types of stimulation in clinical neurophysiology are cathodal, anodal stimulation in TcMEP is more effective, because the cell body and axon hillock, the sites of stimulation for TES, are more sensitive to anodal stimulation. In using TcFMEP, the TcFMEP must be excluded from analysis if the onset latency of TcFMEP is shorter than 10 ms, because they can be thought to be contaminated by direct current spread to the extracranial facial nerve [27]. By using the TcFMEP study, the threshold for FMEP as well as the latency and amplitude of FMEP is usually analyzed. The threshold of FMEP is defined as the minimum voltage required to elicit an FMEP of ≥30 μV in at least 50% of a minimum of several consecutive stimulation trials.

Result and Interpretation

Though not commonly performed, some have proposed that the myogenic facial motor evoked potentials elicited via transcranial electrical stimulation can be used to monitor the functional integrity of the corticobulbar tract, facial motor nucleus, and facial nerve during MVD surgery. Kaufmann et al. measured FMEP with LSR during MVD in 10 HFS patients and conducted FMEP study during MVD for 17 patients with trigeminal neuralgia (TN) [13]. They analyzed latency, amplitude, and duration of the FMEP before and after MVD. They suggested that the amplitude and durations of FMEP significantly decreased on the symptomatic side of HFS patients after MVD, whereas these changes were not observed from the asymptomatic side of HFS patients or TN patients. Also, they presented a dramatic reduction in amplitude and duration of FMEP with disappearance of LSR when decompression of the offending vessel. Otherwise the latency of FMEP revealed no significant change before and after MVD. In other study, they ana-

lyzed retrospectively the threshold of FMEP and the incidence of FMEP to the single pule TcMEP during surgery in 65 patients with HFS and 29 patients with skull base tumors [28]. In the study, the threshold of FMEP is significantly lower in HFS compared to skull base tumor patients. Also, FMEP to the single pulse stimulation were observed in 87% of HFS patients, whereas only 10% in patients with skull base tumor showed FMEP response to single pulse stimulation. Recently, Kaufmann et al. prospectively compared FMEP under total intravenous anesthesia (TIVA) with or without desflurane during MVD for HFS patients to define the hyperexcitability of FMN in HFS [14]. As inhalational anesthetics such as desflurane are well known for their suppressive effects on the level of the alpha motor neuron, they expected that there would be a difference in effect of desflurane on FMEP from symptomatic and asymptomatic side of HFS patients. By this study, they suggested that the suppressive effects of desflurane were less on the symptomatic side than on the asymptomatic side (59% vs. 79%, $p = 0.03$), although desflurane (1 minimum alveolar concentration) suppressed FMEPs on both sides. While showing that M-waves recorded from the mentalis muscle remained unchanged together, they also demonstrated that desflurane had no effect on the peripheral facial nerve or neuromuscular junction. Through such a series of research using TcFMEP, they suggested that the hyperexcitability of FMN might be the main pathogenesis for HFS.

Conclusion

There has been a long debate on the main pathogenesis of HFS: ephaptic transmission/ectopic excitation between individual nerve fiber vs. the Hyperexcitability of the FMN. To elucidate the pathogenesis of HFS, many electrophysiological studies have been conducted, including LSR, blink reflex test, facial F-wave, and TcMEP, so far (Table 1). Much progress about the pathogenesis for HFS has been made due to accumulation of knowledge and development of research meth-

Table 1 The summary of the main electrophysiological studies for hemifacial spasm

Researchers	Year	Subjects	The main findings	The pathogenesis[a]
Lateral spread response				
Nielson	1984	62 HFS, 14 TN	LSR was recorded on symptomatic side in all patients with HFS. Also, after-activity and late-activity were recorded on symptomatic side in HFS patients	Peripheral
Nielsen and Jannetta	1984	59 HFS	After MVD, LSR disappeared in 23% and changed from bidirectional to unidirectional in 45% patients in 1 week	Peripheral
Møller and Jannetta	1984	7 HFS	The latency of the LSR from the orbicularis oculi muscle by stimulating the mandibular branch was larger than the sum of the conduction time from the points of stimulation of the mandibular branch to the REZ of facial nerve and from REZ of the facial nerve to the orbicularis oculi muscle	Central
Yamashita et al.	2002	12 HFS	By using double stimulation, the second LSR presented after a fixed refractory period without facilitation or depression in a recovery curve of latency and amplitude regardless of the inter-stimulus intervals	Peripheral
Wilkinson et al.	2014	22 HFS	During MVD for HFS, desflurane with TIVA significantly decreased only LSR amplitude, not LSR latency than under TIVA	Central
Blink reflex test				
Nielson	1984	62 HFS	The latency and amplitude of the R1 response on symptomatic side were increased as compared with the asymptomatic side and controls	Peripheral
Esteban and Molina-Negro	1986	53 HFS, 20 HC	The latency of R2 response was shortened and the duration of R2 response was greater on symptomatic side in HFS	Central
Møller and Jannetta	1986	4 HFS	Under anesthesia using inhalational anesthetics, the R1 response was observed only on the symptomatic side of HFS patients	Central
Valls-Sole and Tolosa	1989	17 HFS	With double stimulation, the inhibitory effect of the conditioning stimuli upon the test stimuli R2 response was significantly less pronounced at short ISIs in HFS patients	Central
Eekhof et al.	2000	23 HFS, 10 PFPS, 22 HC	Both R1 and R2 response from the orbicularis oris muscle occurred significantly more often on the symptomatic side in HFS patients than HC	Central
Facial F-wave				
Ishikawa et al.	1996	20 HFS, 10 HC	On symptomatic side of HFS, F-wave duration, F/M amplitude ration, and frequency of F-wave significantly increased. However, minimum latency and chronodispersion had no difference between groups	Central
Ishikawa et al.	1996	10 HFS	On symptomatic side of HFS, the facial F-wave duration tended to increase, and a lineal correlation between the facial F-wave duration and the afterdischarge duration of LSR was observed	Central
Ishikawa et al.	1997	20 HFS	F-waves and LSR were still recorded in some patients after the HFS had disappeared completely, and then F-waves and LSR disappeared subsequently	Central
Hai and Pan	2007	10 HFS rabbits	There was a linear correlation between the amplitude ratio of LSR/M-waves and F-waves/M-waves and between the duration of LSR and F-waves	Central
Transcranial facial motor evoked potential				
Wilkinson and Kaufmann	2005	10 HFS, 17 TN	The amplitude and durations of Facial MEP significantly decreased on the symptomatic side of HFS patients after MVD. However, the latency of Facial MEP revealed no significant change before and after MVD	Central

Table 1 (continued)

Researchers	Year	Subjects	The main findings	The pathogenesis[a]
Wilkinson and Kaufmann	2014	65 HFS, 29 skull base tumors	The threshold of Facial MEP are significantly lower in HFS compared to skull base tumor patients. Also, FMEP to the single pulse stimulation were observed in 87% of HFS patients and only 10% in patients with skull base tumor	Central
Wilkinson et al.	2016	31 HFS	A significantly lower threshold of facial MEPs on the symptomatic side. Under desflurane during MVD, more less suppressive effects on facial MEPs of the symptomatic side	Central

HFS Hemifacial spasm, *TN* Trigeminal neuralgia, *PFPS* Post-facial palsy synkinesis, *HC* Healthy control, *MVD* Microvascular decompression, *LSR* Lateral spread response, *MEP* Motor-evoked potential
[a]Pathogenesis that the results of the studies favor more between the peripheral mechanism and the central mechanism

ods. Taken all the previous studies together, the hyperexcitability of the FMN is thought as the main pathogenesis of HFS.

References

1. Gardner WJ. Concerning the mechanism of trigeminal neuralgia and hemifacial spasm. J Neurosurg. 1962;19:947–58.
2. Jannetta PJ, Abbasy M, Maroon JC, Ramos FM, Albin MS. Etiology and definitive microsurgical treatment of hemifacial spasm. Operative techniques and results in 47 patients. J Neurosurg. 1977;47:321–8.
3. Nielsen VK. Pathophysiology of hemifacial spasm: I. Ephaptic transmission and ectopic excitation. Neurology. 1984;34:418–26.
4. Nielsen VK. Electrophysiology of the facial nerve in hemifacial spasm: ectopic/ephaptic excitation. Muscle Nerve. 1985;8:545–55.
5. Nielsen VK. Pathophysiology of hemifacial spasm: II. Lateral spread of the supraorbital nerve reflex. Neurology. 1984;34:427–31.
6. Nielsen VK, Jannetta PJ. Pathophysiology of hemifacial spasm: III. Effects of facial nerve decompression. Neurology. 1984;34:891–7.
7. Kameyama S, Masuda H, Shirozu H, Ito Y, Sonoda M, Kimura J. Ephaptic transmission is the origin of the abnormal muscle response seen in hemifacial spasm. Clin Neurophysiol. 2016;127:2240–5.
8. Esteban A, Molina-Negro P. Primary hemifacial spasm: a neurophysiological study. J Neurol Neurosurg Psychiatry. 1986;49:58–63.
9. Ishikawa M, Ohira T, Namiki J, Gotoh K, Takase M, Toya S. Electrophysiological investigation of hemifacial spasm: F-waves of the facial muscles. Acta Neurochir. 1996;138:24–32.
10. Moller AR, Jannetta PJ. On the origin of synkinesis in hemifacial spasm: results of intracranial recordings. J Neurosurg. 1984;61:569–76.
11. Moller AR. The cranial nerve vascular compression syndrome: II. A review of pathophysiology. Acta Neurochir. 1991;113:24–30.
12. Ferguson JH. Hemifacial spasm and the facial nucleus. Ann Neurol. 1978;4:97–103.
13. Wilkinson MF, Kaufmann AM. Monitoring of facial muscle motor evoked potentials during microvascular decompression for hemifacial spasm: evidence of changes in motor neuron excitability. J Neurosurg. 2005;103:64–9.
14. Wilkinson MF, Chowdhury T, Mutch WA, Kaufmann AM. Analysis of facial motor evoked potentials for assessing a central mechanism in hemifacial spasm. J Neurosurg. 2017;126:379–85.
15. Auger RG, Piepgras DG, Laws ER Jr, Miller RH. Microvascular decompression of the facial nerve for hemifacial spasm: clinical and electrophysiologic observations. Neurology. 1981;31:346–50.
16. Yamashita S, Kawaguchi T, Fukuda M, et al. Lateral spread response elicited by double stimulation in patients with hemifacial spasm. Muscle Nerve. 2002;25:845–9.
17. Wilkinson MF, Chowdhury T, Mutch WA, Kaufmann AM. Is hemifacial spasm a phenomenon of the central nervous system? The role of desflurane on the lateral spread response. Clin Neurophysiol. 2015;126:1354–9.
18. Kugelberg E. Facial reflexes. Brain. 1952;75:385–96.
19. Kimura J. Electrodiagnosis in diseases of nerve and muscle: principles and practice. New York: Oxford University Press; 2001.
20. Eekhof JL, Aramideh M, Speelman JD, Devriese PP, Ongerboer De Visser BW. Blink reflexes and lateral spreading in patients with synkinesia after Bell's palsy and in hemifacial spasm. Eur Neurol. 2000;43:141–6.
21. Valls-Sole J. Facial nerve palsy and hemifacial spasm. Handb Clin Neurol. 2013;115:367–80.
22. Moller AR, Jannetta PJ. Blink reflex in patients with hemifacial spasm. Observations during microvascular decompression operations. J Neurol Sci. 1986;72:171–82.

23. Ishikawa M, Ohira T, Namiki J, Ajimi Y, Takase M, Toya S. Abnormal muscle response (lateral spread) and F-wave in patients with hemifacial spasm. J Neurol Sci. 1996;137:109–16.
24. Hai J, Pan QG. Experimental study on the correlation between abnormal muscle responses and F waves in hemifacial spasm. Neurol Res. 2007;29:553–6.
25. Rasminsky M. Ephaptic transmission between single nerve fibres in the spinal nerve roots of dystrophic mice. J Physiol. 1980;305:151–69.
26. Ishikawa M, Ohira T, Namiki J, et al. Electrophysiological investigation of hemifacial spasm after microvascular decompression: F waves of the facial muscles, blink reflexes, and abnormal muscle responses. J Neurosurg. 1997;86:654–61.
27. Dong CC, Macdonald DB, Akagami R, et al. Intraoperative facial motor evoked potential monitoring with transcranial electrical stimulation during skull base surgery. Clin Neurophysiol. 2005;116:588–96.
28. Wilkinson MF, Kaufmann AM. Facial motor neuron excitability in hemifacial spasm: a facial MEP study. Can J Neurol Sci. 2014;41:239–45.

Magnetic Resonance Imaging Evaluation of Hemifacial Spasm

Hyung-Jin Kim and Minjung Seong

Hemifacial spasm (HFS) is one of the most common disease entities in the spectrum of neurovascular compression syndrome (NVCS) which is defined as a direct contact with mechanical irritation of the cranial nerves (CNs) by the blood vessels [1–5]. Although a wide variety of diseases are categorized as NVCS according to the affected CNs and the resulting symptoms, the evidence-based firm cause-and-effect relationship was generally recognized in only three conditions, including trigeminal neuralgia (TN) for the trigeminal nerve, HFS for the facial nerve, and vago-glossopharyngeal neuralgia for the vagus and glossopharyngeal nerves [3, 6].

After the advent of magnetic resonance imaging (MRI), it had rapidly replaced the role of conventional angiography and computed tomography (CT) for evaluation of the patients with HFS and those with other categories of NCVS such as TN. By virtue of its superb contrast resolution, MRI can simultaneously demonstrate both the CNs and blood vessels, and thus inform us of the detailed anatomic relationship between them [7–14]. Now, MRI with the aid of magnetic resonance angiography (MRA) has become the imaging modality of choice in the diagnosis and treatment planning of the patients with HFS. With

the various state-of-the-art three-dimensional (3D) fast spin-echo (FSE) or fast gradient-echo (FGE) imaging techniques, we can decode the detailed neurovascular relationship that might be responsible for the clinical symptoms [4–6].

In this chapter, we will discuss the role of MRI in patients with HFS: how to image and what to look. Special attention will be paid to the importance of the anatomy of so-called "root exit zone (REZ)" of the facial nerve on MRI. The imaging issues in patients with failed microvascular decompression (MVD) will be addressed as well.

MR Imaging Techniques Used for Hemifacial Spasm

High-resolution 3D MRI has proved useful and highly accurate in the preoperative evaluation of both primary and secondary types of HFS. Recent advances in hardware and software of MRI enabled us to obtain images of high quality in a reasonably short time. Transversely oriented axial images obtained with 3D heavily T2-weighted fast imaging sequence, so-called magnetic resonance cisternography (MRC), is well suited for evaluation of the complex relationship between the nerves and vessels. On these images, the low signal intensity of the nerves and vessels contrasts with the bright signal intensity of background cerebrospinal fluid (CSF) (Fig. 1a). Based on the isotropic images obtained with the 3D technique, the images can be

H.-J. Kim (✉) · M. Seong
Department of Radiology, Samsung Medical Center, Sungkyunkwan University School of Medicine, Seoul, Republic of Korea
e-mail: hyungjin1219.kim@samsung.com; m.seong@samsung.com

© Springer Nature Singapore Pte Ltd. 2020
K. Park, J. S. Park (eds.), *Hemifacial Spasm*, https://doi.org/10.1007/978-981-15-5417-9_7

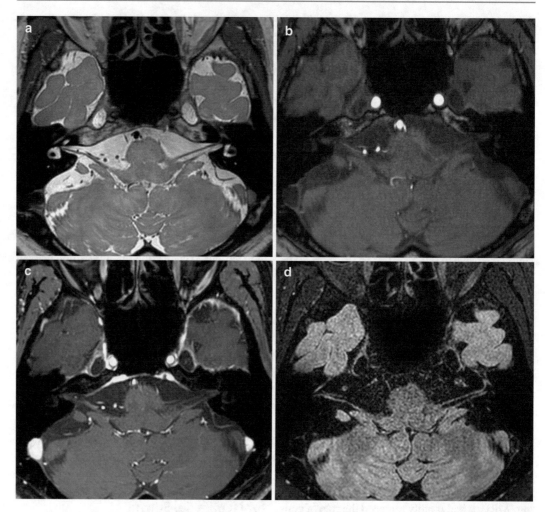

Fig. 1 Various 3D MRI sequences used for evaluation of hemifacial spasm. (**a**) MR cisternography (MRC) using 3D volumetric isotropic turbo spin-echo acquisition (T2 VISTA). (**b**) MR angiography (MRA) using 3D time-of-flight (TOF). (**c**) Contrast-enhanced T1-weighted imaging using 3D balanced turbo field echo (bTFE). (**d**) Fluid-attenuated inversion recovery (FLAIR) image using 3D FLAIR-VISTA

displayed in any planes by multiplanar reformation (MPR) to improve the precision of evaluation [7–14]. Various 3D steady-state FGE and 3D FSE sequences have been used among the different vendors, with the former including FIESTA-C, CISS, and bTFE and the latter including CUBE, SPACE, and VISTA [4–6].

3D time-of-flight (TOF) MRA is another useful MRI technique to visualize the offending arteries (Fig. 1b) [4, 5]. However, small arteries are difficult to see consistently with this technique. It is also useful for detection of high-flow vascular malformation [5]. Image fusion that combines MRC and 3D TOF MRA has been reported to display the neurovascular relationship more vividly in patients with HFS [15, 16].

3D contrast-enhanced T1-weighted imaging sequence is useful in the evaluation of HFS, especially for exclusion of secondary HFS caused by tumor, inflammation, and demyelinating disease (Fig. 1c) [4, 5]. Venous lesions that may cause HFS can also be demonstrated much better with this technique.

Fluid-attenuated inversion recovery (FLAIR) sequence is useful to evaluate the patients with secondary HFS, as seen in the patients with demyelinating diseases, such as multiple sclerosis. By using an inversion recovery pulse, the signal intensity of fluid such as CSF is nulled and the lesion other than fluid is demonstrated as high signal intensity on the images (Fig. 1d) [5]. The superiority of the fusion images generated by combined 3D FLAIR and 3D TOF MRA has been reported for more clear delineation of neurovascular relationship in cases where the REZ of the facial nerve is difficult to trace on MRC (Fig. 2) [17]. The value of image fusion of 3D

FLAIR and MRC has been reported in patients with TN as well [18].

Diffusion-weighted imaging (DWI) is also helpful in the evaluation of patients with HFS. The lesions showing restricted diffusion are seen as an area of high signal intensity on DWI and best exemplified by the diseases causing secondary HFS such as brainstem infarction and epidermoid cyst in the cerebellopontine angle (CPA) cistern (Fig. 3) [5]. Recently, diffusion tensor imaging with tractography has been

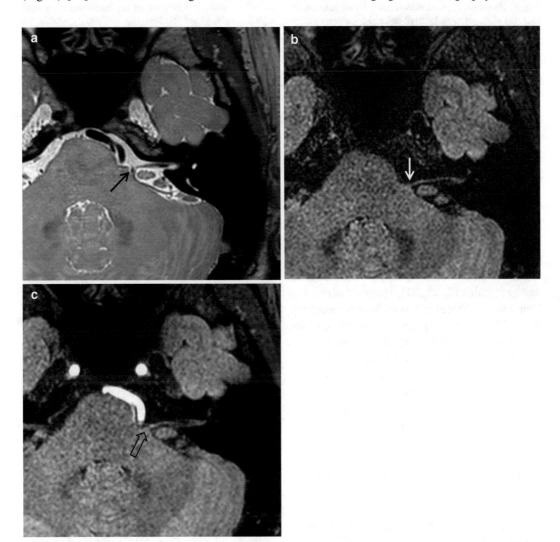

Fig. 2 Usefulness of 3D FLAIR for better demonstration of the proximal course of the facial nerve. (**a**) On MRC, the proximal portion of the facial nerve overlaps with the vascular structure, making it difficult to trace (arrow). (**b**) On 3D FLAIR, the proximal portion of the facial nerve (arrow) is well traced medially thanks to suppression of the signal from the vessels. (**c**) Fusion image of MRA and 3D FLAIR demonstrates more clear relationship between the nerve and offending vessel (open arrow)

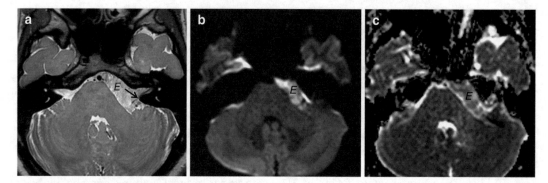

Fig. 3 Hemifacial spasm associated with an epidermoid cyst. (**a**) MRC shows a large ill-defined hyperintense cystic mass (*E*) in the left cerebellopontine angle, displacing the facial nerve laterally (arrow). (**b**) and (**c**) Diffusion-weighted image obtained with *b* = 1000 (**b**) and apparent diffusion coefficient map image (**c**) demonstrate the high signal intensity and the low signal intensity within the mass (*E*), respectively, which is related to restricted diffusion, characteristic of the epidermoid cyst

reported to be useful, especially for patients with TN [6, 19].

Anatomy of the Facial Nerve on MRI and Its Clinical Implication

The most appealing theory as to the pathogenesis of NVCS is that NVC (neurovascular compression) initiates demyelination of the nerve at the REZ, which then causes ephaptic transmission of neural impulse [2, 20, 21]. However, there has been a considerable confusion about the anatomy of the REZ [3, 22, 23]. Unclear definition of the REZ has led to a misconception of the term as synonymous with the term "transition zone (TZ)", also known as Obersteiner–Redlich zone, where a transition occurs between the oligodendrocyte-derived central myelin and the Schwann cell-derived peripheral myelin [6, 22, 23]. Anatomically, however, TZ is only one part of the REZ complex which encompasses the central myelin portion of the nerve root and the subpial portion of the nerve fascicles within the brain stem [4].

Compared to other CNs, the facial nerve is unique for its distinctive course of the REZ. Based on 75 facial nerves in 44 cadaveric brains, Tomii et al. [23] found that unlike the usual cranial nerves, the facial nerve had a long segment of the central glial portion that strongly adhered to the ventral surface of the pons after exiting the brain stem from the pontomedullary sulcus. Based on the works by Tomii et al. [23] and later on by Compos-Benitez and Kaufmann [22], the REZ of the facial nerve can be divided into four parts according to the presence of the central myelin components within the nerve: the root exit point (RExP), where the facial nerve emerges from the brain stem at the upper edge of the supraolivary fossette; the attached segment (AS), where the facial nerve adheres to the ventral surface of the pons; the root detachment point (RDP), where the facial nerve enters the prepontine cistern, separating from the pons; and finally the TZ, where the central myelin transitions into the peripheral myelin (Fig. 4a). Beyond the TZ lies the cisternal portion (CP) of the facial nerve, where the myelin is entirely derived from the Schwann cell. The reported lengths of the AS and the TZ are 8–10 mm and 1–4 mm, respectively [22–24].

Because of this unique anatomy of the REZ of the facial nerve, the oblique coronal reformatted MR images with the plane parallel to the facial nerve on the axial MRC images may be the best plane to appreciate the entire course of the REZ of the facial nerve (Fig. 4b). On the axial images, the portion of the REZ, which is localized at the junction of the facial nerve and the outer surface of the brain stem, may actually represent the RDP and the more proximally located AS cannot

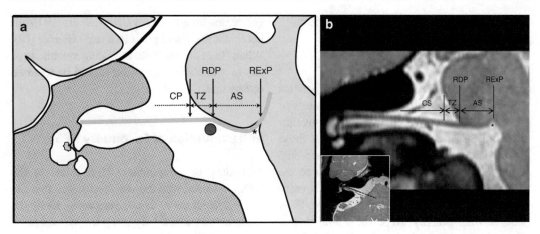

Fig. 4 Anatomy of the root exit zone of the facial nerve. (**a**) and (**b**) A schematic drawing (**a**) and corresponding MRC in oblique coronal plane (**b**) demonstrate the subdivisions of the facial nerve from the root exit point (RExP) at the supraolivary fossette (asterisk) to the cisternal portion (CP). The attached segment (AS) is a relatively long segment that tightly adheres to the ventral surface of the pons before the nerve emerges into the prepontine cistern at the root detachment point (RDP). The transition zone (TZ) is a short segment where the central myelin transitions into the peripheral myelin. The image reformation in oblique coronal plane is obtained with the plane parallel to the facial nerve on the axial MRC as shown in the inset in (**b**)

be well seen as a separate structure. This happens because the tight attachment of the AS to the brain stem frequently makes the distinction between the two structures difficult. A separation of the AS from the brain stem is still not easy even on the oblique coronal reformatted images. However, one can easily imagine the course of the AS underneath the brain stem from the RExP to the RDP on these images (Fig. 4b).

The most important implication of this peculiar anatomy of the REZ of the facial nerve in clinical practice may be that the more the proximal part of the REZ, the greater the responsibility for the site of NVC causing HFS. In a cadaveric study, Tomii et al. [23] demonstrated that NVC occurred proximal to the TZ in more than 80% of the nerves showing vascular compression and suggested that this area might also correspond to the TZ. In their study on 115 patients with MVD for HFS, Campos-Benitez and Kaufmann [22] supported the findings of the study of Tomii et al. They found that 74% of patients with HFS had NVC in the proximal parts of the REZ including AS in 64% and RExP in 10%. In contrast, in only 25% of cases, NVC involved the more distal parts including RDP/TZ in 22% and CP in 3% [22].

The anatomy of the REZ of the facial nerve is also important to evaluate the patients who have persistent symptoms of HFS after MVD. In the study on 18 patients with a failed MVD, Hughes et al. [24] reported that seven of 12 patients (58%), in whom persistent vascular compression was identified on MRI, had NVC at the AS. They also pointed out that the unaddressed vascular compression was typically proximal to the previously placed surgical pledgets in 10 of 12 patients (83%). They recommended a careful scrutiny of the AS on high-resolution MRI to identify a persistent NVC [24].

With the oblique coronal images, one can see the site and severity of NVC at the proximal portion of the REZ of the facial nerve more instantaneously than on the axial images. In a patient with a failed MVD, the NVC at the AS of the facial nerve is also well demonstrated on this oblique coronal image.

MRI Evaluation of Hemifacial Spasm

MRI Evaluation of Primary HFS

The first step of MRI interpretation in patients with HFS is to exclude the secondary causes of HFS, such as tumors, vascular lesions, demyelinating processes, and ischemic changes. If those lesions are not found on MRI, the next step should be carefully focused on the path of the facial nerve to identify the site of vascular compression, if any. To characterize the NVC in patients with primary HFS, there are several things that should be included in the radiologic report on MRI: the type of the offending (culprit) vessels, the site of NVC, and the severity of NVC.

In terms of the type of the offending vessels, NVC in primary HFS is caused by the arteries with the anterior inferior cerebellar artery being the most common vessel in 43–53.2%, followed by the posterior inferior cerebellar artery in 30.9–31% and the vertebral artery in 1.1–23% [22, 25]. HFS can be caused by multiple arteries (Fig. 5). A review of 1174 patients by Hyun et al. [25] identified multiple offending arteries in various combinations in 14.1%. In the study by Campos-Benitez and Kaufmann [22], 38% of patients had multiple vessels compressing the nerve. The venous cause is infrequent with the incidence being reported in 0.3–3% of the cases (Fig. 6) [22, 25].

The site of NVC should be addressed whether it involves the REZ or more distal portion of the facial nerve, such as CP. The reported incidence of the REZ involved in primary HFS nearly approaches 97–100% [13, 22]. To provide more detailed information on the site of NVC to the referring surgeons, it is preferred to comment on the points of neural compression according to the subdivisions of the REZ of the facial nerve, as described previously: the more the proximal site of NVC, the greater the likelihood that the vessel is a real offender (Fig. 7). This approach may help guide surgeons to perform a more optimal surgical management.

The severity of NVC can be classified as simple contact, indentation, and displacement, as the severity increases (Fig. 8). As expected, the greater the severity of NVC, the greater the likelihood that the vessel plays as a real offender [12]. When there are multiple offending vessels, each site of vascular compression should be reported one by one.

MRI Evaluation of Secondary HFS

Secondary or symptomatic HFS refers to the HFS associated with the various diseases that can cause a facial nerve damage anywhere along the facial nerve pathway. By using the various pulse sequences, MRI plays a very important role in the management of secondary HFS not only for the diagnosis of the lesions but also for a precise assessment of the extent of the lesions to help clinicians plan the best treatment for the patients. It is beyond scope of this chapter to deal with the detailed MRI findings of the various lesions that cause secondary HFS, and so only several examples are presented here.

Different kinds of pathology have been implicated as the cause of secondary HFS [5]. In causing HFS, these underlying diseases are collectively thought to induce a neural dysfunction and/or irritation of the facial nerve pathway. The incidence of secondary HFS is approximately one-fourth of that of primary HFS [15]. HFS caused by CPA tumors is rare with the reported incidence in 0.3–2.5% (Fig. 9) [5, 26]. In the study of 2050 patients with HFS by Lee et al. [26], only nine patients (0.44%) had HFS attributable to CPA tumors, including vestibular schwannoma ($n = 2$), meningioma ($n = 5$), and epidermoid cyst ($n = 2$). As previously shown in Fig. 2, DWI is very useful for the diagnosis of the epidermoid cyst. Although HFS associated with the demyelinating diseases, typified by multiple sclerosis, has been reported in the literature, its real incidence is not clear because of its rare occurrence [27]. The vascular lesions can also cause HFS and include the vertebrobasilar artery dolichoectasia, developmental venous anomaly (Fig. 6), arteriovenous malformation, and pial arteriovenous fistula [5]. Other uncommon causes of secondary HFS include the vascular insult, trauma, and infection/inflammation affecting the facial nerve.

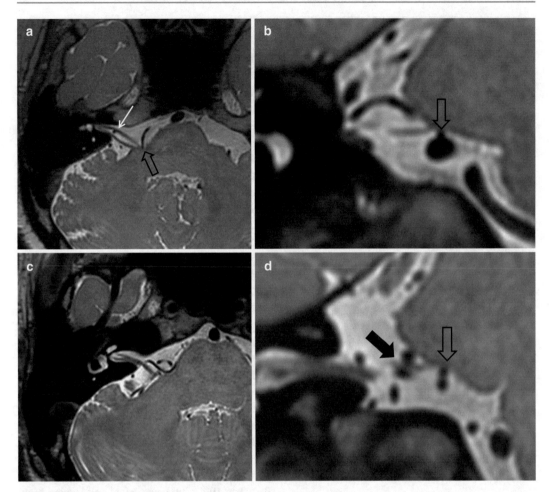

Fig. 5 Examples of various offending vessels in two patients with hemifacial spasm. (**a**) and (**b**) MRCs in axial (**a**) and oblique coronal (**b**) plane demonstrate two separate sites of neurovascular compression (NVC) along the course of the right facial nerve. The proximal site of NVC occurs in the attached segment near the root detachment point with two vessels, the anterior inferior cerebellar artery (upper) and the vertebral artery (lower), showing a tandem type of NVC (open arrow). The distal site of NVC takes place in the cisternal portion near the porus acusticus by the anterior inferior cerebellar artery, causing a mild anterior displacement of the facial nerve (arrow). (**c**) and (**d**) MRCs in axial (**c**) and oblique coronal (**d**) plane demonstrate two separate sites of NVC along the course of the right facial nerve. Proximally, a tandem type of NVC occurs in the attached segment by the anterior inferior cerebellar artery (upper) and the posterior inferior cerebellar artery (lower) in combination (open arrow). More distally, another site of NVC is found at the transition zone where the facial nerve is compressed between the branch of the superior petrosal vein (upper) and the anterior inferior cerebellar artery (lower), showing a sandwich type of NVC (thick arrow)

NVC in Asymptomatic Subjects

It is well known that a vascular contact is not infrequent on MRI in subjects with no clinical signs of NVC. In their works on MRI, Tash et al. [7] and Fukuda et al. [12] reported the incidence of the asymptomatic vascular contact on the facial nerve in 21% and 15%, respectively.

According to Kakizawa et al. [28], the incidence was much higher, being reported in 78.6%. In their series, however, there was no severe deviation of the facial nerve. It may be that it is not the existence of NVC itself but the severity of NVC that determines a provocation of the symptoms. Sometimes, as demonstrated in Fig. 10, MRI shows the facial nerve that is compressed more

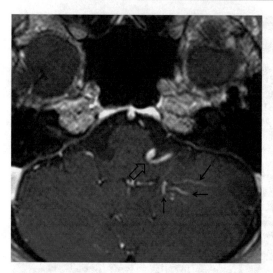

Fig. 6 Hemifacial spasm caused by developmental venous anomaly. 3D axial contrast-enhanced bTFE MRI shows multiple small collecting veins (arrows) in the left cerebellum that drain into a single dilated transcortical vein (open arrow) at the root exit zone of the facial nerve, characteristic of developmental venous anomaly

severely in the asymptomatic side than in the symptomatic side. It is critical to always refer to the clinical information to avoid such pitfalls during image interpretation, because the imaging findings are significant only in patients with the symptoms in the affected side [4].

MRI Evaluation of Persistent or Recurrent HFS After MVD

MVD is a well-established method of treatment for HFS with success rates of 90–95% for initial operation [6, 24, 25]. Despite a successful surgical decompression, however, the recurrence has been reported in up to 25% of the patients [6]. The causes responsible for the persistent or recurrent symptoms post MVD include the missed offending vessels, insufficient decompression, malposition of the surgical implant (Teflon), newly developed vascular compression, Teflon-related adhesion, arachnoid adhesion, and Teflon granuloma [29–32].

In patients with a persistent HFS after MVD, the unaddressed vascular compression is typically proximal to the previously placed surgical material. Bigder and Kaufmann [33] reported that persistent NVC was found proximal to the prior implant material in 11 of 12 patients who underwent a repeat MVD due to persistent NVC. In all 12 patients, NVC involved the REZ, including the AS in 11 patients and the RExP in three patients [33]. In 21 patients who underwent a repeat MVD for recurrent HFS, Lee et al. [30] reported that NVC was found at the REZ in 15 patients and at the cisternal segment in six patients. Reimaging with high-resolution MRI usually identifies the culprit vessels in patients with failed MVD. The predominant proximal location of NVC can be accurately depicted on MRI (Fig. 11). In 12 of 14 patients with evidence of persistent NVC after MVD on MRI, Hughes et al. [24] reported that the locations of NVC were the AS in seven (58%), RDP in one (8%), and TZ in four (33%). In 10 of 12 patients (83%), the contact occurred proximal to the existing surgical implant.

Among others, due to its histocompatibility and absorption resistance, Teflon is currently used as the material of choice in MVD for patients with NVCS including TN and HFS. Since the early 1990s, however, giant cell foreign body reaction induced by Teflon, which causes a granuloma formation, was reported in patients with TN after MVD, with an incidence ranging from 1.1 to 7.3% [34–36]. In contrast, the occurrence of Teflon granuloma after MVD for HFS has been reported much less commonly. As suggested by Chen et al. [36], the greater overall prevalence of Teflon granulomas in MVD for TN may be explained by the fact that due to a longer REZ of the trigeminal nerve, there is a greater chance of Teflon that comes in contact with the tentorium and dura, which then provokes an inflammatory reaction.

Radiologically, Teflon granuloma may present as an enhancing CPA mass on CT and MRI many years or even decades after surgery. It can be con-

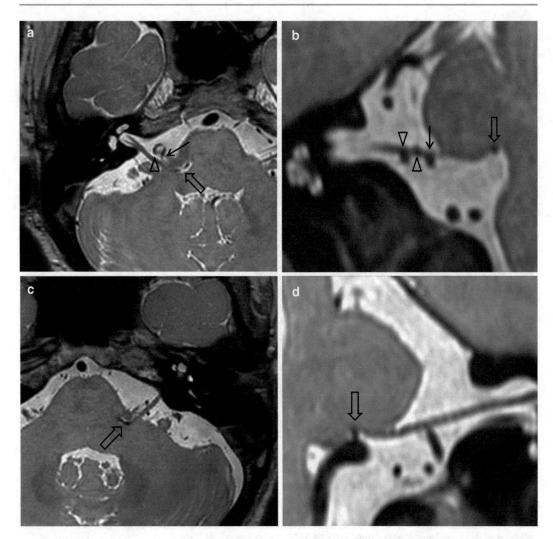

Fig. 7 Examples of various sites of neurovascular compression in two patients with hemifacial spasm. (**a**) and (**b**) MRCs in axial (**a**) and oblique coronal (**b**) plane demonstrate multiple sites of neurovascular contact by the anterior inferior cerebellar artery including the root exit point (open arrows), transition zone (arrows), and cister- nal portion (arrowheads). (**c**) and (**d**) MRCs in axial (**c**) and oblique coronal (**d**) plane demonstrate a tandem type of neurovascular compression at the attached segment by the anterior inferior cerebellar artery (upper) and vertebral artery (lower) in combination (open arrow)

fused with other CPA tumors (Fig. 12) [34, 35, 37]. MRI demonstrates an oval to round heterogeneous mass that is hypointense on both T1- and T2-weighted images. Early, actively growing gran- uloma may demonstrate more avid enhancement, whereas older, quiescent granuloma shows mini- mal or no enhancement. On CT, focal calcification is often noticed. Although Teflon granuloma fre- quently continues to grow, associated malignant transformation has not been reported [35].

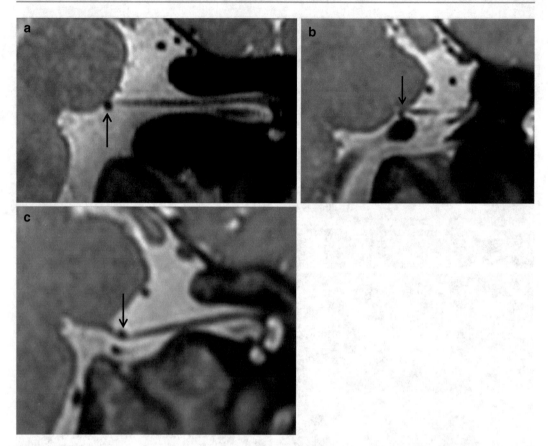

Fig. 8 Examples of the severity of neurovascular compression in two patients with hemifacial spasm. (**a**) Simple contact. MRC in oblique coronal plane demonstrates the anterior inferior cerebellar artery which is in contact with the root detachment point of the facial nerve with no evidence of pressure effect (arrow). (**b**) Indentation. MRC in oblique coronal plane demonstrates a focal indentation of the root detachment point of the facial nerve by the ante-rior inferior cerebellar artery (upper) and the vertebral artery (lower) in a tandem type of neurovascular compression (arrow). (**c**) Displacement. MRC in oblique coronal plane demonstrates a severe compression on the transition zone of the facial nerve by the anterior inferior cerebellar artery, causing an angulation and displacement of the nerve inferiorly (arrow)

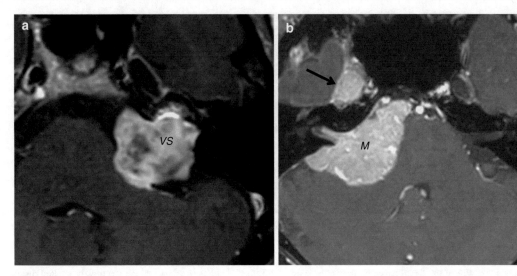

Fig. 9 Examples of secondary hemifacial spasm. (**a**) Vestibular schwannoma. Axial contrast-enhanced T1-weighted MRI shows a large lobulated soft tissue mass (*VS*) with heterogeneous enhancement in the left cerebellopontine angle. The internal auditory canal is replaced and marked widened by the mass. The brain stem is compressed by the mass as well. (**b**) Meningioma. Axial contrast-enhanced T1-weighted MRI shows a large well-enhancing dural-based soft tissue mass (*M*) in the right cerebellopontine angle. The mass partly grows into the internal auditory canal and causes significant compression on the brain stem and cerebellum. The ipsilateral Meckel's cave is also involved by the mass (arrow)

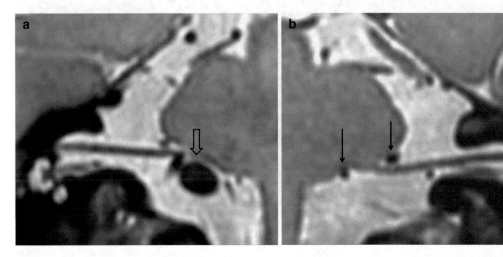

Fig. 10 Asymptomatic NVC. (**a**) and (**b**) MRCs in oblique coronal plane in the same patient demonstrate neurovascular compression in the root exit zone of bilateral facial nerves, more conspicuous on the right (open arrow in **a**) than on the left (arrows in **b**). However, the patient complained of a twitch only on her left face. The image interpretation should always be based on clinical information, because the imaging features are significant only in the symptomatic side

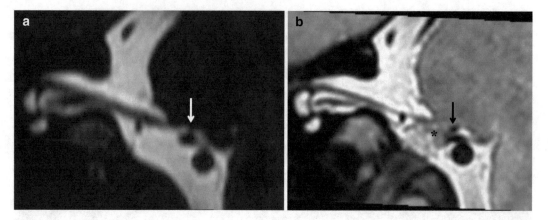

Fig. 11 Recurrent hemifacial spasm after microvascular decompression (MVD). (**a**) Preoperative MRC in oblique coronal plane demonstrates a focal indentation of the attached segment of the facial nerve by the anterior inferior cerebellar artery (arrow). (**b**) MRC in oblique coronal plane obtained 4 years after MVD demonstrates persistent neurovascular compression at the same site by the same offending artery as before (arrow). Note the position of Teflon pledgets (asterisk) which are mostly located distal to the site of neurovascular compression

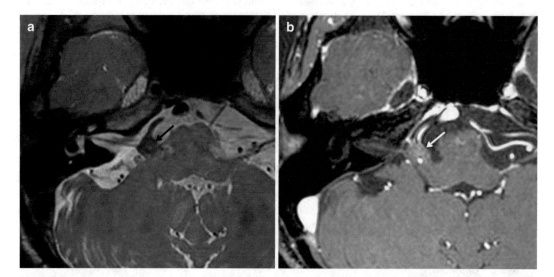

Fig. 12 Recurrent hemifacial spasm caused by Teflon granuloma after microvascular decompression (MVD). (**a**) and (**b**) Axial MRC (**a**) and 3D contrast-enhanced T1-weighted MRI (**b**) obtained 15 years after MVD dem-onstrate a mass at the root exit zone of the right facial nerve (arrows). The lesion shows low signal intensity on T2-weighted image (**a**) and mild homogenous enhancement after contrast injection (**b**)

Conclusion

In primary HFS, NVC initiates demyelination of the REZ of the facial nerve, which can be divided into four segments: RExP, AS, RDP, and TZ. High-resolution 3D MRI is the imaging modality of choice in the preoperative evaluation of HFS. 3D heavily T2-weighted imaging (MRC), aided by 3D TOF MRI and other pulse sequences, is well suited for evaluation of the complex anatomy at the site of NVC. On MRC, the oblique coronal plane, reformatted parallel to the facial nerve, is the best plane to appreciate the AS where the facial nerve adheres to the ventral surface of the pons and also where NVC is frequently missed during MVD. The roles of MRI in HFS are to exclude other secondary causes such as tumors and demyelinating diseases and to identify the offending vessels in terms of the type (artery or vein), the site of NVC, and the severity of NVC. Because vascular contact is frequently seen in asymptomatic subjects, it is critical to refer to the clinical information to avoid mistakes during MRI interpretation. By correctly depicting the proximal location of NVC, high-resolution MRI is also effective in patients with HFS who suffer from the persistent or recurrent symptoms after MVD. Rarely, Teflon granuloma can cause recurrent symptoms and may present as an enhancing CPA mass on MRI that should be differentiated from true CPA tumors.

References

1. Jannetta PJ. Neurovascular compression in cranial nerve and systemic disease. Ann Surg. 1980;192:518–25.
2. Devor M, Amir R, Rappaport ZH. Pathophysiology of trigeminal neuralgia: the ignition hypothesis. Clin J Pain. 2002;18:4–13.
3. Guclu B, Sindou M, Meyronet D, Streichenberger N, Simon E, Mertens P. Cranial nerve vascular compression syndromes of the trigeminal, facial and vagoglossopharyngeal nerves: comparative anatomical study of the central myelin portion and transitional zone: correlations with incidences of corresponding hyperactive dysfunctional syndromes. Acta Neurochir. 2011;153:2365–75.
4. Donahue JH, Ornan DA, Mukherjee S. Imaging of vascular compression syndromes. Radiol Clin N Am. 2017;55:123–38.
5. Chen SR. Neurological imaging for hemifacial spasm. Int Ophthalmol Clin. 2018;58:97–109.
6. Haller S, Etienne L, Kovari E, et al. Imaging of neurovascular compression syndromes: trigeminal neuralgia, hemifacial spasm, vestibular paroxysmia, and glossopharyngeal neuralgia. AJNR Am J Neuroradiol. 2016;37:1384–92.
7. Tash R, DeMerritt J, Sze G, et al. Hemifacial spasm: MR imaging features. AJNR Am J Neuroradiol. 1991;12:839–42.
8. Du C, Korogi Y, Nagahiro S, et al. Hemifacial spasm: three-dimensional MR images in the evaluation of neurovascular compression. Radiology. 1995;197:227–31.
9. Mitsuoka H, Tsunoda A, Okuda O, et al. Delineation of small nerves and blood vessels with three-dimensional fast spin-echo MR imaging: comparison of presurgical and surgical findings in patients with hemifacial spasm. AJNR Am J Neuroradiol. 1998;19:1823–9.
10. Yamakami I, Kobayashi E, Hirai S, et al. Preoperative assessment of trigeminal neuralgia and hemifacial spasm using constructive interference in steady state-three-dimensional Fourier transformation magnetic resonance imaging. Neurol Med Chir (Tokyo). 2000;40:545–56.
11. Lee MS, Kim MS, Hong IS, et al. Clinical usefulness of magnetic resonance cisternography in patients having hemifacial spasm. Yonsei Med J. 2001;42:390–4.
12. Fukuda H, Ishikawa M, Okumura R, et al. Demonstration of neurovascular compression in trigeminal neuralgia and hemifacial spasm with magnetic resonance imaging. Comparison with surgical findings in 60 consecutive cases. Surg Neurol. 2003;59:93–100.
13. Naraghi R, Tanrikulu L, Troescher-Weber R, et al. Classification of neurovascular compression in typical hemifacial spasm: three-dimensional visualization of the facial and the vestibulocochlear nerves. J Neurosurg. 2007;107:1154–63.
14. Tanrikulu L, Scholz T, Nikoubashman O, et al. Preoperative MRI in neurovascular compression syndromes and its role for microsurgical considerations. Clin Neurol Neurosurg. 2015;129:17–20.
15. Lu AY, Yeung JT, Gerard JL, et al. Hemifacial spasm and neurovascular compression. Sci World J. 2014;2014:349319. https://doi.org/10.1155/2014/349319.
16. Satoh T, Onoda K, Date I. Fusion imaging of three-dimensional magnetic resonance cisternograms and angiograms for the assessment of microvascular decompression in patients with hemifacial spasm. J Neurosurg. 2007;106:82–9.
17. Yim YJ. Value of fusion imaging with multiplanar reconstruction generated by 3D FLAIR and 3D TOF MR angiography in the evaluation of the patients with hemifacial spasm. Thesis for the degree of Doctor of Philosophy submitted to the Graduate School of Sungkyunkwan University School of Medicine. 2011.
18. Cha J, Kim ST, Kim H-J, et al. Trigeminal neuralgia: assessment with T2 VIST and FLAIR VISTA fusion imaging. Eur Radiol. 2011;21:2633–9.
19. Lutz J, Linn J, Mehrkens JH, et al. Trigeminal neuralgia due to neurovascular compression: high-spatial-resolution diffusion-tensor imaging reveals microstructural neural changes. Radiology. 2011;258:524–30.
20. De Ridder D, Møller A, Verlooy J, Cornelissen M, De Ridder L. Is the root entry/exit zone important in microvascular compression syndromes? Neurosurgery. 2002;51:427–34.
21. Prasad S, Galetta S. Trigeminal neuralgia. Historical notes and current concepts. Neurologist. 2009;15:87–94.

22. Campos-Benitez M, Kaufmann AM. Neurovascular compression findings in hemifacial spasm. J Neurosurg. 2008;109:416–20.

23. Tomii M, Onoue H, Yasue M, et al. Microscopic measurement of the facial nerve root exit zone from central glial myelin to peripheral Schwann cell myelin. J Neurosurg. 2003;99:121–4.

24. Hughes MA, Branstetter BF, Taylor CT, et al. MRI findings in patients with a history of failed prior microvascular decompression for hemifacial spasm: how to image and where to look. AJNR Am J Neuroradiol. 2015;36:768–73.

25. Hyun S-J, Kong D-S, Park K. Microvascular decompression for treating hemifacial spasm. lessons learned from a prospective study of 1174 operations. Neurosurg Rev. 2010;33:325–34.

26. Lee SH, Rhee BA, Choi SK, et al. Cerebellopontine angle tumors causing hemifacial spasm: types, incidence, and mechanism in nine reported cases and literature review. Acta Neurochir. 2010;152:1901–8.

27. Marin Collazo IV, Tobin WO. Facial myokymia and hemifacial spasm in multiple sclerosis: a descriptive study on clinical features and treatment outcomes. Neurologist. 2018;23:1–6.

28. Kakizawa Y, Seguchi T, Kodama K, et al. Anatomical study of the trigeminal and facial cranial nerves with the aid of 3.0-tesla magnetic resonance imaging. J Neurosurg. 2008;108:483–90.

29. Jiang C, Xu W, Dai Y, et al. Failed microvascular decompression surgery for hemifacial spasm: a retrospective clinical study of reoperations. Acta Neurochir. 2017;159:259–63.

30. Xu X-L, Zhen X-K, Yuan Y, et al. Long-term outcome of repeat microvascular decompression for hemifacial spasm. World Neurosurg. 2018 Feb;110:e989–97. https://doi.org/10.1016/j.wneu.2017.11.144.

31. Park CK, Lee SH, Park BJ. Surgical outcomes of revision microvascular decompression for persistent or recurrent hemifacial spasm after surgery: analysis of Radiologic and intraoperative findings. World Neurosurg. 2019;131:e454–9. https://doi.org/10.1016/j.wneu.2019.07.191.

32. Lee S, Park SK, Lee JA, et al. Missed culprits in failed microvascular decompression surgery for hemifacial spasm and clinical outcomes of redo surgery. World Neurosurg. 2019 Sep;129:e627–33. https://doi.org/10.1016/j.wneu.2019.05.231.

33. Bigder MG, Kaufmann AM. Failed microvascular decompression surgery for hemifacial spasm due to persistent neurovascular compression: an analysis of reoperations. J Neurosurg. 2016;124:90–5.

34. Oda K, Higuchi T, Murai Y, et al. Teflon granuloma after microvascular decompression for hemifacial spasm: a case report and literature review. Neurosurg Rev. 2017;40:513–6.

35. Deep NL, Graffeo CS, Copeland WR, et al. Teflon granulomas mimicking cerebellopontine angle tumor following microvascular decompression. Laryngoscope. 2017;127:715–9.

36. Chen J-F, Lee S-T, Lui T-N, et al. Teflon granuloma after microvascular decompression for trigeminal neuralgia. Surg Neurol. 2005;53:281–7.

37. Capelle H-H, Brandis A, Tschan C, et al. Treatment of recurrent trigeminal neuralgia due to Teflon granuloma. Headache Pain. 2010;11:339–44.

Surgical Principles of Hemifacial Spasm: How We Do Microvascular Decompression

Seunghoon Lee and Kwan Park

Indication of Microvascular Decompression

The diagnosis of hemifacial spasm (HFS) is made largely based on clinical manifestations of the patient. Recurrent, paroxysmal, involuntary twitching of the facial muscles is the characteristic features of HFS and is manifested mostly unilaterally and less frequently bilaterally [1, 2]. For all patients who show typical symptoms and have offending vessel compressing the facial nerve identified on the MRI, microvascular decompression (MVD) can be performed when general condition permits brain surgery under general anesthesia. If the patient has minimal or atypical symptom and the diagnosis is uncertain, electrophysiologic studies such as electromyography (EMG) or nerve excitability test (abnormal muscle response; AMR) are helpful.

Secondary HFS caused by, for instance, tumors at cerebellopontine angle (CPA) can be identified on the preoperative magnetic resonance image (MRI). Majority of the secondary HFS has its own offending vessel between tumor and the facial nerve. Therefore, offending vessel underneath the brain tumor should be verified and decompressed completely from the facial nerve after tumor resection [3].

It is controversial when to perform MVD in HFS patients. Longer preoperative period might influence on the nucleus of the facial nerve and result in facial nucleus degeneration, which may explain in part the MVD failure cases comprising about 10% of the patients. However, there have been no evidence for this, and no prognostic difference between early versus late MVD was acknowledged to date. We usually recommend the MVD surgery when the patient has been experiencing this disorder to be progressive, and the tonic-clonic spasm is so severe that the patient starts to avoid social interaction and finally is willing to take the risks of brain surgery which has possible complications.

Preoperative Evaluation

The patient is proceeded to review of general condition for general anesthesia, and consulted to relevant department if there is any abnormality. We perform preoperative hearing function review by otologists in all HFS patients. Temporal bone computed tomography (CT), Brain MRI with MR angiography including 3D PD TSE (three-dimensional proton density-weighted turbo spin-echo) images are taken so that the bony structure of the skull, neurovascular relationship, and other

S. Lee (✉)
Department of Neurosurgery, Samsung Medical Center, Sungkyunkwan University School of Medicine, Seoul, Republic of Korea
e-mail: shben.lee@samsung.com

K. Park
Department of Neurosurgery, Konkuk University Medical Center, Seoul, Korea (Republic of)
e-mail: kwanpark@skku.edu

© Springer Nature Singapore Pte Ltd. 2020
K. Park, J. S. Park (eds.), *Hemifacial Spasm*, https://doi.org/10.1007/978-981-15-5417-9_8

brain conditions are evaluated. Coronal images which are angled parallel to the facial nerve as well as axial images are of great help to understand the neurovascular relationship. Furthermore, preoperative EMG, nerve conduction study, and nerve excitability test, i.e. AMR, are routinely evaluated in every patient. We published our data that preoperative identification of AMR is helpful for better detection of AMR and higher disappearance of AMR during and after MVD [4].

Patient Position for Lateral Suboccipital Retrosigmoid Approach

We prefer a park-bench position or three quarter prone position. Compared to supine or prone position, a park-bench position requires less neck flexion in any direction, which is beneficial in patients especially with obesity, short neck, or muscular neck. After the patient is in position, we perform three-pin skull fixation with Mayfield® skull clamps; one pin is located on ipsilateral forehead and two are on occipital bone near occipital sinus above transverse sinus. Finally, head is immobilized with the head being rotated 5°–10° to the contralateral side that makes the mastoid tip on top. And a lateral tilt of neck using gravity without further force and an anterior flexion of neck allowing about two-finger breadth space between neck and jaw can make the lateral suboccipital area be widened and adequate for a retrosigmoid approach. Excessive flexion of neck should be avoided for possible airway obstruction, blockage of contralateral jugular venous return, or postoperative neck pain. At last, but not least, immobilization of the body and placing positioners at each pressure position is important to avoid any potential complication related to misposition or malposition such as compressive neuropathy, vascular compromise, and even fall-down during movement of surgical table. Ipsilateral shoulder is pulled in the caudal direction and fixated with medical plaster tape bounded to the surgical table tightly so that that the ipsilateral lateral suboccipital area can be easy to access. The contralateral arm is dropped naturally below the table and wrapped with an arm sling. Body positioners made of soft foam are placed at each point where the body and the surgical bed or instruments are touched. In the end, any additional hazard to the patient should be assessed while the surgical bed is moving; being tilted or in reverse Trendelenburg position (Fig. 1).

From Skin to Dural Opening

Using landmarks of skull such as mastoid notch, zygomatic arch, and inion, the locations of transverse and sigmoid sinus are presumed and the location of skull opening is determined. Considering the thickness of neck and hair line, a curvilinear lazy "S"-shaped skin incision line is marked. Vertical skin incision is also feasible, but it leaves surgical wound on the neck outside of the hairline and medio-caudal skin tag may block the microscopic view when a neurosurgeon tries to inspect the root exit zone (REZ) of the facial nerve in a caudal-to-cephalic direction. Skin incision starts with the depth to the fascia layer (Fig. 2a), and muscle dissection of sternocleidomastoid muscle and splenius capitis muscle in order is performed. White coarse connective tissue layer is exposed and an occipital artery can be identified below it (Fig. 2b). After coagulation and cut of an occipital artery, dissection of the scalp continues down to the skull, and profuse venous bleeding from mastoid emissary vein at mastoid foramen can be encountered. Hemostasis of bleeding from mastoid emissary vein can be accomplished using monopolar cautery or bone wax according to the size of the vein. When the vein is small, monopolar cautery alone is sufficient to control the bleeding. However, if the vein is bigger and the mastoid foramen is large not to be controlled with monopolar cautery, bone wax is used to stop the bleeding. If the foramen is so large that bone wax keeps being pushed into the foramen, the obstruction of the sinus can be anticipated when the large amount of bone wax is pushed in. Scalp dissection over the skull continues to the posterior margin of the mastoid process

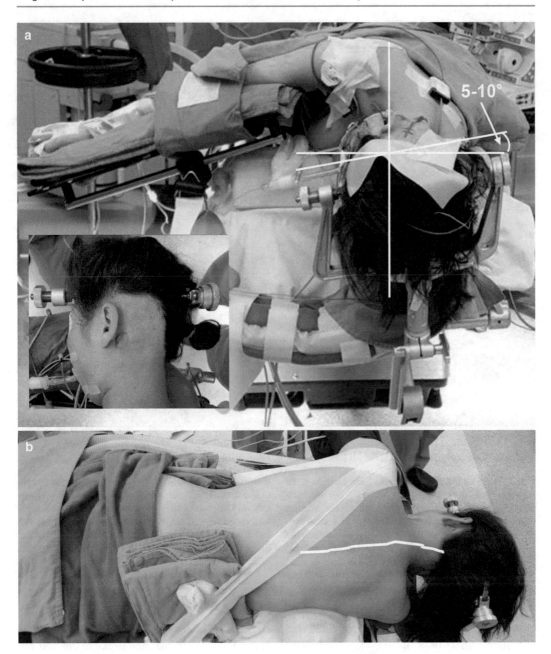

Fig. 1 Positioning of the patient for microvascular decompression. (**a**) The patient is in a park-bench position and the head is fixated with three-pin skull clamps. The head is rotated 5°–10° to the contralateral side that makes the mastoid tip on top, (**b**) and a lateral tilt of neck using gravity without further force and an anterior flexion of neck allowing about two-finger breadth space between neck and jaw can make the lateral suboccipital area be widened and adequate for a retrosigmoid approach. See text for further details

laterally, transverse sinus superiorly, and just below inferior nuchal line inferiorly. And medial limit of dissection depends on the thickness of the scalp, which enables the visualization of CPA without being obstructed by skin tag. Monopolar cautery should be used carefully below inferior nuchal line, or instead, blunt dissection with Penfield dissector #1 in a peel-off fashion is

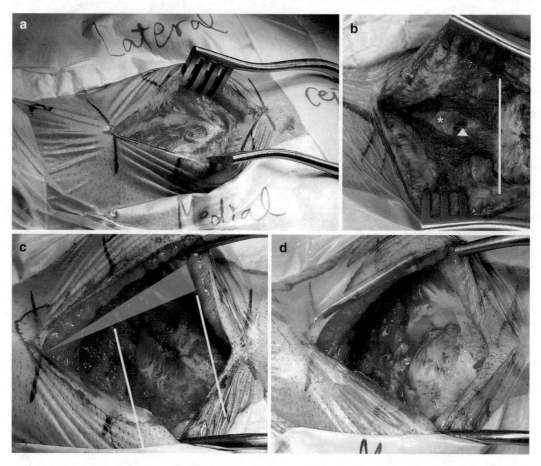

Fig. 2 Scalp and skull opening. (**a**) Skin incision starts with the depth to the fascia layer (asterisk), (**b**) and muscle dissection of sternocleidomastoid muscle and splenius capitis muscle in order is performed. White coarse connective tissue layer (filled star) is exposed and an occipital artery can be identified below it. Sometimes, mastoid foramen emissary vein (filled triangle) is so large that can be seen before an occipital artery is identified. (**c**) Scalp dissection over the skull continues to the posterior margin of the mastoid process laterally, transverse sinus superiorly, and just below inferior nuchal line inferiorly. A mastoid process is located in blue triangle, and superior and inferior nuchal lines are depicted in yellow lines. (**d**) After the scalp dissection, a craniectomy with the average size of 2.5cm × 3.5 cm is performed. (See text)

effective and safe without worrying about damage of vertebral artery or occipital condylar emissary vein (Fig. 2c).

After the scalp dissection and exposure of the adequate extent of skull, a cranial opening is performed. The size of a cranial opening is also customized according to the patient's anatomical characteristics. A 2.5 cm × 3.5 cm of a craniectomy usually gives sufficient operating space. Lateral margin of the craniectomy is to the very edge of sigmoid sinus, and superior margin does not necessarily meet transverse sinus in MVD for

HFS; around 0.5 cm above the mastoid notch is enough to explore the cranial nerve 7th and 8th complex. Inferior margin confines to inferior nuchal line, and undercutting of the inner table of the skull at the inferior nuchal line is sufficient for approaching to the cisterna magna in case of difficulty in cerebrospinal fluid (CSF) drainage at cerebellopontine cistern (Fig. 2d).

Management strategy of venous sinus injury is composed of elevation of head and application of hemostatic agents. If there is profuse bleeding from the sinus tear, the position of head needs to be higher and high-pressure suction of blood clot is required to estimate the size of injury. Use the hemostatic agent such as Surgicel® or TachoSil® as small amount as possible over the injured site and cottonoid patties are used to cover the site and press it gently. If there is larger amount of hemostatic agents used or too much pressure is applied, the size of tear can be increased. After the bleeding is reduced, surgical adhesive is applied over the injury area. Venous bleeding is bound to be ceased, that is why the neurosurgeon should remain calm and use as small amount of hemostatic agent, and as small pressure on the site as possible to control the venous sinus injury.

After cleansing out the bone dusts with saline irrigation, a dural opening is followed. A curvilinear durotomy is performed parallel to the sigmoid sinus and the dura is reflected laterally. It is easy to damage the cortical vessel of the cerebellum during durotomy, and placing a small cottonoid patty is helpful to avoid the injury. A cerebellopontine cistern is punctured and CSF drainage is pursued while the cerebellum is gently retracted using brain spatula. If the CSF is not drained enough, cisterna magna is punctured at the inferomedial direction. If the cerebellum is adequately sunken down, rubber and cottonoid patties are placed over the cerebellum to avoid damaging it when the instruments are placed in and out of posterior fossa.

Exploration Along the CPA to the Neurovascular Compression Site

Approaching to the jugular foramen is performed in the first place. Usually, there is a thick and touch arachnoid membrane covering lower cranial nerves (LCNs), which tightly anchors nerves to the adjacent tissues. When the dissection is insufficient, bradycardia can be induced and cerebellar retraction becomes difficult. Therefore, dissection around LCNs using micro-bayonet or micro-scissors needs to be thorough to the very medial and superior part of LCNs (Fig. 3a). And then, root exit and entry zones of cranial nerve (CN) 7th and 8th are gradually visualized in the

cephalic direction. The corridor between the complex of CN 7th and 8th and LCNs are the route to the REZ or neurovascular compression site (Fig. 3b). Usually, broad arachnoid membrane covered over REZ needs to be dissected so that there is no hindrance of visualization. As Jannetta mentioned, retraction of cerebellum using retractor blade or suction should be directed perpendicular to the axis of the CN 7th and 8th complex, not the longitudinal direction of the complex [5]. Frequently, the CN 7th is hard to be visualized even with excessive cerebellar retraction. Head position change by surgical bed movement can be tried and helpful. Under the continuous brainstem auditory evoked potentials (BAEPs) monitoring, frequent release of retraction is required. Prolonged retraction may cause changes in BAEPs and hearing loss. And branches from anterior inferior cerebellar artery or posterior inferior cerebellar artery near internal auditory canal need to be checked repeatedly to see if there is a vasospasm. Stretching causes shrinkage of the vessel diameter, and warm saline or topical vessel dilator such as papaverine needs to be applied. A direct visualization of neurovascular compression site with indentation on the facial nerve is known to be the single-most important factor to determine the prognosis of HFS patients after MVD [6, 7]. Try hard to find the culprit near REZ of the facial nerve, and exploration can be extended to the cisternal portion or medial side of the facial nerve if there is no significant indentation on the REZ. Angled endoscope is helpful to observe the deep neurovascular compression site without excessive cerebellar retraction, especially if the patient has a prominent flocculus or complex vessels blocking the REZ.

Decompression of the Facial Nerve

We have used an interposition method in every HFS patient. Although a transposition method may be, theoretically, more definitive way of decompression of the nerve, there is no direct comparison study of the MVD outcomes between two methods. Moreover, the transposition method cannot be applied in some types of HFS, espe-

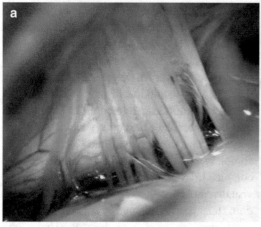

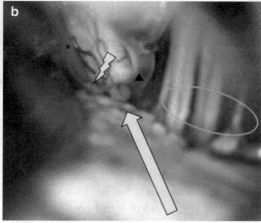

Fig. 3 Approaching to the neurovascular compression site. (**a**) The dissection around LCNs using micro-bayonet or micro-scissors needs to be thorough to the very medial and superior part of LCNs. (**b**) Then, root exit and entry zones of CN 7th (asterisk) and 8th and an offending vessel (filled triangle) are gradually visualized in the cephalic direction. The corridor (yellow arrow) between the complex of CN 7th and 8th and LCNs (circle) are the route to the neurovascular compression site (lightning bolt). *LCN* lower cranial nerve, *CN* cranial nerve

cially HFS with offending vessel having many perforators. And we think that the interposition method is reasonable in the aspect of clinical outcome; overall, 90% chance of spasm-free rate is reported using the interposition method without risk-taking of complications during struggling around narrow CPA [2, 7, 8].

Teflon felts (Teflon felt (BARD® PTFE Felt Pledget, Bard Peripheral Vascular Inc., Tempe, Arizona, USA) are prepared in three sizes (Fig. 4) and used according to the size of the affected offending vessel or anatomical working space. Adequate amount of Teflon felt should be used to decompress the facial nerve completely and to avoid possible Teflon granuloma. After complete decompression, Teflon felt is immobilized using surgical adhesive. Although surgical adhesive is going to be melting away in the end, it will work in effect during the period of brisk change in CSF dynamics during and right after the surgery. And there may be an inflammatory reaction upcoming before the adhesive is melting away, and which can prevent the Teflon felt from slippage.

CNs should not be manipulated directly in principle. However, some degree of manipulation cannot be avoided. A slight stretching of the CN 7th and 8th is easily observed during MVD even with the infrafloccular approach as Jannetta rec-

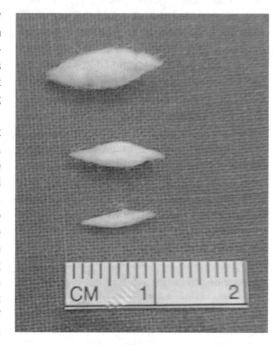

Fig. 4 Preparation of Teflon felts. Teflon felts are prepared in three sizes, and used according to the size of the affected offending vessel or anatomical working space

ommended [5]. And prominent flocculus or medially displaced REZ sometimes needs further retraction than the retraction in typical MVD surgeries. Inevitable manipulation of CNs is anticipated in HFS with offending vessels at medial side, in the middle of the facial nerve, or between CN 7th and 8th. For the better recovery of probably damaged CNs during manipulations, prolonged retraction should be avoided, and normal vasculature, even the small arteriole and vein, should be preserved. If there is a warning sign during monitoring of BAEPs or free-running EMG, which implies nerve irritation or damage, the cerebellar retraction should be released. Other measures which can be applied are as follows: head position is to be lowered, warm saline is applied in posterior fossa, and topical papaverine or intravenous steroid can be used.

From Dura to Skin Closure

After the sufficient decompression of the facial nerve, posterior fossa is to be filled with warm saline. If there is any active bleeding from the irrigation, exploration in posterior fossa to find bleeding point should be sought and managed properly. Usually, continuous saline irrigation or gentle compression with cottonoid patty or Teflon felt can manage the bleeding. Thereby, dural closure is initiated. Previously we used muscle plugs between the dural sutures to prevent CSF leakage [9]. Recently we are using DuraGen®, TachoSil®, and surgical adhesive to support the dural closure for the cosmetic problems, "neck depression", which is a common complaint from postoperative patients at the outpatient clinic. Dural sutures are performed with the plugged DuraGen®, TachoSil® is used as an overlay on dura, and finally, surgical adhesive is applied on it. Mastoid air cell sealing needs to be completed in the first place for sure. Then, a cranioplasty using polymethyl methacrylate bone cement is performed to fill the bony defect from a craniectomy, and artificial bone flap is fixed to the nearby skull using plate and screw. One or two surgical knots from dura are tied over the fixating plate which was bridging between the artificial bone and the surrounding skull to suspend the dura and allow the dura to adhere to the overlying bone [10] (Fig. 5).

Massive saline irrigation at lateral suboccipital area is performed to cleanse out the bone dust and to find the bleeding point of whole layers of scalp. After the irrigation and hemostasis of scalp, scalp closure starts with two or three neck muscle sutures. Fascia sutures need to be done in compact intervals. Subcutaneous layer and skin sutures are performed orderly without a surgical drain.

Postoperative Management

After the surgery, the patient is moved to the intensive care unit for immediate postoperative management. If the routine brain CT taken on postoperative day (POD) 1 shows no significant abnormality, the patient is sent to the general ward and continues to be taken care of. We routinely use postoperative steroid, methylprednisolone, for the first 24 h after MVD. The patient is discharged on POD 5, and postoperative hearing function is examined before the discharge. Regular outpatient clinic is scheduled on postoperative 1 month, 3 months, 6 months, and 12 months. After that, follow-up is performed in annual or biannual intervals. We do not have regular postoperative medication.

Use of Endoscope

Facial palsy and hearing loss are the major complications after MVD. Possible causes of them include stretching of CNs, direct trauma, and vasospasm (e.g. labyrinthine artery in case of hearing loss) during cerebellar retraction or nerve manipulation [11]. Moreover, hearing loss after MVD surgery is mainly due to the cerebellar retraction during MVD [12, 13], and we have observed that the risk of hearing loss is increased when the patient had large flocculus or medially deviated brainstem. In those cases, further retraction is required to observe and explore the neurovascular compression site. In case of facing a warning sign [14] in monitoring of BAEPs during exploration at CPA, neurovascular relationship cannot be understood in detail. Teflon felt is inserted to the point where the vessel is visualized and the instrument is reached at best. Surgical

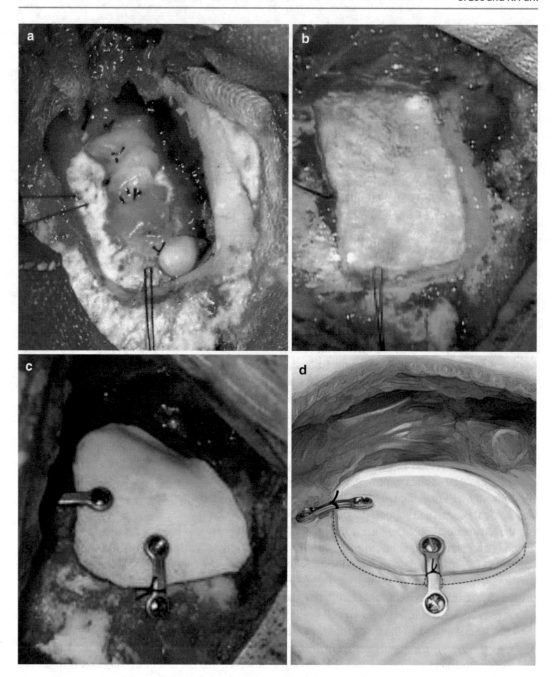

Fig. 5 Dura and skin closure. To prevent CSF leakage, we use three types of barriers. (**a**) Dural sutures are performed with the plugged DuraGen®, (**b**) TachoSil® is used as an overlay on dura, and finally surgical adhesive is applied on it. (**c**) A cranioplasty using polymethyl methacrylate bone cement is performed to fill the bony defect from a craniectomy, and artificial bone flap is fixed to the nearby skull using plate and screw. One or two surgical knots from dura are tied over the fixating plate which was bridging between the artificial bone and the surrounding skull to suspend the dura. (**d**) Illustration demonstrates how knots suspend the dura and allow the dura to adhere to the overlying bone. *CSF* cerebrospinal fluid

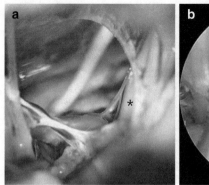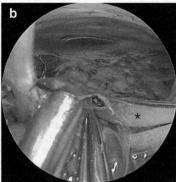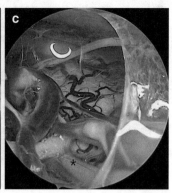

Fig. 6 Endoscopic assistance. Endoscope-assisted MVD has a great advantage for understanding the neurovascular relationship, and also for checking the decompressed facial nerve after releasing retractor or suction device. (**a**) Vessels and prominent flocculus are blocking the root exit zone of the facial nerve (asterisk) from the microscopic view. (**b**) 30°-angled endoscope shows clear neurovascular relationship in the same patient. (**c**) After placement of one Teflon felt, further exploration can be considered for further decompression using more Teflon felts. *MVD* microvascular decompression

outcome from this unclear decompression without identifying the culprit vessel can be inferior [15]. Moreover, the neurovascular relationship can be distorted when the CSF is drained and the cerebellum is retracted either by retractor blade or suction device, so that the offending vessel can be misinterpreted in some cases.

Therefore, initial observation of neurovascular compression site without excessive arachnoid dissection or cerebellar shifting is important to find the culprit offending vessel. It is the endoscope that helps in this respect, especially 30°-angled endoscope. Endoscope-assisted MVD has a great advantage for understanding the neurovascular relationship, and also for checking the decompressed facial nerve after releasing retractor or suction device (Fig. 6). If the neurosurgeon is an expert in endoscope, fully endoscope MVD can be performed, too. Small cranial opening and better visualization can minimize the operation time and maximize the surgical outcome. Although the corridor between the complex of CN 7th and 8th and the LCNs is so narrow that heat from endoscope may cause nerve injury, development of instruments and skillful hands can make up for its present short-

comings. Basic surgical skills are prerequisite for the endoscope users for the possible difficulty in CSF drain, hemostasis, or cerebellar swelling.

References

1. da Silva Martins WC, de Albuquerque LAF, de Carvalho GTC, Dourado JC, Dellaretti M, de Sousa AA. Tenth case of bilateral hemifacial spasm treated by microvascular decompression: review of the pathophysiology. Surg Neurol Int. 2017;8:225.
2. Miller LE, Miller VM. Safety and effectiveness of microvascular decompression for treatment of hemifacial spasm: a systematic review. Br J Neurosurg. 2012;26:438–44.
3. Zhang X, Wang XH, Zhao H, Tang YD, Zhou P, Yuan Y, et al. Surgical treatment of secondary hemifacial spasm: long-term follow-up. World Neurosurg. 2019;125:e10–5.
4. Lee S, Park SK, Lee JA, Joo BE, Kong DS, Seo DW, et al. A new method for monitoring abnormal muscle response in hemifacial spasm: a prospective study. Clin Neurophysiol. 2018;129:1490–5.
5. McLaughlin MR, Jannetta PJ, Clyde BL, Subach BR, Comey CH, Resnick DK. Microvascular decompression of cranial nerves: lessons learned after 4400 operations. J Neurosurg. 1999;90:1–8.
6. Kim HR, Rhee DJ, Kong DS, Park K. Prognostic factors of hemifacial spasm after microvascular decompression. J Korean Neurosurg Soc. 2009;45:336–40.

7. Lee JA, Kim KH, Kong DS, Lee S, Park SK, Park K. Algorithm to predict the outcome of microvascular decompression for hemifacial spasm: a data-mining analysis using a decision tree. World Neurosurg. 2019;125:e797–806.

8. Sindou M, Mercier P. Microvascular decompression for hemifacial spasm: outcome on spasm and complications. A review. Neurochirurgie. 2018;64:106–16.

9. Park JS, Kong DS, Lee JA, Park K. Intraoperative management to prevent cerebrospinal fluid leakage after microvascular decompression: dural closure with a "plugging muscle" method. Neurosurg Rev. 2007;30:139–42.

10. Lee S, Park SK, Joo BE, Lee JA, Kong DS, Park K. A surgical strategy to prevent delayed epidural hematoma after posterior fossa surgery using lateral suboccipital retrosigmoid approach. J Clin Neurosci. 2018;52:156–8.

11. Bartindale M, Kircher M, Adams W, Balasubramanian N, Liles J, Bell J, et al. Hearing loss following posterior fossa microvascular decompression: a systematic review. Otolaryngol Head Neck Surg. 2018;158:62–75.

12. Lee MH, Lee HS, Jee TK, Jo KI, Kong DS, Lee JA, et al. Cerebellar retraction and hearing loss after microvascular decompression for hemifacial spasm. Acta Neurochir. 2015;157:337–43.

13. Li N, Zhao WG, Pu CH, Yang WL. Quantitative study of the correlation between cerebellar retraction factors and hearing loss following microvascular decompression for hemifacial spasm. Acta Neurochir. 2018;160:145–50.

14. Park SK, Joo BE, Lee S, Lee JA, Hwang JH, Kong DS, et al. The critical warning sign of real-time brainstem auditory evoked potentials during microvascular decompression for hemifacial spasm. Clin Neurophysiol. 2018;129:1097–102.

15. Lee S, Park SK, Lee JA, Joo BE, Park K. Missed culprits in failed microvascular decompression surgery for hemifacial spasm and clinical outcomes of redo surgery. World Neurosurg. 2019;129:e627–33.

Technical Difficulties of Microvascular Decompression Surgery for Hemifacial Spasm

Kwan Park and Seunghoon Lee

Offending Vessel with Perforators into Brainstem

Branching vessels from anterior inferior cerebellar artery (AICA), posterior inferior cerebellar artery (PICA), or vertebral artery (VA) at ponto-medullary junction can go into the brainstem and supply it. These perforating vessels usually anchor the root vessel to the brainstem and prevent the root vessel from being mobilized (Fig. 1). If there is an offending vessel anchored to the brainstem right over the root exit zone (REZ) of the facial nerve, MVD surgery gets difficult. High elevation of the offending vessel as in a typical MVD surgery is dangerous due to brainstem infarction and hemorrhage in the cistern caused by perforator rupture. Therefore, careful inspection of the offending vessel if there is perforator anchoring the offending vessel to the brainstem is advised before elevating it up to insert a Teflon felt. During surgery, careful detachment of the offending vessel from the REZ using dissector, and the Teflon felt is inserted without elevating the offending vessel. Place the

K. Park (✉)
Department of Neurosurgery, Konkuk University
Medical Center, Seoul, Korea (Republic of)
e-mail: kwanpark@skku.edu

S. Lee
Department of Neurosurgery, Samsung Medical
Center, Sungkyunkwan University School of
Medicine, Seoul, Republic of Korea
e-mail: shben.lee@samsung.com

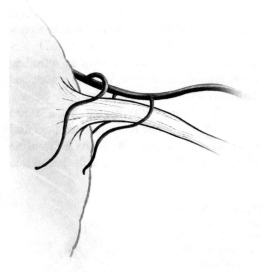

Fig. 1 Offending vessel with perforators. Perforating vessels anchor the root vessel or the offending vessel to the brainstem and prevent the root vessel from being mobilized. High elevation of the offending vessel as in a typical MVD surgery is dangerous due to brainstem infarction and hemorrhage in the cistern caused by perforator rupture. Insertion of the Teflon felt to the space between the nerve and the vessel as shown in the illustration is applicable. *MVD* microvascular decompression

Teflon felt first, and then thrust it in between the vessel and the facial nerve. According to the location of the perforators, several small Teflon felt may be required. In this type of HFS, we can hardly observe indentation or discoloration on the facial nerve, and therefore, no other conflicts of neurovascular compression along the whole

© Springer Nature Singapore Pte Ltd. 2020
K. Park, J. S. Park (eds.), *Hemifacial Spasm*, https://doi.org/10.1007/978-981-15-5417-9_9

axis of the facial nerve should be looked into before decompressing the facial nerve from the vessel with perforators.

Offending Vessel Encircling the Facial Nerve

Encircling offending vessel type in HFS is the offending vessel rounding around the facial nerve more than 270° in coil (Fig. 2). It will be the best decompression of the facial nerve if the facial nerve which is caught in the vessel is pulled out of the vessel. However, we could not perform a MVD in this manner in most cases with encircling offending vessel type. Rather, circumferential insertion of the Teflon felt between the nerve and the vessel is the usual method of decompression. Just as how the Teflon felt is inserted in perforator type, careful detachment of the offending vessel from the facial nerve using dissector, and the Teflon felt is inserted piece by piece circumferentially; placing the Teflon felt first, and then thrusting it in between the vessel and the facial nerve. It is better to insert the Teflon felt from the medial side of the facial nerve to the lateral side

of it for avoiding the hindrance of visualization of the medial side if the insertion starts with the lateral side first (Fig. 3). The offending vessel in encircling type usually has a small diameter and vasospasm of it can be observed during MVD. Warm saline or papaverine irrigation may help to relieve the spasm.

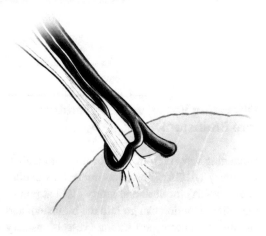

Fig. 2 Encircling offending vessel. Encircling offending vessel is rounding around the facial nerve more than 270° in coil, and compressing the facial nerve

Fig. 3 Illustration of the steps of decompressing the facial nerve from the encircling offending vessel. Careful detachment of the offending vessel from the facial nerve using dissector, and a small Teflon felt is inserted piece by piece circumferentially; placing the Teflon felt first, and then thrusting it in between the vessel and the facial nerve. It is better to insert the Teflon felt from the medial side of the facial nerve to the lateral side of it for avoiding the hindrance of visualization of the medial side if the insertion starts with the lateral side first

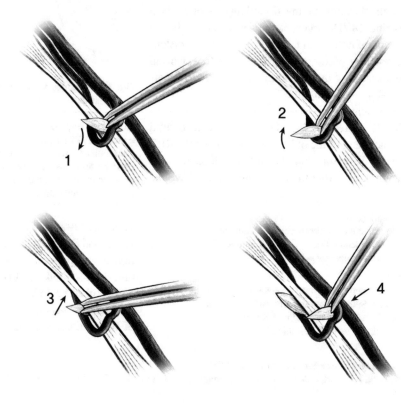

Vertebral Artery

The VA as an offending vessel has a potential surgical difficulty during MVD due to its large diameter and stiffness. MVD for VA-associated HFS has difficulties in mobilization of itself and in visualization of the neurovascular compression site near the REZ of the facial nerve. This difficulty accounts for the decreased efficacy of MVD surgery in patients with VA-associated HFS relative to those with non-VA-associated HFS [1, 2]. If the VA has a dolichoectatic change, full REZ visualization can be hardly achieved and surgery gets more difficult. Several previous studies have demonstrated various surgical techniques to completely mobilize the VA away from the facial nerve (transposition method) [3–9]. However, those surgical methods are not universally applicable to all cases and are not easily performed by all neurosurgeons. Moreover, there is no evidence that the transposition method outdoes the interposition method in respect of clinical outcome.

We perform the MVD using the interposition method in every case, the "Fulcrum Teflon method". After dural opening, arachnoid dissection around the lower cranial nerves (LCNs) followed by further cerebellar retraction usually facilitated the visualization of the REZ of the facial nerve at the ventromedial portion of the pontomedullary junction. However, the VA was identified instead in the microscopic surgical field by pushing the LCNs toward the neurosurgeon and obstructing the REZ in VA-associated HFS. Attempts to lift up the VA toward the

petrous bone high enough to expose the REZ using a microsurgical instrument often failed due to its large size and stiffness. A large piece of Teflon felt functioning as the fulcrum is placed between the proximal VA and the ventromedial brainstem near the LCNs. With the help of the fulcrum Teflon, the VA could be elevated away from the brainstem and maintained at the height of the Teflon's thickness. A surgical space could then be widened by either moving the VA distal to the fulcrum or pushing the fulcrum together with the overlying VA to further elevate the VA (Fig. 4). As shown in previous reports, majority of the VA-associated HFS had multiple offending vessels including VA [10–12]. The co-offending vessel underneath the VA and the neurovascular compression site could eventually be observed by slightly changing the angle of the microscope. Complete MVD was achieved by inserting one or two more Teflon pieces without inducing further neurovascular damages. Although we technically used the interposition method, we placed the Teflon felt between the VA and the facial nerve proximal or distal to the neurovascular compression site; hence, the site was free of both the offending vessel and Teflon felt.

Offending Vessel Located Medial to or at the Cisternal Segment of the Facial Nerve

In most HFS cases, neurovascular compression site is found at the REZ of the facial nerve. If there is no definitive offending vessel at the REZ,

Fig. 4 Fulcrum Teflon method. Illustration of widened surgical space with the help of the fulcrum Teflon felt. Attempts to lift up the VA to expose the REZ using a microsurgical instrument often failed due to its large size and stiffness. A large piece of Teflon felt functioning as the fulcrum is placed between the proximal VA and the ventromedial brainstem near the LCNs. Then, the VA

could be elevated away from the brainstem and maintained at the height of the Teflon's thickness. A surgical space could be widened by either moving the VA distal to the fulcrum or pushing the fulcrum together with the overlying VA to further elevate the VA. *VA* vertebral artery, *REZ* root exit zone, *LCN* lower cranial nerve exploration should be extended to the cisternal segment and medial side of the facial nerve. In our previous report, we could encounter atypical locations of neurovascular compression site, which may be the cause of MVD failure. Those locations include the medial side and cisternal segment of the facial nerve instead of typical locations such as lateral side and REZ of the facial nerve [13]. Neurovascular compression sites at both locations are not easily visualized, and the chances of manipulating of the cranial nerve (CN) 7th and 8th are high during vessel mobilization and decompression of the facial nerve. Therefore, thrusting the Teflon felt between the vessel and the nerve without elevating the vessel or the nerve is the basic surgical skill used in this type of HFS for there being no large space that allows lifting-up. While Teflon felt between the vessel and the medial side of the facial nerve tends to be fixed in position (Fig. 5), it can be slipped out of the neurovascular compression site at cisternal segment if the axis of the nerve and the vessel are perpendicular to each other and conflict area is small. The Teflon felt

should be placed more broadly and slippage of the Teflon felt can be prevented.

The offending vessel located at the cisternal segment of the facial nerve sometimes makes the wide-spaced CN 7th and 8th. CN 7th and 8th are located just next to each other and exit or enter through the internal auditory canal. Usually, CN 7th is almost blocked by CN 8th with the routine infrafloccular approach via the retrosigmoid approach because of this proximity of the two CNs. Moving the head position or microscope angle enables the visualization of the lateral portion of CN 7th. However, if the offending vessel is located between CN 7th and 8th, the CN 7th can be moved for, up or downward direction. Preoperative MRI showed elongated and bent CN 7th and large space is shown between CN 7th and 8th (Fig. 6). Unusual location of the CNs may mislead the identification of each CN, and REZ of the facial can also be deviated to the direction where the facial nerve is pushed. During MVD, CN identification can be aided by facial nerve direct stimulation. If the course of the offending vessel can be changed out of the com-

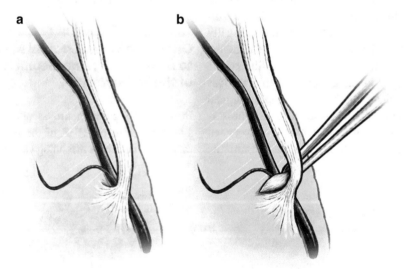

Fig. 5 Offending vessel located at the medial side of the facial nerve. (**a**) Neurovascular compression site is hardly visualized, decompression of the facial nerve cannot be performed without manipulating it, and the chances of postoperative facial palsy increase in this type of HFS. (**b**) Therefore, thrusting the Teflon felt between the vessel and the nerve without elevating the vessel or the nerve is the basic surgical skill for there being no large space that allows lifting-up. *HFS* hemifacial spasm

Fig. 6 Wide-spaced CN 7th and 8th due to the offending vessel at the cisternal segment. The offending vessel located at the cisternal segment of the facial nerve sometimes makes the wide-spaced CN 7th and 8th. (**a**) Preoperative MRI showed elongated and bent CN 7th and wide space is shown between CN 7th and 8th. (**b**) During operation, a medially located PICA (asterisk) was identified and it has moved the CN 7th anteriorly and made a wide space between the two CNs. *CN* cranial nerve, *MRI* magnetic resonance image, *PICA* posterior inferior cerebellar artery

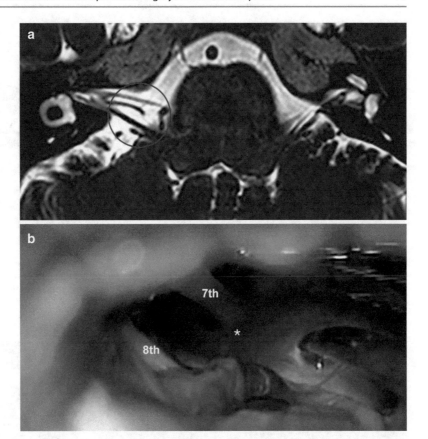

plex of CN 7th and 8th, the vessel is moved away out of the complex. However, most of the cases need to be decompressed in situ. As other medial or cisternal segment offending vessels, high chance of CN dysfunction can be anticipated due to possible manipulation of them.

Penetrating Offending Vessel Through the Facial Nerve

In this type of HFS, the offending vessel literally penetrates the facial nerve (Fig. 7). There are cases where blood vessels penetrate the middle of the facial nerve or divide the nerve into main thick portion and thin small portion. It is a very rare type of HFS and only six patients were revealed to have penetrating offending vessel in our more than 4000 MVD surgical cases.

MVD in HFS patients with penetrating offending vessel through the facial nerve is thought to be the most surgically challenging and demands delicate hands. All surface of the facial nerve is theoretically blocked off from the vessel, which is not easy without damaging the facial nerve. Therefore, high possibility of insufficient decompression of the facial nerve may lead to less favorable clinical outcome, and unavoidable facial nerve manipulation has higher chance of facial palsy. Interposition of Teflon felt between the facial nerve and the vessel is pursued, and neurectomy should be avoided as much as possible. Intraoperative monitoring of free-running electromyography and abnormal muscle response is helpful to decide the extent of surgery. As in other types with narrow space between the nerve and the vessel, dissection at the plane of neurovascular conflict, followed by interposition of the Teflon felt is performed.

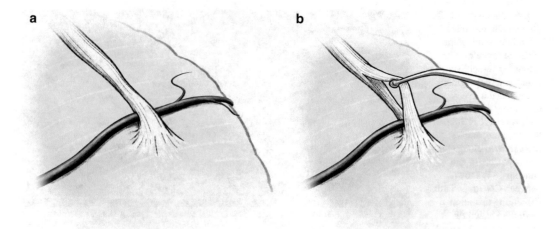

Fig. 7 Penetrating offending vessel. (**a**) Illustration shows the offending vessel literally penetrating the facial nerve. Manipulation of the facial nerve is unavoidable and this yields a high chance of the facial palsy. (**b**) The offending vessel penetrates the middle of the facial nerve

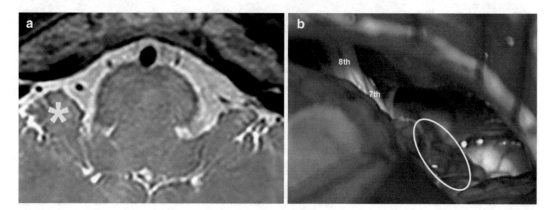

Fig. 8 Flocculus blocking the visualization of the neurovascular compression site. (**a**) Preoperative MRI showed a larger flocculus on the right (asterisk) than left side. (**b**) While approaching the neurovascular compression site, the flocculus blocked the visualization of the site and further retraction of the cerebellum including flocculus was needed. Identification of the indentation on the facial nerve failed in this patient, and frequent changes of the BAEPs were noted without resultant hearing loss after the surgery. *MRI* magnetic resonance image, *BAEP* brainstem auditory evoked potential

Prominent Flocculus Blocking the Neurovascular Compression Site

As Jannetta [14] mentioned earlier, most of the MVD for HFS is performed in infrafloccular corridor. After the dissection of the arachnoid membrane at the level of jugular foramen, we follow the corridor between the complex of CN 7th and 8th and LCNs. At this point, large flocculus can block the visualization of the complex of CN 7th and 8th (Fig. 8). Inevitably, retraction of the cerebellum including flocculus needs to be performed further, which can sometimes result in changes of brainstem auditory evoked potentials and hearing loss. Moreover, difficulty in identification of the neurovascular compression site in detail gives insufficient decompression of the facial nerve and less favorable clinical outcome.

In this regard, angled-endoscopic view provides the neurovascular relationship in situ without retraction, and reduces the retraction time during exploration of the neurovascular compression site. After the neurovascular relationship is fully understood, the culprit vessel is moved away and the Teflon felt is inserted quickly while retracting the flocculus. Flocculus can be resected to visualize REZ and perform MVD. But severe nystagmus or dizziness can be troublesome for patients, and we avoid resecting it.

Redo Surgery

Due to adhesion all along the cerebellopontine angle to the REZ and the possible higher complication rate from it, decision to perform a redo surgery tends to be made reluctantly for all neurosurgeons in general. Wider cranial opening is necessary, cautious arachnoid dissection should be guaranteed before cerebellar retraction, and it is unnecessary to remove all Teflon felt. Unusual location where neurovascular compression occurs should be inspected thoroughly. Further details will be discussed in the chapter "Redo surgery for failed microvascular decompression for hemifacial spasm" of redo surgery for failed MVD.

References

1. Kim JP, Park BJ, Choi SK, Rhee BA, Lim YJ. Microvascular decompression for hemifacial spasm associated with vertebrobasilar artery. J Korean Neurosurg Soc. 2008;44(3):131–5.
2. Kurokawa Y, Maeda Y, Toyooka T, Inaba K. Microvascular decompression for hemifacial spasm caused by the vertebral artery: a simple and effective transposition method using surgical glue. Surg Neurol. 2004;61(4):398–403.
3. Attabib N, Kaufmann AM. Use of fenestrated aneurysm clips in microvascular decompression surgery. Technical note and case series. J Neurosurg. 2007;106(5):929–31.
4. Bejjani GK, Sekhar LN. Repositioning of the vertebral artery as treatment for neurovascular compression syndromes. Technical note. J Neurosurg. 1997;86(4):728–32.
5. Ichikawa T, Agari T, Kurozumi K, Maruo T, Satoh T, Date I. "Double-stick tape" technique for transposition of an offending vessel in microvascular decompression: technical case report. Neurosurgery. 2011;68(2 Suppl Operative):377–82.
6. Kyoshima K, Watanabe A, Toba Y, Nitta J, Muraoka S, Kobayashi S. Anchoring method for hemifacial spasm associated with vertebral artery: technical note. Neurosurgery. 1999;45(6):1487–91.
7. Lee SH, Park JS, Ahn YH. Bioglue-coated Teflon sling technique in microvascular decompression for hemifacial spasm involving the vertebral artery. J Korean Neurosurg Soc. 2016;59(5):505–11.
8. Masuoka J, Matsushima T, Kawashima M, Nakahara Y, Funaki T, Mineta T. Stitched sling retraction technique for microvascular decompression: procedures and techniques based on an anatomical viewpoint. Neurosurg Rev. 2011;34(3):373–9.
9. Zaidi HA, Awad AW, Chowdhry SA, Fusco D, Nakaji P, Spetzler RF. Microvascular decompression for hemifacial spasm secondary to vertebrobasilar dolichoectasia: surgical strategies, technical nuances and clinical outcomes. J Clin Neurosci. 2015;22(1):62–8.
10. Mikami T, Minamida Y, Akiyama Y, et al. Microvascular decompression for hemifacial spasm associated with the vertebral artery. Neurosurg Rev. 2013;36(2):303–8.
11. Masuoka J, Matsushima T, Nakahara Y, et al. Outcome of microvascular decompression for hemifacial spasm associated with the vertebral artery. Neurosurg Rev. 2017;40(2):267–73.
12. Yang DB, Wang ZM. Microvascular decompression for hemifacial spasm associated with the vertebral artery. Acta Neurol Belg. 2017;117(3):713–7.
13. Lee S, Park SK, Lee JA, Joo BE, Park K. Missed culprits in failed microvascular decompression surgery for hemifacial spasm and clinical outcomes of redo surgery. World Neurosurg. 2019;129:e627–33.
14. McLaughlin MR, Jannetta PJ, Clyde BL, Subach BR, Comey CH, Resnick DK. Microvascular decompression of cranial nerves: lessons learned after 4400 operations. J Neurosurg. 1999;90(1):1–8.

Various Applications of Microvascular Decompression Other than for Hemifacial Spasm

Min Ho Lee and Jae Sung Park

Neurovascular compression syndrome (NVC) refers to a group of disorders that are caused by a direct compression on a cranial nerve by a blood vessel or vessels, and hemifacial spasm also belongs to it. After his first microvascular decompression (MVD) for trigeminal neuralgia (TN), Jannetta suggested that MVD might also be applied to other forms of NVC. Indeed, MVD is now established as the treatment of choice for HFS. This chapter presents other disease entities for which MVD may provide cure or improvement.

Trigeminal Neuralgia

Trigeminal neuralgia (TN) is a syndrome characterized by paroxysmal facial pain in the somatosensory distribution of the trigeminal nerve. Studies in the general population show that the prevalence of TN is around 0.3%, peaking in the 1960s, with some preference for female [1]. There are many treatment modalities for TN patients that are unable to obtain sufficient relief with medication. TN is the first case of neurovascular

M. H. Lee (✉)
Department of Neurosurgery, Uijeongbu St. Mary's Hospital, The Catholic University of Korea, Seoul, Republic of Korea
e-mail: minho919.lee@catholic.ac.kr

J. S. Park
Department of Neurosurgery, Konkuk University School of Medicine, Chungju Hospital, Chungju, Korea (Republic of)

compression syndrome in which Jannetta applied MVD for treatment, and MVD is still the preferred treatment modality [2–4]. There are several large series with long-term follow-up studies have been reported, with pain free rates of 70~80%, with 5~10 years follow-up [5–9]. In addition to MVD, stereotactic radiosurgery (SRS) has also been performed, but according to recent meta-analysis, MVD was associated with greater rate of short- and long-term pain freedom, lower incidences of facial numbness and dysesthesia, and pain recurrence compared to SRS [10].

MVD for TN is quite analogous to that for HFS. With the park-bench position (3/4 lateral prone decubitus), retromastoid suboccipital craniectomy can be performed as described by McLaughlin and colleagues [11]. In general, to gain access to the trigeminal nerve, it is relatively easy to approach via superolateral aspect of the cerebellum. In the process, one should be cautious not to injure the petrosal vein, for the iatrogenic obliteration of it may cause complications including venous infarction, sigmoid thrombosis, cerebellar hemorrhage, etc., and the incidence of the complications was reported up to 6.2% [12].

TN is the condition for which MVD is most frequently performed, followed by HFS. According to a recent study [8], the patients in the TN group were older than those in the HFS group. There was no predilection as to which side of the face was affected by HFS; however, TN was reported to be present more frequently on the right side. The offending vessels were mainly the

AICA and/or PICA in the HFS group, as opposed to the SCA in the TN group. The initial response to MVD for TN did not differ from that for HFS, although the recurrence rate following the former was significantly lower than the latter. MVD is an effective treatment for both HFS and TN. MVD is a very promising intervention for HFS but is associated with a risk of recurrence when used to treat TN. The application of MVD surgery should be carefully considered in the context of these specific conditions.

Glossopharyngeal Neuralgia

Glossopharyngeal neuralgia (GPN) is a rare pain syndrome in the sensory distribution of the IX cranial nerve with a brief episodic unilateral, sharp, and stabbing pain. The glossopharyngeal nerve distribution includes the angle of the jaw, ear, tonsillar fossa, and the tongue base, and GPN accounts for 0.2–1.3% of all types of cranial neuralgias. The prevalence of GPN is estimated to be approximately 0.8/100,000 populations/year, which is far less common than TN (4.7/100,000) [13]. GPN shows a predilection for the left side of the body, whereas TN is more commonly observed on the right side [14]. The etiology of glossopharyngeal neuralgia is not entirely elucidated, but the neurovascular compression appears to be one of the main causes [15, 16].

The glossopharyngeal nerve is a mixture of cranial nerves that contain the somatic sensory fibers from the oropharynx, mastoid, middle ear, and Eustachian tube, as well as the posterior third of the tongue. The middle ear and mastoid areas are also innervated by the glossopharyngeal nerve via the tympanic branch or Jacobson's nerve [15, 17]. Like TN, carbamazepine, gabapentin, and pregabalin are first-line pharmacological treatments for glossopharyngeal neuralgia. Nerve block may be an option for GPN when refractory to medications; however, one must not trivialize the possibility of complications including dysphagia or hoarseness, not to mention arrhythmia and syncope [18, 19].

Jannetta performed MVD for six GPN patients in 1977 and reported the pain control rate of 80% [20]. Zhao et al. reported 94.3% of pain control rate based on their performance of MVD for 35 GPN patients [21]. MVD has been established as an effective and safe treatment option for glossopharyngeal neuralgia [22].

Geniculate Neuralgia

Geniculate neuralgia, also known as nervus intermedius neuralgia, is a pain in the ear triggered by sensory or mechanical stimuli at the posterior wall of the auditory canal without any pathology. The causative nerve is intermediate nerve (Wrisberg's nerve). It is a sensory component of the CN VII, containing sensory and parasympathetic fibers. The intermediate nerve joins the motor root of the facial nerve in the internal auditory canal (meatal or intracanalicular segment) to form a common trunk. The sensory auricular branch of the facial nerve arises from the vertical segment of the facial nerve, between the second genu and the chorda tympani nerve origin. This auricular sensory nerve usually arcs laterally and courses inferiorly to supply the posterior and inferior external ear canal at the region of the osseous–cartilaginous junction, as well as the inferior portions of the pinna [23].

In addition to the aforementioned TN and glossopharyngeal neuralgia, geniculate neuralgia may also be treated with medications such as carbamazepine.

Jannetta reported in 1997 that 13 patients had good results in 14 patients by sectioning intermediate nerves [24]. However, Holste et al. also had good results in 13 of 15 patients, but recurred in 6 patients with 4.8 year of median pain-free interval, which was interpreted somewhat negatively [25]. Other researchers tried surgical treatments, but geniculate neuralgia seems to have performed transection rather than MVD [26, 27]. Other researchers reported at the case report level that MVD had good results [28–31]. On the other hand, there was a case series of chorda tympani was identified and preserved through postauricular mastoidectomy and sensory branches of the facial nerve were identified and transected [32]. Taken all together, the initially, transection of intratemporal division of the cutaneous branches of the facial nerve can be tried. If this is unsuccessful, resection of the nervus intermedius is advised; simultaneous microvascular decompression should be performed if there is also vascular compression [23].

Hemimasticatory Spasm

Hemimasticatory spasm is a rare clinical entity characterized by involuntary and paroxysmal contractions of the masseteric muscles on one side of the face. Reported results show hemimasticatory spasm more commonly presents in females in the third and fourth decade [33, 34]. The cause is the motor branch of the trigeminal nerve. Hemimasticatory spasm involves the masseter and the temporalis muscles, with the medial pterygoid muscle also rarely being involved. There is usually no involvement of the jaw-opening muscles. Motor branch travels inferior to the trigeminal ganglion and inferomedial to the mandibular branch when reaching the foramen ovale. Pathophysiologic examination can be used to distinguish oromandibular dystonia, etc. The characteristic EMG findings of hemimasticatory spasm include irregular bursts of motor unit potentials that correlate with the involuntary masseter spasms [33, 34].

It is relatively effective for botulinum toxin type A injection. However, this is not as fundamental a treatment as HFS is. Several studies with relatively good results have been reported [35–39]. Among them, Wang et al. reported a case series. Of the six cases, four cases showed improvement immediately after surgery and no recurrence [38]. If there is contraindication to MVD, or if the patient is hesitant about surgical treatment, botulinum toxin injection may be tried first.

Tinnitus, Vertigo

Tinnitus is a very common disease, reported by 10–20% in adults. In most cases, mild symptoms are controlled by medication, but in some cases, they interfere with daily life, which greatly affects the quality of life. The American Academy of Otolaryngology–Head and Neck Surgery (AAO-HNS) recently issued a clinical practice guideline for tinnitus in 2014. But, use of MVD in the treatment for tinnitus is not addressed, yet [40]. However, some groups have performed MVD on medically refractory tinnitus, and report relatively good results. In 1975, Jannetta attempted to explain hyperactive CN VIII symptoms such as intractable tinnitus and/or vertigo [41]. Since then, MVD has been attempted by many surgeons [42–44]. Recently, there was a systemic literature review with previously reported 43 cases. The authors reported MVD for CN VIII showed 70% success rate, which is lower than that of HFS or TN. The tinnitus-only patients had an even lower success rate of 60%, suggesting the importance of accurate diagnosis. The shorter the duration of symptoms, 5-year or less, showed the better the results.

Hypertension

Refractory hypertension is believed to affect 5–30% of the general population [45]. In the 1970s, Jannetta mentioned the possibility for neurogenic etiology in refractory hypertension [46]. They discovered a relationship between arterial compressions of the brainstem, in particular the rostral ventrolateral medulla (RVLM) and its impact on the regulation of cardiovascular activities. Since then, the idea has been established through anatomy studies. RVLM contains sympathoexcitatory bulbospinal neurons that play an important role in the control of blood pressure [47, 48]. And effectiveness of MVD has been demonstrated in several case series [49–52]. Compression was due to dolicoectatic vertebral artery or basilar artery at the medulla, which showed improvement after decompression. Sandou et al. reported the result of MVD for HFS with hypertension [53]. They mobilized the low cranial nerve in patients with HTN while operating MVD for HFS patients. At last follow-up, blood pressure was normalized in 28 patients

(58.33%), and medication is withdrawn in 14 patients (50%).

There must be some patients with refractory hypertension who could potentially benefit from MVD operation to normalize their BP. However, the long-term impact of this procedure remains debatable. Despite questionable long-term outcomes, the observed positive short-term outcomes indicate potential for future interventions to improve hypertension in this population.

Spasmodic Torticollis

Spasmodic torticollis is atonic and clonic neuromuscular disorder characterized by continuous or intermittent involuntary spasm of the cervical musculature. The exact cause of spasmodic torticollis has not yet been identified. Various hypotheses are suggested, such as abnormal neurotransmitters concentration [54], or maybe genetics [55]. Various surgical treatments are suggested—myotomy, neurectomy, selective peripheral denervation, MVD, and even deep brain stimulation. In 1995, Jannetta and Jho performed MVD on accessory nerves in 20 patients and reported cure in 13 patients and symptom improvement in 4 patients [56].

Compared to other diseases mentioned in this article, relatively many operations were performed. Li et al. [57] compared MVD (80 patients) and neurectomy (41 patients), and reported better improvement of symptoms in the MVD group than neurectomy. And they reported improvement of symptoms by 6 months.

The American Academy of Neurology recommends the Botulinum toxin as first-line treatment for primary spasmodic torticollis [58]. However, the limitation of Botulinum toxin is also obvious. Thus, MVD could be chosen in spasmodic torticollis patients with confirmed accessory nerve compression.

Others

In addition, MVD has been tried for a wide variety of disease entities. Table 1 describes the neurovascular conflict syndromes that can occur depending on the each cranial nerve. Some of these are hypoactivity. This is far from the general definition of neurovascular conflict. However, it may be also caused by neurovascular compression with the cranial nerve, which is included in the table.

Olfactory nerve has not been reported with disease due to neurovascular conflict. In the case

Table 1 Periodic table of neurovascular compression syndromes

Cranial nerve		Pathology	References
I	Olfactory nerve		none
II	Optic nerve	Paroxysmal phosphenes	[59, 60]
III	Oculomotor nerve	Oculomotor nerve palsy[a] Ocular neuromyotonia[a]	[61–64]
IV	Trochlear nerve	Superior oblique myokymia	[65–67]
V	Trigeminal nerve	Trigeminal neuralgia	[b]
	Mandibular branch	Hemimasticatory spasm	[35–39]
VI	Abducens nerve	Abducens nerve palsy[a]	[68, 69]
VII	Facial nerve	Hemifacial spasm	[b]
	Intermediate nerve	Geniculate neuralgia	[23, 70–72]
VIII	Cochlear, Vestibular nerve	Tinnitus, Vertigo	[41, 42, 44, 73]
IX	Glossopharyngeal nerve	Glossopharyngeal neuralgia	[20, 22, 74–77]
X	Vagus nerve	Essential hypertension	[46, 49–51]
XI	Accessory nerve	Spasmodic torticollis	[56, 57]
XII	Hypoglossal nerve	Hypoglossal palsy[a] Hemilingual spasm	[78–80]

[a]Hypoactivity is not hyperactivity. This is far from the general definition of neurovascular conflict
[b]Too many references and not mentioned

of olfactory nerve, it is unlikely to cause any special symptoms because there is no main vessel around the olfactory bulb. In addition, olfaction is a very subjective symptom, which may be one reason why the patient does not actively complain about it.

In the case of optic nerve, MVD was reported to be effective when paroxysmal phosphenes occurred in some case reports [59, 60]. The patient, reported by De Ridder D [59], complained light flash with visual field deterioration. An MRI scan showed an ectatic distal right internal carotid artery (ICA) abutting the undersurface of the right optic nerve and a compression of the optic chiasm by the anterior communicating artery. They were approached by orbito-zygomatic craniotomy and Teflon was inserted between the A1 and optic chiasm. After surgery, symptoms improved mostly. It is a very rare case, but it is a good example of how MVD can work in a disease we overlook.

The cranial nerves (oculomotor and abducens nerves), involved in extra ocular movement, developed palsy by compression [61–64]. In the case of trochlear nerves, there were reports of superior oblique myokymia treated with MVD [65–67]. Superior oblique myokymia is a rare condition of unclear etiology. Since 1906, about 116 cases have been reported. Although meta-analysis has been published, there is no established treatment [81]. Bringewald [82] first presented the possibility of neurovascular compression and was confirmed in some cases. However, some studies have not found compression. Yet, superior oblique myokymia cannot be truly defined as a neurovascular compression syndrome.

Hypoglossal nerve has been reported with hypoactivity (hypoglossal palsy) and hyperactivity (hemilingual spasm) [78–80, 83, 84]. De Ridder et al. reported a case of hemilingual spasm caused by an arachnoid cyst [83]. Subsequently, it is suggested that hemilingual spasm could be caused by the tortuosity of the extracranial internal carotid artery [85]. Afterward, treatment of hemilingual spasm with MVD by decompression of cranial nerve XII at the lower brainstem was reported [78].

Conclusion

Even with preliminary results, MVD has been tried in a wide variety of diseases. From cranial nerve I to XII (although not found in CN I), neurovascular compression syndrome for each cranial nerve has been discovered and treated. This neurovascular compression syndrome can occur not only in the cranial nerve, but also in other autonomic nervous systems. Much research is still needed.

MVD surgery, started from Jannetta, is not an old technique but is still changing and developing. It is expected that the experiences and efforts of neurosurgeons will help more patients in the future.

References

1. De Toledo IP, Réus JC, Fernandes M, et al. Prevalence of trigeminal neuralgia: a systematic review. J Am Dent Assoc. 2016;147:570–576.e2.
2. Kaufmann AM, Price AV. A history of the Jannetta procedure. J Neurosurg. 2019;132:1–8.
3. Jannetta PJ. Arterial compression of the trigeminal nerve at the pons in patients with trigeminal neuralgia. J Neurosurg. 1967;26(Suppl):159–62.
4. Jannetta PJ. Trigeminal neuralgia and hemifacial spasm—etiology and definitive treatment. Trans Am Neurol Assoc. 1975;100:89–91.
5. Barker FG 2nd, Jannetta PJ, Bissonette DJ, Larkins MV, Jho HD. The long-term outcome of microvascular decompression for trigeminal neuralgia. N Engl J Med. 1996;334:1077–83.
6. Sarsam Z, Garcia-Fiñana M, Nurmikko T, Varma T, Eldridge P. The long-term outcome of microvascular decompression for trigeminal neuralgia. Br J Neurosurg. 2010;24:18–25.
7. Tyler-Kabara E, Kassam A, Horowitz M, Urgo L, Hadjipanayis C, Levy E. Predictors of outcome in surgically managed patients with typical and atypical trigeminal neuralgia: comparison of results following microvascular decompression. J Neurosurg. 2002;96:527–31.
8. Lee MH, Lee JA, Park K. Different roles of microvascular decompression in hemifacial spasm and trigeminal neuralgia. J Neurol Surg B Skull Base. 2019;80:511–7.
9. Wang DD, Raygor KP, Cage TA, et al. Prospective comparison of long-term pain relief rates after first-time microvascular decompression and stereotactic radiosurgery for trigeminal neuralgia. J Neurosurg. 2017;128:68.

10. Lu VM, Duvall JB, Phan K, Jonker BP. First treatment and retreatment of medically refractive trigeminal neuralgia by stereotactic radiosurgery versus microvascular decompression: a systematic review and Meta-analysis. Br J Neurosurg. 2018;32:355–64.

11. McLaughlin MR, Jannetta PJ, Clyde BL, Subach BR, Comey CH, Resnick DK. Microvascular decompression of cranial nerves: lessons learned after 4400 operations. J Neurosurg. 1999;90:1–8.

12. Narayan V, Savardekar AR, Patra DP, et al. Safety profile of superior petrosal vein (the vein of Dandy) sacrifice in neurosurgical procedures: a systematic review. Neurosurg Focus. 2018;45:E3.

13. Koopman JS, Dieleman JP, Huygen FJ, de Mos M, Martin CG, Sturkenboom MC. Incidence of facial pain in the general population. PAIN®. 2009;147:122–7.

14. Rey-Dios R, Cohen-Gadol AA. Current neurosurgical management of glossopharyngeal neuralgia and technical nuances for microvascular decompression surgery. Neurosurg Focus. 2013;34:E8.

15. Singh PM, Kaur M, Trikha A. An uncommonly common: glossopharyngeal neuralgia. Ann Indian Acad Neurol. 2013;16:1–8.

16. Teixeira MJ, de Siqueira SR, Bor-Seng-Shu E. Glossopharyngeal neuralgia: neurosurgical treatment and differential diagnosis. Acta Neurochir. 2008;150:471–5.

17. Franzini A, Messina G, Franzini A, et al. Treatments of glossopharyngeal neuralgia: towards standard procedures. Neurol Sci. 2017;38:51–5.

18. Cosgrove MA, Towns DK, Fanciullo GJ, Kaye AD. Interventional pain management. In: Vadivelu N, Urman RD, Hines RL, editors. Essentials of pain management. New York: Springer; 2011. p. 237–99.

19. Benzon H, Rathmell JP, Wu CL, Turk D, Argoff CE, Hurley RW. Practical management of pain. Philadelphia: Elsevier Health Sciences; 2013. E-Book

20. Laha RK, Jannetta PJ. Glossopharyngeal neuralgia. J Neurosurg. 1977;47:316–20.

21. Zhao H, Zhang X, Zhu J, Tang YD, Li ST. Microvascular decompression for glossopharyngeal neuralgia: long-term follow-up. World Neurosurg. 2017;102:151–6.

22. Patel A, Kassam A, Horowitz M, Chang YF. Microvascular decompression in the management of glossopharyngeal neuralgia: analysis of 217 cases. Neurosurgery. 2002;50:705–10.

23. Tang IP, Freeman SR, Kontorinis G, et al. Geniculate neuralgia: a systematic review. J Laryngol Otol. 2014;128:394–9.

24. Lovely TJ, Jannetta PJ. Surgical management of geniculate neuralgia. Am J Otol. 1997;18:512–7.

25. Holste KG, Hardaway FA, Raslan AM, Burchiel KJ. Pain-free and pain-controlled survival after sectioning the nervus intermedius in nervus intermedius neuralgia: a single-institution review. J Neurosurg. 2018;131:1–8.

26. Sachs E. The role of the nervus intermedius in facial neuralgia: report of four cases with observations on the pathways for taste, lacrimation, and pain in the face. J Neurosurg. 1968;28:54–60.

27. Tubbs RS, Mosier KM, Cohen-Gadol AA. Geniculate neuralgia: clinical, radiologic, and intraoperative correlates. World Neurosurgery. 2013;80:e353–7.

28. Saers S, Han K, De Ru J. Microvascular decompression may be an effective treatment for nervus intermedius neuralgia. J Laryngol Otol. 2011;125:520–2.

29. Sakas DE, Panourias IG, Stranjalis G, Stefanatou MP, Maratheftis N, Bontozoglou N. Paroxysmal otalgia due to compression of the intermediate nerve: a distinct syndrome of neurovascular conflict confirmed by neuroimaging: case report. J Neurosurg. 2007;107:1228–30.

30. Bellotti C, Medina M, Oliveri G, et al. Neuralgia of the intermediate nerve combined with trigeminal neuralgia: case report. Acta Neurochir. 1988;91:142–3.

31. Younes WM, Capelle H-H, Krauss JK. Microvascular decompression of the anterior inferior cerebellar artery for intermediate nerve neuralgia. Stereotact Funct Neurosurg. 2010;88:193–5.

32. Eshraghi AA, Buchman CA, Telischi FF. Sensory auricular branch of the facial nerve. Otol Neurotol. 2002;23:393–6.

33. Cruccu G, Inghilleri M, Berardelli A, et al. Pathophysiology of hemimasticatory spasm. J Neurol Neurosurg Psychiatry. 1994;57:43–50.

34. Christie C, Rodriguez-Quiroga SA, Arakaki T, Rey RD, Garretto NS. Hemimasticatory spasm: report of a case and review of the literaturedagger. Tremor Other Hyperkinet Mov (N Y). 2014;4:210.

35. Sun H, Wei Z, Wang Y, Liu C, Chen M, Diao Y. Microvascular decompression for hemimasticatory spasm: a case report and review of the literature. World Neurosurg. 2016;90:703.e705–10.

36. Chon KH, Lee JM, Koh EJ, Choi HY. Hemimasticatory spasm treated with microvascular decompression of the trigeminal nerve. Acta Neurochir. 2012;154:1635–9.

37. Yan KK, Wei JB, Lin W, Zhang YH, Zhang M, Li M. Hemimasticatory spasm with a single venous compression treated with microvascular decompression of the trigeminal motor rootlet. World Neurosurg. 2017;104:1050.e1019–22.

38. Wang YN, Dou NN, Zhou QM, et al. Treatment of hemimasticatory spasm with microvascular decompression. J Craniofac Surg. 2013;24:1753–5.

39. Dou NN, Zhong J, Zhou QM, Zhu J, Wang YN, Li ST. Microvascular decompression of trigeminal nerve root for treatment of a patient with hemimasticatory spasm. J Craniofac Surg. 2014;25:916–8.

40. Tunkel DE, Bauer CA, Sun GH, et al. Clinical practice guideline: tinnitus. Otolaryngol Head Neck Surg. 2014;151:S1–s40.

41. Jannetta PJ. Neurovascular cross-compression in patients with hyperactive dysfunction symptoms of the eighth cranial nerve. Surg Forum. 1975;26:467–9.

42. Brandt T, Strupp M, Dieterich M. Vestibular paroxysmia: a treatable neurovascular cross-compression syndrome. J Neurol. 2016;263(Suppl 1):S90–6.

43. Sindou M, Mercier P. Hemifacial spasm associated with other cranial nerve syndromes: literature review. Neurochirurgie. 2018;64:101–5.

44. van den Berge MJC, van Dijk JMC, Posthumus IA, Smidt N, van Dijk P, Free RH. Microvascular decompression of the cochleovestibular nerve for treatment of tinnitus and vertigo: a systematic review and meta-analysis of individual patient data. J Neurosurg. 2017;127:588–601.

45. Roberie DR, Elliott WJ. What is the prevalence of resistant hypertension in the United States? Curr Opin Cardiol. 2012;27:386–91.

46. Jannetta PJ, Gendell HM. Clinical observations on etiology of essential hypertension. Surg Forum. 1979;30:431–2.

47. Smith PA, Meaney JF, Graham LN, et al. Relationship of neurovascular compression to central sympathetic discharge and essential hypertension. J Am Coll Cardiol. 2004;43:1453–8.

48. Nicholas JS, D'Agostino SJ, Patel SJ. Arterial compression of the retro-olivary sulcus of the ventrolateral medulla in essential hypertension and diabetes. Hypertension. 2005;46:982–5.

49. Levy EI, Scarrow AM, Jannetta PJ. Microvascular decompression in the treatment of hypertension: review and update. Surg Neurol. 2001;55:2–10.

50. Barley J, Ellis C. Microvascular decompression: a surgical option for refractory hypertension of neurogenic etiology. Expert Rev Cardiovasc Ther. 2013;11:629–34.

51. Barley J, Ellis C. Microvascular decompression surgery for refractory hypertension of neurogenic causes. J Clin Hypertens (Greenwich). 2013;15:217.

52. Sindou M. Is there a place for microsurgical vascular decompression of the brainstem for apparent essential blood hypertension? a review. Adv Tech Stand Neurosurg. 2015;42:69–76.

53. Sindou M, Mahmoudi M, Brinzeu A. Hypertension of neurogenic origin: effect of microvascular decompression of the CN IX-X root entry/exit zone and ventrolateral medulla on blood pressure in a prospective series of 48 patients with hemifacial spasm associated with essential hypertension. J Neurosurg. 2015;123:1405–13.

54. Scheidt C, Rayki O, Nickel T, et al. Psychosomatic aspects of idiopathic spasmodic torticollis. Results of a multicenter study. Psychother Psychosom Med Psychol. 1998;48:1–12.

55. Shin M, Douglass LM, Milunsky JM, Rosman NP. The genetics of benign paroxysmal torticollis of infancy: is there an association with mutations in the CACNA1A gene? J Child Neurol. 2016;31:1057–61.

56. Jho HD, Jannetta PJ. Microvascular decompression for spasmodic torticollis. Acta Neurochir. 1995;134:21–6.

57. Li X, Li S, Pu B, Hua C. Comparison of 2 operative methods for treating laterocollis and torticollis subtypes of spasmodic torticollis: follow-up of 121 cases. World Neurosurg. 2017;108:636–41.

58. Simpson DM, Hallett M, Ashman EJ, et al. Practice guideline update summary: botulinum neurotoxin for the treatment of blepharospasm, cervical dystonia, adult spasticity, and headache: report of the Guideline Development Subcommittee of the American Academy of Neurology. Neurology. 2016;86:1818–26.

59. De Ridder D, Sime MJ, Taylor P, Menovsky T, Vanneste S. Microvascular decompression of the optic nerve for paroxysmal phosphenes and visual field deficit. World Neurosurg. 2016;85:367.e365–9.

60. Miyazawa T, Uozumi Y, Tsuzuki N, Shima K. "Phosphene": early sign of vascular compression neuropathy of the optic nerve. Acta Neurochir. 2009;151:1315–7.

61. Inoue T, Hirai H, Shimizu T, et al. Ocular neuromyotonia treated by microvascular decompression: usefulness of preoperative 3D imaging: case report. J Neurosurg. 2012;117:1166–9.

62. Suzuki K, Muroi A, Kujiraoka Y, Takano S, Matsumura A. Oculomotor palsy treated by microvascular decompression. Surg Neurol. 2008;70:210–2.

63. Kheshaifati H, Al-Otaibi F, Alhejji M. Microvascular decompression for oculomotor nerve palsy: a case report and literature review. World Neurosurg. 2016;88:695.e691–3.

64. Fukami S, Akimoto J, Fukuhara H, Kohno M. Microvascular decompression for oculomotor nerve palsy due to compression by infundibular dilatation of posterior communicating artery. World Neurosurg. 2018;119:142–5.

65. Scharwey K, Krzizok T, Samii M, Rosahl SK, Kaufmann H. Remission of superior oblique myokymia after microvascular decompression. Ophthalmologica. 2000;214:426–8.

66. Mikami T, Minamida Y, Ohtsuka K, Houkin K. Resolution of superior oblique myokymia following microvascular decompression of trochlear nerve. Acta Neurochir. 2005;147:1005–6.

67. Fam MD, Scott C, Forster A, Kamel MH. Microvascular decompression for superior oblique myokymia: case report. Br J Neurosurg. 2014;28:552–5.

68. De Ridder D, Menovsky T. Neurovascular compression of the abducent nerve causing abducent palsy treated by microvascular decompression. Case Report J Neurosurg. 2007;107:1231–4.

69. Yamazaki T, Yamamoto T, Hatayama T, et al. Abducent nerve palsy treated by microvascular decompression: a case report and review of the literature. Acta Neurochir. 2015;157:1801–5.

70. Peris-Celda M, Oushy S, Perry A, et al. Nervus intermedius and the surgical management of geniculate neuralgia. J Neurosurg. 2018;131:1–9.

71. Kenning TJ, Kim CS, Bien AG. Microvascular decompression and nervus intermedius sectioning for the treatment of geniculate neuralgia. J Neurol Surg B Skull Base. 2019;80:S316–s317.

72. Nguyen VN, Basma J, Sorenson J, Michael LM 2nd. Microvascular decompression for geniculate neural-

gia through a retrosigmoid approach. J Neurol Surg B Skull Base. 2019;80:S322.

73. Zhang L, Yu Y, Yuan Y, Xu J, Xu X, Zhang J. Microvascular decompression of cochleovestibular nerve in patients with tinnitus and vertigo. Neurol India. 2012;60:495–7.

74. Lu VM, Goyal A, Graffeo CS, Perry A, Jonker BP, Link MJ. Glossopharyngeal neuralgia treatment outcomes after nerve section, microvascular decompression, or stereotactic radiosurgery: a systematic review and meta-analysis. World Neurosurg. 2018;120:572–582.e577.

75. Teton ZE, Holste KG, Hardaway FA, Burchiel KJ, Raslan AM. Pain-free survival after vagoglossopharyngeal complex sectioning with or without microvascular decompression in glossopharyngeal neuralgia. J Neurosurg. 2019;132:1–7.

76. Langmayr JJ, Russegger L. [Surgical treatment of glossopharyngeal neuralgia]. Wien Med Wochenschr 1992;142:281–83.

77. Resnick DK, Jannetta PJ, Bissonnette D, Jho HD, Lanzino G. Microvascular decompression for glossopharyngeal neuralgia. Neurosurgery. 1995;36:64–8.

78. Osburn LL, Moller AR, Bhatt JR, Cohen-Gadol AA. Hemilingual spasm: defining a new entity, its electrophysiological correlates and surgical treatment through microvascular decompression. Neurosurgery. 2010;67:192–5.

79. Cheong JH, Kim JM, Yang MS, Kim CH. Resolution of isolated unilateral hypoglossal nerve palsy following microvascular decompression of the intracranial vertebral artery. J Korean Neurosurg Soc. 2011;49:167–70.

80. Kuroi Y, Tani S, Ohbuchi H, Kasuya H. Microvascular decompression for hypoglossal nerve palsy secondary to vertebral artery compression: a case report and review of the literature. Surg Neurol Int. 2017;8:74.

81. Zhang M, Gilbert A, Hunter DG. Superior oblique myokymia. Surv Ophthalmol. 2018;63:507–17.

82. Bringewald PR. Superior oblique myokymia. Arch Neurol. 1983;40:526.

83. De Ridder D, Alessi G, Lemmerling M, Fransen H, De Waele L. Hemilingual spasm: a new neurosurgical entity? Case report. J Neurosurg. 2002;97:205–7.

84. Moller AR, Osburn LL, Cohen-Gadol AA. Hemilingual spasm: pathophysiology. Neurosci Lett. 2009;461:76–9.

85. Heckmann JG, Marthol H, Bickel A, Dorfler A, Neundorfer B. Hemilingual spasm associated with tortuosity of the extracranial internal carotid artery. Cerebrovasc Dis. 2005;20:208–10.

Intraoperative Neurophysiological Monitoring in Microvascular Decompression for Hemifacial Spasm

Sang-Ku Park

Introduction

Globally, HFS is not a common disease. However, it is a relatively common disease in Korea, China, and Japan. We have had many experiences and many thoughts in touching more than 4000 cases. 'How can I proceed more safely?', 'How can I detect nerve damage at a faster rate?' Etc. In this chapter, we will look at the tests performed during HFS surgery. There are so many opinions on the warning criteria that determine the BAEPs test and hearing loss. There is also a question about whether the lateral spread response measured only in patients with HFS is satisfactory if it is lost during surgery, and if it does not disappear during surgery. We hope that you will have a closer look at the tests used in MVD surgeries and understand them correctly to help you become safer and more successful.

Conventional Intraoperative Neurophysiological Monitoring

Brainstem Auditory Evoked Potentials (BAEPs)

BAEPs is an electrical stimulus induced by stimulation from cochlear to brainstem by sound stimuli, which normally form 5–7 waves [1] and can evaluate the auditory pathway function. Jewett and Williston reported this noninvasive method for the first time in 1971 [2].

The Origin of Waveforms

The origin of each waveform is as follows: the I wave is the distal portion of the cochlear nerve; the II wave is the frequency, intensity, and phase of the wave sound originating from the proximal portion of the cochlear nerve. It plays a role of transferring information about the phase. Wave III performs the information processing of time and frequency, the control role of sending the wave neuron information from the cochlear nucleus to the appropriate part of the auditory nerve system. The wave IV functions to analyze the difference in the time and the intensity of the wave generated in the superior olivary complex. Therefore, it plays an important role in locating the sound source. Wave V is the most important path of the auditory system in the lateral lemniscus. It responds sensitively to the change of the negative stimulus by time and stimulus size. Wave VI and wave VII occur in the inferior

S.-K. Park (✉)
Department of Neurosurgery, Konkuk University Medical Center, 120-1 Neungdong-ro, Gwangjin-gu, Seoul, Republic of Korea
e-mail: heydaum@hanmail.net

© Springer Nature Singapore Pte Ltd. 2020
K. Park, J. S. Park (eds.), *Hemifacial Spasm*, https://doi.org/10.1007/978-981-15-5417-9_11

colliculus, which also deal with auditory and somatosensory information. Wave VI and wave VII are not all observed, so they are generally excluded from the waveform evaluation [3].

Filtering

The BAEPs waveform is a very small waveform, so it is strong at 150 Hz to block the waveform from shaking even a little. HFF 3 KHz using a filter to make fine and fast waveforms visible, Use a notch filter to prevent the incorporation of electrical artifacts.

Electrode Position

Surface disc electrodes are used for recording and are placed on both mastoid processes and the vertex. A two-channel montage is used: Cz-Ai (vertex to ipsilateral mastoid) and Cz-Ac (vertex to contralateral mastoid). The auditory stimulus is a broadband click delivered through a tubal insert phone sponge. The operative ear is stimulated with an intensity of 85 dB normal hearing level (nHL), and a masking stimulus of 55 dB nHL or intensity of 120 dB sound pressure level (SPL), and a masking stimulus of 80 dB SPL are delivered to the contralateral ear.

Stimulation Rate and Averaging Time

Several hospitals around the world are doing a little different test for BAEPs. This is because the test has not yet been fully established, and experience with the surgery is different. Based on the paper, Thirumala [4], the stimulation rate was 17.5 Hz, averaging 256 (at least) times, and Amagasaki [5], the stimulation rate 10 Hz and averaging 1000 times. In Damaty [6], the stimulation rate was 11.1 Hz, averaging 1000 times, and so on, it is slightly different. James [7] reported that the test time was longer than 1 min, but the results should be reported within 15 s, depending on the stimulation rate 31.1 Hz, averaging 1000 and 4000 times.

During BAEPs monitoring, investigating patterns of BAEPs changes, analyzing correlations with surgical techniques, and investigating possible technical factors can identify the cause of BAEPs changes and provide appropriate information to the rest of the surgical team [8].

However, if the test method is different, the analysis of the information that can be provided will be different, so further research will be carried out in the future.

Cause of Waveform Changes

Sindou and Fernández-Conejero explain the causes of BAEPs waveform changes: (1) Extension of the VIIIth nerve caused by excessive cerebellar traction. (2) Manipulation of the labyrinthine artery or the motorized cerebellar artery that can cause a vasospastic reaction. (3) Direct trauma due to coagulation or heating due to appliances. (4) Classification of new CN compression of the eighth CN at the end of surgery by material inserted between VIIth and VIIIth complexes [9, 10].

Legatt studied the waveforms I, III, and V in detail. He concluded that cochlear ischemia or infarction during infarction or infarction and temporal bone drilling due to internal auditory artery injury affects all BAEPs components including wave I, delayed and attenuated direct mechanical or thermal trauma to the eighth nerve, and removed III and V waves, but waves I from the cochlea of the eighth nerve can be preserved. And extending the I–III interpeak interval during the retraction of the cerebellum and brainstem reflects stretching of the eighth nerve and is often reversible [8].

The origin of wave I is the cochlear junction at the distal end of the cochlea, and vasospasm in the eighth nerve may cause potentially reversible BAEPs changes. The changes in BAEPs, which all waveforms disappear without Wave I, may reflect direct mechanical or thermal damage of the brainstem, brainstem compression, or ischemia due to vascular injury [4, 9].

Post Operation Hearing Loss

In general, postoperative hearing loss is 2.3~21.2% [11–13]. This postoperative hearing disorder is a good test for hearing conservation in the BAEPs test in MVD surgery, and loss of wave V during MVD is a specific indicator of postoperative hearing loss. Current alarm criteria used to alert surgeons are a sensitive indicator of impending hearing loss after surgery [4].

Other studies have also reported that hearing impairment occurs when the offending vessel is present between 7 and 8 CN [14, 15].

According to several studies, the high-frequency (4–8 kHz) hearing loss (HFHL) associated with the cochlea, such as the impaired appearance of the cellular dysfunction caused by the acoustic trauma created by the drill, also occurs [16, 17].

Warning Criteria

In the past, the BAEPs warning criteria showed an interest in the latency change of wave V. After a while, both the latency and the amplitude of the V waveform are considered to be important. Recently, it has no relation with latency, the decrease or loss of amplitude is reported to be closely related to postoperative hearing.

In the case of classification based on the latency of V waveform, a study defined warning criteria by dividing by 0.4 ms (watching) signal, 0.6 ms (risk "warning" signal), and 1 ms [18, 19], and some studies have focused on the latency of the wave V [20–22].

The American Clinical Neurophysiology Society reported that V latency longer than 1 ms or 10% increase in latency and/or with a 50% reduction in amplitude as a warning criterion [23, 24].

James studied hearing loss in cerebellopontine angle (CPA) surgery, where the 1 ms prolongation of the wave V or a 50% amplitude reduction was considered to be an indicator of influencing [7]. Although the BAEPs waveform is poor in patients with normal hearing, the BAEPs loss during surgery may be normal after hearing, so further studies are needed.

Hatayama was not associated with postoperative hearing loss when the latency period of the wave V was extended, and all patients who showed a decrease in amplitude more than 40% of the wave V were accompanied by prolongation of the latency and it was reported to be highly related to postoperative hearing loss [25].

Legget [8], Jo [26], Jung [27], and Park [28] have little association with the latency of wave V and hearing loss but the reduction of amplitude by more than 50% was highly related to hearing loss.

Thirumala [4] and Park [28] reported that the disappearance of wave V was an indicator of postoperative hearing loss.

Damaty showed a New Score that predicted postoperative hearing impairment by analyzing the scores of the changes of amplitude of BAEPs wave I, amplitude of wave V, change of latency, and IPL change of wave I–III [6].

The optimal stimulation rate and averaging time should be studied in the future, and the optimum warning criteria will be derived if the optimal test method is established based on these studies.

Lateral Spread Response (LSR)

In hemifacial spasm, the LSR test was performed on all surgeries, beginning with the study of ephaptic transmission and ectopic excitation of Nielsen [29].

Moller and Jannetta played a major role in the study of MVD surgery in earnest. LSR tests were made with bandpass from 0.3 to 3 KHz and latency was observed at an average of 11.03 + 0.66 ms [30].

Electrode Position

A dermis needle was inserted at the junction of the facial nerve at the midpoint between the trauma of the ipsilateral side and the outer side of the eyeball. Electrical stimulation is performed with two electrodes spaced 3 cm apart from the temporal or zygomatic branch, and the direction of the stimulation is toward the brainstem and a 0.2–0.3 ms stimulus with an intensity of 5–25 mA [31]. It is also very effective to stimulate the lower branch (buccal or mandibular), such as stimulating the upper branch of the facial nerve, to measure the LSR in the upper branch. Most of the reactions in both LSRs are measured, but only in one place.

Meaning of LSR

It is difficult to explain precisely whether LSR is a central mechanism or a peripheral mechanism.

But there are more claims of central mechanism [10, 32, 33]. If the origin of the facial nerve is affected by the arterial blood vessels, an anomaly occurs between the facial nerve strands. It is hypothesis that this phenomenon affects the facial motor nuclei in the brainstem and that the brainstem function becomes hyperexcitic state, resulting in hemifacial spasm.

A specific electrophysiologic study has shown that the disappearance of the LSR signifies good separation of facial nerve and blood vessels [10, 34–40]. Thirumala predicted that if LSR remains in operation in 2011, it may be possible to have spasm after surgery. LSR and the outcome are not statistically related, but long-term effects are predictable [31]. In addition, the case reported that the LSR did not disappear until the end of the operation, and that the HFS disappeared a year later [33], adding to the expectation and concentration of LSR measurement during surgery. However, questions are now being raised about what LSR measurements mean. Because we have learned from experience that LSR and spasm are not necessarily proportional.

Prognosis of LSR

LSR is not measured in normal persons [34, 35, 37, 41]. The relationship between long-term prognosis and LSR that does not disappear even after adequate decompression. There are many studies that LSR is useful during surgery, and studies have shown that the loss of LSR during surgery is a good surgical outcome after surgery [10, 32, 33, 35, 38, 42], showed that patients who did not disappear during surgery had a good outcome after surgery. Even if the LSR does not disappear during surgery, surgery is terminated when a change in shape is observed [34–37]. Because of these various studies, the LSR has been controversial. LSR was reported to be associated with spasm at 1 week postoperatively, but not at 1 year [43], and reported a different result from that reported in association with long-term outcome [31].

Some researchers believe LSR monitoring during surgery is useful for predicting facial outcomes of MVD-reported surgeries for HFS. It has not completely disappeared, but some

research has shown that a reduction in LSR amplitude is associated with favorable outcomes [35, 44, 45]. However, some reports claim that the prognostic value of LSR monitoring for predicting the long-term outcome of facial seizures is still questionable [34, 46, 47]. In addition, some authors have reported that the loss of LSR after decompression is not always correlated with favorable postoperative results [44, 48].

Hatem et al. [47] reported complete improvement of spasm at the last follow-up for 10 of 10 patients who continued until the end of the operation. They suggested that uninjured LSRs are not always a bad prognosis. This is because motor nuclei and excitability can take months or years to normalize.

Persistence (Residual) of LSR

If the LSR response remains unchanged despite all surgical interventions, additional irradiation and/or decompression may be performed during surgery to relieve all vascular compression of the facial nerve [35, 44, 49]. However, this method is not recommended because he or she may have additional hearing loss or facial paralysis after extensive surgery for the purpose of LSR loss. Eckardstein et al. reported that 90% or more postoperative spasm is resolved if LSR is lost during surgery. However, it is not a bad thing to remain, and even if it does not disappear, it takes no further action [50].

Damaty reported that the LSR did not predict whether the operation was successful or not, and reported that LSR did not cause bad results even if it did not disappear during surgery. In particular, it has been suggested that endoscopy may be a good way to confirm decompression of peripheral blood vessels [51].

Lee et al. reported that the disappearance of LSR is helpful in the operation. Patients with symptoms less than 1 year had a significantly higher LSR loss rate after decompression than those with symptoms for 10 years or more. EMG monitoring was useful in confirming proper insertion of Teflon, especially when multiple vessels were involved. The LSR was suitable for short-term follow-up but not long-term follow-up for more than 1 year [52].

Reoperation

Park suggests that it is better to judge after 1 year of reoperation if more than 1 year of observation is needed because some people have improved after 10 months [53].

Thirumala reported on residual LSRs, suggesting that long-term outcomes may be seen even if LSR is not lost if decompression is appropriate, and intraoperative communication between surgeons and testers is important in producing adequate decompression [54]. It was advised to consider reoperation if the LSR did not disappear during surgery but there were signs of trembling for more than 1 year after surgery [31].

Optimal Method of LSR Test

LSR does not necessarily disappear after decompression, but when the CSF flows out after the dura open, the brain is sunken down and naturally the facial nerve and offender are separated and disappear [42]. It is therefore very important to accurately assess at which point the LSR is lost. Therefore, it is recommended that the LSR measurement be divided into at least six sections. (1) Before the dura open (2) After the dura open (3) Before the Teflon insertion (4) After the Teflon insertion (5) After the dura closed (6) After the operation is done. If the LSR is lost after dura open, it may be lost by retraction before Teflon insertion. It may slowly disappear after MVD, and may be measured again, so the measurement should be checked in at least six steps [42].

The pattern of loss of LSR varies greatly depending on the patient and most is suddenly lost. Sometimes it gets smaller and then disappears, In some cases, the small amplitude does not disappear and the surgery is terminated. Some studies have categorized these different cases [55], In our experience, it is rather a larger measure, there was a case where it was maintained without being reduced at all. There is a need for further studies on various aspects of these LSR.

It is very important to detect the moment when the LSR disappears during surgery. Knowing quickly what has been lost in any operation gives a lot of confidence in surgery, so informing the moment when it disappears gives a very useful help to the flow of surgery. Although LSR has a very significant role as an indicator during surgery, the meaning of the residual LSR still needs further research. It is still uncertain to predict LSR symptoms during surgery over the long term of more than 1 year.

Use a filter such as free-running EMG with LFF 10 Hz, HFF 3 KHz, and notch filter off so that the LSR waveform similar to 50~60 Hz will not be measured by the filter.

When the test is performed during the operation rather than the preoperative waveform, the amplitude of the LSR is measured as small as 100 µV due to anesthesia [56]. Therefore, it should be taken into consideration that LSR, which is measured very small in preoperative examination, may not be measured during surgery.

Zhong-Lee Response (ZLR)

Measurements should be made at the same site as the LSR measurement site, and stimulation should be performed prior to decompression of the nerve and blood vessels within 5 mm of the nerve, followed by stimulation after decompression. Stimulate with a bipolar stimulator (0.2 ms, 2 mA). Concurrent surgical monitoring of LSR + ZLR monitoring provides more valuable neurosurgical guidance than LSR for HFS, regardless of whether the facial nerve compression vessel is simple or complex [57].

There is LSR that is lost before decompression or LSR that is not lost after decompression. If this is the case, ZLR can be used effectively to check for contact between the vessel and the facial nerve [41]. However, other studies have shown that MVD was successful in 20 patients without HFS, before decompression ZLR was observed in 19 patients, 13 patients were lost after decompression, 6 patients were measured, three of these six were weak, while the other three had waveforms similar to those before decompression. The reason for this is that, in addition to being attached to the distal part of the facial nerve, if there is more part of the distal part of the facial nerve, it can be induced by electrical stimulation [58].

A monopolar stimulator is effective for mapping localization of the facial nerve. However, if the intensity of the stimulation is too high, facial neuro-EMG responses are measured in a wide area around the facial nerve, and therefore should be examined with low electrical stimulation. When direct electrical stimulation is applied to the facial nerve, it is stimulated to less than 1 V or 0.5 mA. However, if electrical stimulation is applied to the facial nerve in the presence of CSF or blood, "current shunting" may occur and a false-positive reaction may occur and so it is advantageous to check with Voltage rather than Current [59].

Facial Motor Neuron Function Test

Facial F-Wave

Facial F-wave is possible to record changes in elicitability of the facial F-wave during MVD. Immediate changes in hyperexcitability of the FMN can be observed by monitoring changes in F-wave elicitability. Facial F-waves can be obtained from the mentalis muscle by stimulating the mandibular branch of the facial nerve (Fig. 1).

The facial F-wave represents the backfiring of the facial motor neurons after being activated antidromically. F-wave activity was shown to be an index of motor neuron excitability [60].

F-wave persistence appearance was reported in patients with HFS [61] and found to decrease

after adequate MVD, albeit with delay as long as 2 years.

Facial MEP

Facial MEP is a hyperexcitability facial motor nuclei in hemifacial spasm patients. Therefore, a large amplitude waveform is formed compared to the normal side. There is no change in the amplitude of the normal side after decompression during surgery but surgical site can be used as an indicator of good surgical outcome with decreased amplitude [62].

However, there are some drawbacks to using Facial MEP as a test.

First, electrical stimulation should be given to the facial motor cortex. However, since patients are slightly different, it is difficult to find the exact location. Therefore, the brain cortex area should be extensively stimulated to include the facial motor cortex site within the stimulus range. In other words, it is difficult to know whether the facial motor cortex is fully excited every time it is inspected to achieve the optimal waveform. It is doubtful whether the facial motor cortex was sufficiently stimulated if a waveform was observed in only one of the Oris or mentalis (Fig. 2c).

Second, the amplitude of the waveform varies with each test. Because the MEP waveform is the response of the muscle twitching, the waveform is observed slightly differently each time it is examined, as the stimulus intensity increases, the amplitude increases, in some cases, the waveform

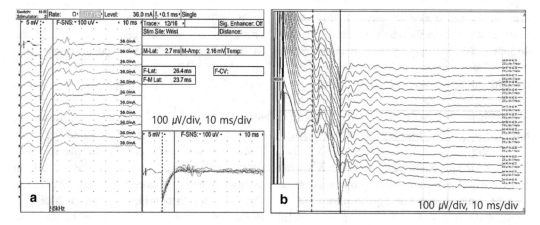

Fig. 1 Preoperative facial F-wave test waveform (**a**) and intraoperative facial F-wave test waveform (**b**)

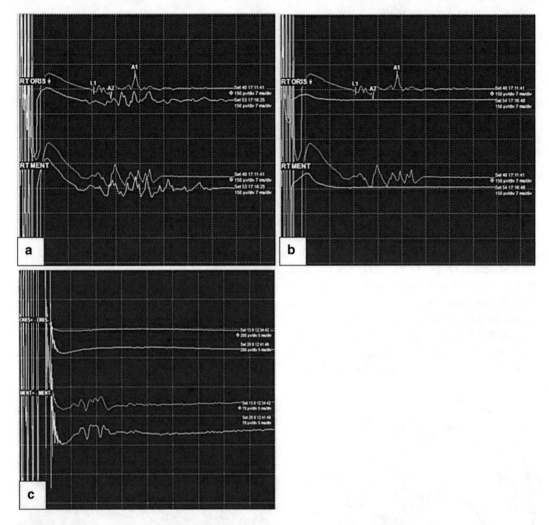

Fig. 2 (**a**) Typical waveforms of facial MEP measured during surgery, (**b**) Facial MEP waveform lost during surgery, (**c**) Waveform measured only in mentalis but not in oris

may disappear even if the muscle relaxant anesthesia is slightly deeper (Fig. 2b).

Third, there is no clear criterion that the amplitude of the waveform should be decreased after the decompression during the operation to make the operation successful. In addition, each patient has a different magnitude of amplitude reduction after decompression and so it is difficult to evaluate whether the operation is clearly done.

Fourth, since it is necessary to compare with the waveform of the normal side as well as the operation site, it is troublesome to further install the electrode.

In this way, it is inconvenient to use four Facial MEPs as an indicator during surgery, in particular, since the presence of a waveform like the LSR is not all or not, it may be helpful for surgery. I think it is difficult to predict the success or failure of surgery with only Facial MEP.

It has the disadvantages as above, but there is a unique advantage. It is very useful for distinguishing facial palsy after surgery due to facial nerve damage during MVD surgery and when facial MEP is performed in a cerebella-pontine angle tumor with a trans-labyrinthine approach, the surgical approach to the distal portion of the facial nerve, it is very useful to distinguish proximal lesions of facial nerve during tumor removal. In vascular surgery, it is very useful to distinguish the damage of the facial motor cortex due to the blood supply problem [62].

New Advanced INM in MVD

Brainstem Auditory Evoked Potentials (BAEPs)

Electrode Position

If the electrode is placed on the back of the ear, it is close to the surgical site and there is a risk of contamination (Fig. 3a). So, there are problems to think about when choosing under earlobe or choosing in front of ear (Fig. 3b, c). Facial nerves are frequently touched during the process of separating facial nerve from blood vessels in surgery. However, under the earlobe or the frontal part of the ear is also affected by facial nerve,

touching the facial nerve during surgery affects the BAEPs waveform, which hinders smooth waveform analysis. Therefore, when electrodes are attached to the tragus or antitragus of the auricle, it is completely separated from the facial nerve, and a very stable waveform can be obtained (Fig. 3d).

BAEPs Ipsilateral and Contralateral Wave Forms

The BAEPs waveform is generally characterized by 5–7 waveforms. Wave I, III, and V are always well seen, wave VI, VII are measured or not measured. A contralateral wave III was observed in the latency between ipsilateral wave II and III, the latency of the contralateral wave IV was observed more rapidly than the ipsilateral wave IV, the latency of the contralateral wave V is later than that of the ipsilateral wave V [63] (Fig. 4a).

In the BAEPs test, only the A1-Cz test is performed in the left operative case, on the right side, only the A2-Cz test is performed. Care

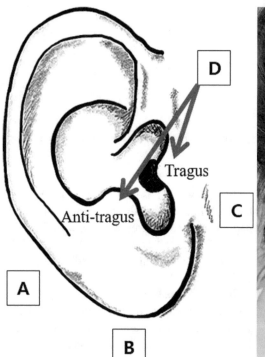

Fig. 3 Photographs of electrodes inserted on the Tagus, anti-tragus. The facial nerve can only be touched by manipulation during surgery. In this case, if the BAEPs electrode is installed in the facial nerve peripheral branch, it can be affected and not measured smoothly

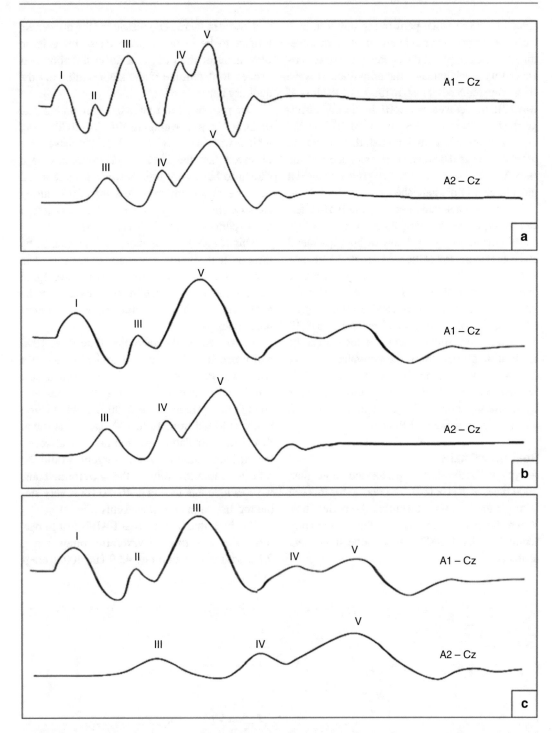

Fig. 4 Conventional BAEPs waveforms (**a**), contralateral waveforms were observed (**b**), and (**c**) judged ipsilateral waveforms differently. If the BAEPs test is carried out on the ipsilateral side only, it may not be possible to correctly judge various changes in the waveform. Changes in the contralateral side waveform can be observed at the same time to help with these problems

should be taken when performing this test, as it can cause errors in correct identification of waveforms. For example, if the cochlear nerve experiences very weak damage, the amplitude of wave III is often reduced or lost. In this case, the loss of wave III or delayed wave III is clearly distinguished. If there is a loss of wave III and the latency of wave V is maintained, there is no big problem, but if the wave III is extended and the wave V is extended, it is very severely damaged it must be clearly distinguished.

When such a situation occurs, the BAEPs test can be helpful in observing the A1-Cz and A2-Cz waveforms in a single window of both ipsilateral and contralateral test of the surgical site. Compared with the contralateral wave, it is easily discernible that the ipsilateral wave III disappears and the latency of the wave V is maintained (Fig. 4b). In addition, the determination of whether wave III and wave V are simultaneously extended can be easily distinguished from the contralateral wave (Fig. 4c). Therefore, in the operating room, it is recommended to observe the waveform changes by measuring both ipsilateral and contralateral when performing the BAEPs test.

Real-Time BAEPs

A typical BAEPs test is performed more than 2000 times at 10 Hz [64]. The reason for this test is to get the best waveform that the patient has. When the patient was awake, the examination required at least 2000 s to remove movement artifacts.

However, there is no reason for the movement artifact to be incorporated because there is no movement of the patient, a faster test time was needed to determine if the surgery affected the auditory nerve.

Our previous test required a stimulation rate of 26.9 Hz and averaging time of 1000–2000, which was about 37.1–74.3 s. We have also observed that the BAEPs waveform has been shown to have very little wave V latency and a decrease in amplitude of less than 50% during surgery, immediately following the next test, I had experienced BAEPs loss (Fig. 5).

This phenomenon is thought to be caused by the fact that abnormal waveforms are mixed in normal waveform due to too much averaging time, or that the stimulation rate is too slow to be appropriate for discriminating between auditory nerve damage.

So we reduce the averaging time from 1000 and proceed to 100 times. I looked to see if a waveform like 1000 would be maintained at a certain level, and I found that 400 times was enough for averaging time 1000. The stimulation rate was tested from 10 to 100 Hz. It was found that if the stimulation was too fast, the effect of the artifact caused by the surgery would be reflected more severely on the waveform. Thus, we conclude that testing at 40–50 Hz is very stable for discriminating waveforms (Fig. 6).

We think that it is the best BAEPs test to perform with the setting of waveform completion at 9.1 s with stimulation rate 43.9 Hz, 400 averag-

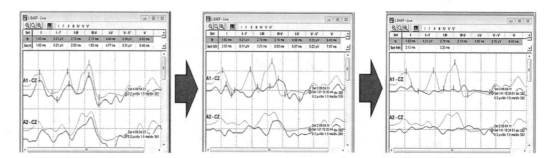

Fig. 5 Each waveform change step at stimulation rate of 26.9 Hz and averaging time of 1000–2000. If the averaging time is prolonged, it is difficult to determine whether the nerve is damaged because the normal waveform and the abnormal waveform are mixed. We have experienced a lot of sudden loss of waveforms during the next test after a fine change in the waveform was observed during the long time average

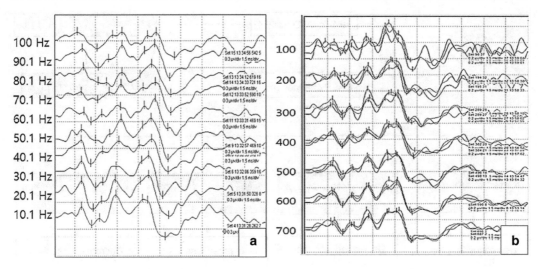

Fig. 6 (**a**) Waveform change according to stimulation rate (with averaging of 1000 trials). (**b**) Averaging trials (with 43.9 Hz/s stimulation rate) from IOM of BAEPs. Although the waveform was formed from 10 to 100 Hz, the faster the test above 50 Hz, the more the waveform was affected by the surgical operation. Thus, when the test was performed at a speed of about 40 Hz, stable waveforms were not affected by the surgical operation. Reproducibility was enough about 400 times to achieve the same effect as 1000 times of averaging time. (These results are from Joo et al. [65])

ing time. With the development of medical equipment, waveforms that were not possible with analog equipment in the past have become possible, and testing in less than 10 s is possible on most equipment. Indeed, a continuous test was repeated every 9.1 s and the amplitude of the waveform was reduced or lost within 10 s [65].

When we used Wave V amplitude loss >50% on an alert basis, we were able to reduce the incidence of postoperative hearing impairment to less than 0.4% (Table 1). We believe that obtaining a stable BAEPs within 10 s is a real-time IOM procedure of BAEPs and very important for preventing CN VIII damage during MVD surgery (Fig. 7).

New Warning Criteria

If the auditory nerve is instantaneously damaged, averaging is performed for a long period of time, and abnormal waveforms are mixed in the conventional normal waveform, so that the waveform is very small. Therefore, we observe the finely changed waveform. That is, the continuous and slowly varying latency will be reflected in the averaging test.

So, sudden decrease or disappearance of amplitude within 10 s will not be reflected.

Table 1 Comparison the protocol for INM of BAEPs and postoperative hearing loss

INM BAEPs	Previous protocol[a]	Real-time protocol[b]	p value
Number of averaging trials	1000–2000 times	400 times	
Stimulation rate	26.9 Hz	43.9 Hz	
Time to obtain BAEPs	About 37.1–74.3 s	about 9.1 s	
Warning criteria (of the wave V)	1 ms latency prolongation or a 50% decrease in amplitude	A 50% decrease in amplitude	
Postoperative hearing loss	4.02%	0.39%	0.002

[a]Means the protocol that used in our previous study (Jo et al. [26])
[b]Means the protocol that used in our recent study (Joo et al. [65])

So far we have been doing BAEPs tests with slightly different settings, but we have not thought that we cannot detect real-time nerve changes because the averaging process is too long.

The real-time BAEPs method with a stimulus frequency of 9.1 s is characterized in that the

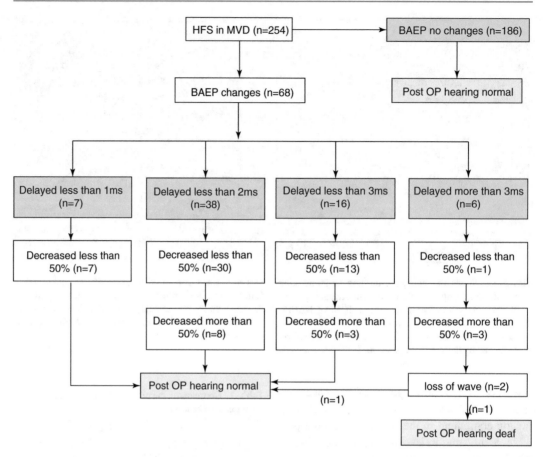

Fig. 7 Examination with real-time BAEPs allows you to determine the state of the auditory nerve in 9.1 s. In particular, we noticed that the latency changes very slowly and continuously, and the amplitude changes very quickly. Even within 9.1 s, the waveforms were lost. The prolonga-tion of latency by about 1 ms was observed very frequently and was not related to postoperative hearing. However, as latency was extended by more than 2 ms, the amplitude decreased more often, leading to waveform loss. (These results are Joo et al. [65])

latency changes gradually and slowly, and the amplitude changes suddenly decrease or disappear within 10 s. Observe clearly. In other words, latency and amplitude change independently, latency changes are observed very often, and amplitude changes are observed very rare. In addition, changes in latency only have no effect on postoperative hearing changes (Fig. 8). Therefore, the change in amplitude should be observed much more important than the change in latency delayed [28].

For warning criteria of BAEPs, we used a "sliding scale" as follows: (1) the observation sign: latency prolongation of 1 ms without an amplitude reduction of at least 50%; (2) the warning sign: latency prolongation of 1 ms with a reduction in amplitude of at least 50%; (3) and the critical sign: loss of wave V. When the neurophysiologist detected the observation sign during MVD surgery, he notified the surgeon immediately, but the surgeon did not perform any corrective maneuvers. However, when the warning or critical sign happened, the neurophysiologist immediately notified the surgeon, the surgeon halted the operation that he was doing, and did surgical corrective measures (Table 2).

However, when the BAEPs wave V latency is prolonged by more than 2 ms, amplitude is often reduced by 50% or more. In particular, prolongation of more than 3 ms may result in more than 50% decrease in amplitude and BAEPs loss more frequently, so prolonged BAEPs wave V latency

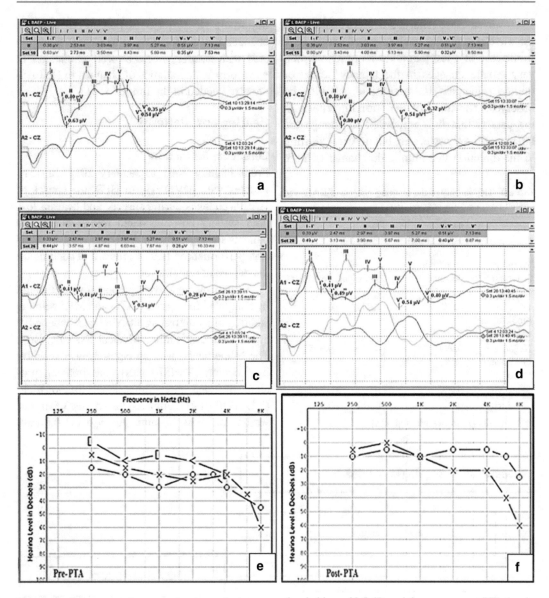

Fig. 8 (**a–d**) representative case showing only latency prolongation (≥1 ms) without a significant change in amplitude. (The latency of wave V was delayed by 3.20 ms from 7.13 to 10.33 ms with a minimal decrease in the amplitude.) The patient in this example did not experience postoperative hearing loss. (The average pre-PTA threshold was 22.5 dB, and the average post-PTA threshold was 6.25 dB.) (**e**) The INM of BAEPs during MVD surgery (white arrow = baseline of wave V; black arrow = wave V showing maximal prolongation in latency), (**f**) Pure tone audiometry of the patients obtained prior to surgery and 7 days postoperatively

should also be observed carefully. Therefore, we want to present the relationship between latency and amplitude as follows (Fig. 9).

If you go through the process in real-time BAEPs in less than 10 s, there are many cases where the amplitude of the wave V suddenly changes within 10 s. Most of the reasons for the decrease in wave V amplitude were due to brain retraction. It is important to quickly detect when the waveform has decreased and notify the neurosurgeon, as soon as the retractor that affects the cochlear nerve is removed, the waveform is observed to recover. We conducted the BAEPs test in real-time in this way. In 2016, 417 patients who underwent surgery did not have any HL after surgery (Fig. 10).

However, these criteria are applicable only if the waveform results can be derived within 10 s. In the case of a test that takes more than 1 min, very minute changes in the latency period or minute changes in amplitude should be observed. The trend of the warning criteria in the past is that it is important to change the latency of the V waveform, but now it is shifting to the direction that amplitude change is more important. It is also important to observe changes in waveform I due to ischemia associated with auditory nerve as well as changes in waveform V, and studies of changes in I-III IPL are also needed in the future [4, 9].

Hearing Loss Patterns

Cerebellar Retraction

From January 2001 to January 2011, in 1518 patients with MVD due to HFS, the causes of BAEPs changes during surgery and hearing loss after surgery are considered retractors used in patients during surgery. An anatomical analysis was performed on how it affected depending on the degree of withdrawal.

All patients underwent preoperative evaluation by computed tomography (CT) and MRI with the addition of a contrast agent. MRI was obtained at 3.0 T (Achieva; Philips Medical Systems, Best, The Netherlands) with an eight-channel sensitivity-encoding (SENSE) head coil. Imaging protocol included three-dimensional time-of-flight (3-D TOF; TR/TE/flip angle, 25 ms/3.5 ms/20 section thickness 1.6 mm, slice spacing 0.8 mm,

Table 2 The validity of the warning criteria of BAEPs to predict postoperative hearing loss

Warning criteria (of the wave V)	Latency prolongation (≥ 1ms) with amplitude decrement (≥ 50%)	Transient loss	Permanent loss
Positive predictable value	0.075	0.211	0.545
Negative predictable value	0.999	0.996	0.995
Sensitivity	0.909	0.727	0.545
Specificity	0.865	0.967	0.994

These results are from Park et al. [28]

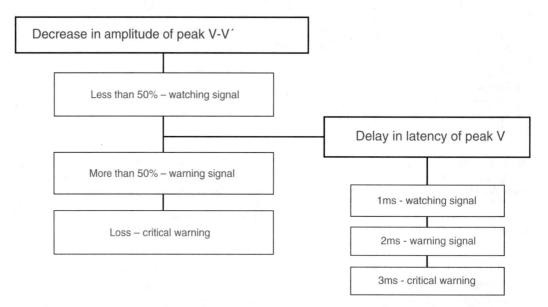

Fig. 9 Warning criteria applied when using real-time BAEPs. The relationship between latency and amplitude is presented separately based on wave V. In real-time BAEPs tests, latency changes were always observed first, followed by amplitude changes. In addition, latency change was observed slowly and continuously, and amplitude change suddenly decreased by more than 50% within 10 s

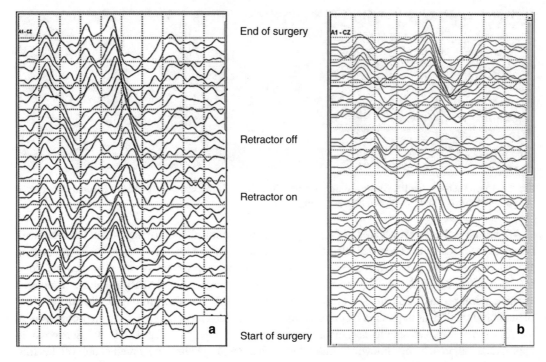

End of surgery

Retractor off

Retractor on

Start of surgery

Fig. 10 If the wave V waveform is delayed (**a**) or lost (**b**) by the retractor, the retractor is removed, immediately testing and reporting the recovered waveform in real time. If you remove the retractor immediately after quickly detecting nerve damage, it will recover even if the waveform is lost

50048 matrix, acquisition time = 5 min 29 s), and 3-D T2 VISTA (volumetric isotropic T2-weighted acquisition; TR/TE/flip angle, 2000 ms/228 ms/20 NSA = 2, mm3, ETL=70, 24850 matrix, voxel size = 0.8.8.8 acquisition time=4 min 10 s). Neurovascular compression was determined by experienced neuroradiologists.

Among the 1518 patients, 106 (6.98%) displayed functional hearing changes. Hearing loss was permanent in 12 patients (0.79%). Of the 1412 patients with stationary hearing compared with preoperative audiometry, 96 patients were selected who were individually matched with respect to sex, age, and degree of spasm. BAEPs changed immediately after cerebellar retraction in 7 of 12 hearing-loss patients, suggesting the importance of retraction on hearing outcomes. The distance from the cerebellar surface of the petrous temporal bone to the neurovascular compression point was measured. The median distance of cerebellar retraction in the hearing loss group was 13.77 mm, which was longer than the median distance in the control group (Fig. 11).

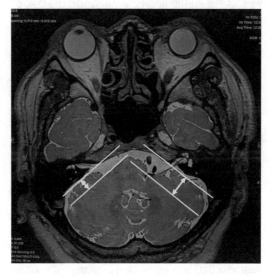

Fig. 11 A preoperative MRI showed the left vertebral artery and posterior inferior cerebellar artery in close proximity to the root exit zone of the ipsilateral facial nerve, and the vertical distance of cerebellar retraction was more than 10 mm. View of measured vertical length from the petrous temporal bone to neurovascular compression point. The greater the amount of retraction, the greater the damage to the cochlear nerve

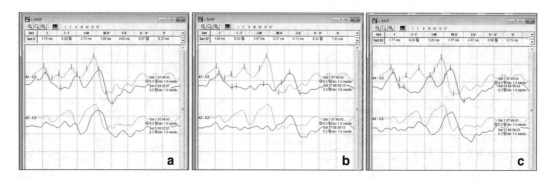

Fig. 12 During surgery, the BAEPs waveform prolonged from wave III (**a**), wave of III was lost (**b**), and unrecovered picture with the procedure extended to the end of surgery (**c**)

We retrospectively investigated attempting to correlate the distance of cerebellar retraction and BAEPs change with the occurrence of hearing loss after MVD for HFS. BAEPs monitoring is a valuable intraoperative indicator for preservation of auditory function, as in previous reported studies [26]. Therefore, we should design the most appropriate approach by knowing individual pathologic anatomy before surgery and the retraction should be done minimally, and in a direction perpendicular, rather than longitudinal, to the axis of the eighth nerve, to minimize the risk of postoperative hearing impairment. Furthermore, according to our experience, individual pathologic anatomy before surgery assessed by MRI is crucial to adapt and design the most appropriate approach for every patient.

Delayed Hearing Loss

We retrospectively reviewed the medical records of 3462 patients who received MVD for reflective spasms from January 1998 to August 2017. Five of these were normal immediately after surgery, but there were five cases of delayed hearing loss. After a while (middle, 22 days, 10–45 days) they suddenly developed hearing problems.

All of these patients showed postoperative sensorineural hearing loss, a common phenomenon was observed, and in the brainstem auditory evoked potentials (BAEPs), the inter-peak latency of waves III was prolonged during surgery, but recovered within a short time.

We have not yet identified the incidence of prolonged inter-peak latency of waves I–III. Therefore, it is difficult to conclude that this is a characteristic feature of delayed hearing loss.

But we think about prolongation of the inter-peak latency of waves I–III seems to be associated with the occurrence of delayed hearing loss. During surgery, the BAEPs waveform prolonged from III. If you stay prolonged without recovery even at the end of surgery, hearing problems may occur after surgery (Fig. 12). Hence, it is possible that BAEPs changes may predict delayed hearing loss, but this issue requires further investigation. Analysis of more cases will be necessary to identify the exact cause of delayed hearing loss and to determine whether BAEPs monitoring can be used to predict delayed hearing loss after MVD for HFS [66].

The Significance of Wave I Loss of Brainstem Auditory Evoked Potentials

We looked closely at the loss of BAEPs waveforms during surgery. We enrolled 670 patients with primary HFS who underwent MVD surgery with IOM of BAEPs from January 2015 to December 2016. We distinguish between the case where wave I is observed when the BAEPs waveform is lost and when the wave I is lost and not all waveforms are observed (Fig. 13).

If the remaining waveforms are lost while wave I is maintained, it is mainly caused by retractor due to direct nerve damage. After retractor removing, we know that the waveform recovers normally. However, the loss of all waveforms without wave I is thought to be caused by a problem in the blood supply, which is observed as a delayed change phenomenon in which the waveform disappears after 10 min without any change in the waveform immediately after damage. In particular, the postopera-

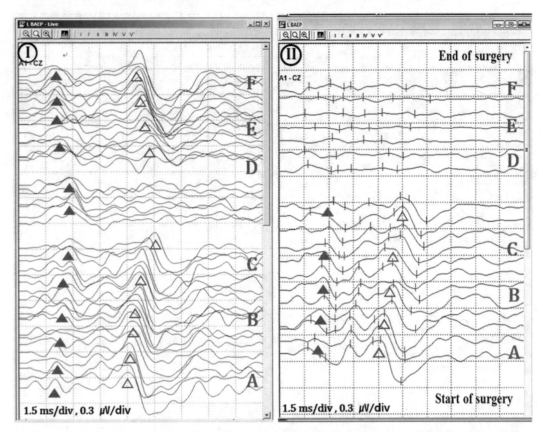

Fig. 13 Example of brainstem auditory evoked potentials (BAEPs) according to persistence of wave I in patients with wave V loss during microvascular decompression surgery. (I: BAEPs with persistence of wave I; II: BAEPs without persistence of wave I; A: Surgery start; B: Dural opening; C: Direct microvascular decompression procedure start; D: Direct microvascular decompression procedure end; E: Dural closure; F: Surgery end; filled triangle: wave I; unfilled triangle: wave V.)

tive patient's condition in which all waveforms are lost and not recovered without wave I is considered to be the result of vasospasm-like artery affecting the vestibular cochlear, as severe damage to the entire vestibular system occurs (Table 3, Fig. 14).

In the case of failure to recover due to loss of waveform during surgery, hearing loss occurs after surgery (Table 4). Especially if all waveforms are lost without wave I, much higher postoperative hearing loss may occur (Fig. 15).

New Method LSR

We have used LSR as an indicator of good surgical outcome during surgery.

Table 3 The proportion of postoperative complications according to persistence of wave I among the patients showing wave V loss

	w/i persistence of wave I	w/o persistence of wave I	*p* value
Patients, *n*	24	12	
Dizziness, *n* (%)	0	5 (41.67%)	0.002
Tinnitus, *n* (%)	0	3 (25.00%)	0.031
Diplopia, *n* (%)	0	1 (8.34%)	0.333
Hoarseness, *n* (%)	0	1 (8.34%)	0.333
Hearing loss, *n* (%)	2 (8.33%)	6 (50.00%)	0.009
Subtype of hearing loss, *n* (Low: High: Total)	2:0:0	0: 0: 6	

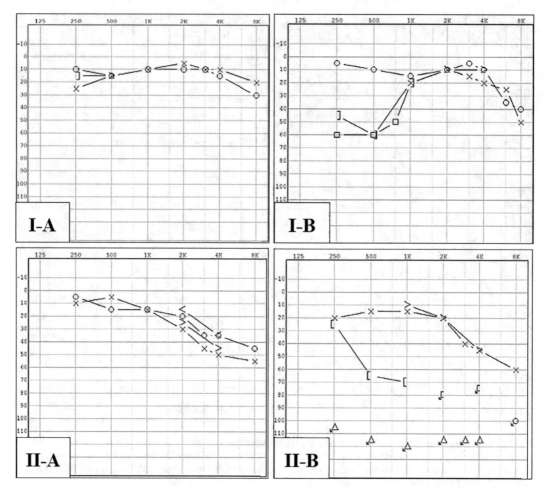

Fig. 14 Example of Pre-/Post-PTA according to timing of wave V loss during microvascular decompression surgery for hemifacial spasm. (I-A/B: Pre/Post-PTA of the immediate-change group; II-A/B: Pre/Post-PTA of the delayed-change; PTA: pure tone audiometry)

Table 4 The comparison of postoperative hearing loss according to BAEPs changes

BAEPs change (Wave V)		Patients, *n* (%)	Postoperative hearing loss, *n*
Only amplitude decrement (≥50%)		12 (1.8%)	0
Only latency prolongation (>1 ms) (the observation sign)		9 (1.3%)	0
Latency prolongation (>1 ms) with amplitude decrement (>50%) (the warning sign)		93 (13.9%)	0
No change		520 (77.6%)	0
Wave V loss (the critical sign)	Transient loss	26 (3.9%)	0
	Permanent loss	10 (1.5%)	8
Total		670	8

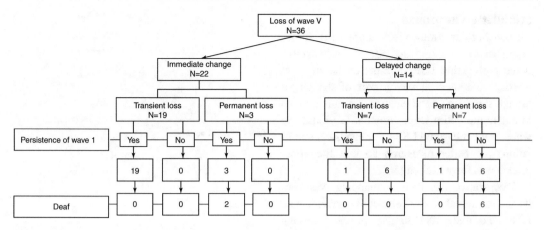

Fig. 15 The differences in postoperative hearing loss according to persistence of wave I in patients with wave V loss during microvascular decompression surgery for hemifacial spasm: Immediate change: Immediate-change group; Delayed change: Delayed-change group; Immediate phase: During the decompressive procedure; Delayed phase: After the decompressive procedure

Most institutions stimulate the temporal or zygomatic branch of the facial nerve about 3 cm lateral to the lateral margin of the orbit during LSR monitoring. The direction of stimulation with paired needles or surface electrodes is centripetal toward the brainstem with the cathode positioned proximally [67, 68].

To our knowledge, there have been no attempts to develop or apply different LSR monitoring methods to increase efficacy. A new LSR monitoring method including facial nerve mapping and centrifugal stimulation of the facial nerve before MVD was studied.

Recording Site

Orbicularis oculi is placed on the eyebrows around the orbit by touching the hand with the electrode so that the electrode does not enter the eye. If you plug in parallel 3~4 cm above the Oculi electrode location, it is in Frontalis position. Orbicularis oris is inserted in the upper lip because it is difficult to plug under the lip because of the intubatio tube. Because the thickness of the upper lip is thin, care should be taken that the electrode does not penetrate through the flesh and into the mouth. If the electrode penetrates into the mouth, noise may be mixed into the electromyogram waveform due to the saliva of the mouth. The mentalis position must be between

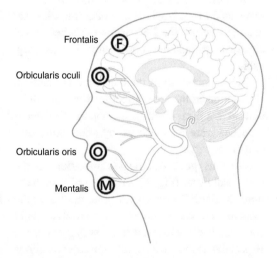

Fig. 16 Illustration of electrode position according to facial nerve branch. The position of the orbicularis oculi is placed in the eye, and the position of the orbicularis oris is located between the nose and the upper lip. The location of the orbicularis oculi is placed at the eye level to eliminate adjacent microscopic connections between the zygomatic branch and the buccal branch. The location of the orbicularis oris is placed between the nose and the upper lip to solve the difficulty of electrode installation because the lower lip is opened due to the intubation tube

the jawbone and the lower lip with the jawbone in the palm of your hand. A person with a lot of flesh is mistaken for the shape of the jaw mixed with the neck, and it is not a mentalis but a mistake of putting the electrode in the neck (Fig. 16).

Stimulation Response

We sought to determine whether electrical stimulation occurs after the facial nerve is fully excited when performing LSR measurements. In other words, we looked at which part of the Upper branch should give the electric stimulus enough to excite the entire Upper branch. We looked at whether the measured LSR after electrical stimulation was better measured so that the upper branch was fully excited [56].

Preoperative facial nerve mapping was conducted for a total of 486 consecutive patients with HFS who underwent surgery at the Samsung Medical Center between February 2015 and August 2016. Patients were monitored for an LSR using centrifugal stimulation of the facial nerve during MVD with the aid of preoperative mapping data.

We observed the response of the muscles that responded to the frontalis and oculi muscles that occurred before the LSR to assess whether the electrical stimulating upper branch was fully excited.

In the same way as the conventional method, stimulation in the stem direction was very difficult to confirm whether or not all the upper branches were fully excited by stimulating several facial parts.

We thus gave electrical stimulation to the peripheral nerve (Fig. 17), as opposed to conventional testing. The response of the muscles in the frontalis and oculi muscles, which are electrically stimulated, was observed to be very responsive. Especially, frontalis and oculi muscular reactions were observed when stimulating various parts of the body by moving the electric stimulating facial region. So when I was stimulating somewhere I was able to clearly see if the Upper branch was fully excited (Fig. 18).

It is also known that the upper branch is fully excited by the well-observed muscle response in both frontalis and oculi muscle reactions, and that LSR measurements in this case also measure a larger amplitude at lower intensities it was.

Stimulation Direction and Intensity

In contrast to the conventional method, we applied electrical stimulation to the frontalis and

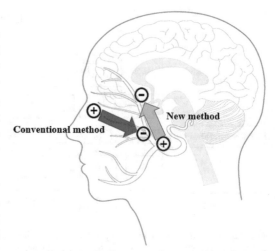

Fig. 17 The direction of stimulation in the conventional and new methods. In the conventional method, electrodes are placed in the temporal or zygomatic branch of the facial nerve, about 3 cm lateral to the lateral margin of the orbit, and centripetal impulses are transmitted toward the brainstem with the cathode positioned proximally. In contrast, electrodes were inserted intradermally with the anode located proximally over the area just anterior to the mandibular fossa and the cathode located distally in the temporal branch of the facial nerve in the new method. The direction of stimulation was centrifugal outward from the brainstem

oculi muscles to detect where the Upper Branch was fully excited.

We used a preoperative test to fix the temporomandibular joint to the anodic electrode in order to use the LSR in the same area as the well-measured area. In addition, the cathode electrode was divided into three large areas and stimulated.

Three directions were designated as 'F' for the frontalis electrode, 'O' for the oculi attached to the oculi, and 'F-O' for the midpoint between the frontalis and oculi (Fig. 19).

LSR was measured in 428 out of 486 patients. Direction F was the most frequent in 325 (66.9%), direction F-O was in 91 (18.7%), and direction O was in 12 (2.5%), and this is not a zygomatic branch but a temporal branch (Table 5).

We have experienced the patient's condition good after surgery even if the LSR is not lost. It is also known that the time of LSR disappears during surgery is very diverse.

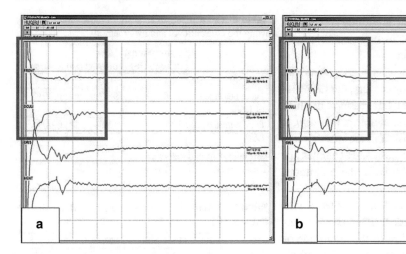

Fig. 18 The response measured in the frontalis and oculi muscles when the upper branch was stimulated by the conventional stimulation method was very small (**a**). The response measured in the frontalis and oculi muscles when the upper branch was stimulated by the new method stimulation was very large, making it easy to assess whether the upper branch was sufficiently excited (**b**)

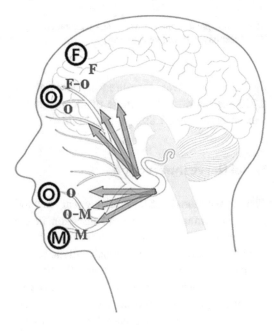

Fig. 19 Facial nerve mapping performed preoperatively. The anode was the reference point or pivot and was placed over the location just anterior to the mandibular fossa. The cathode was first placed in the direction of the frontalis muscle, vertical with respect to the anode, and then moved fanwise toward the direction of the orbicularis oculi muscle while stimulating the facial nerve. The locations of maximal LSR elicitation were divided into three regions: F (the direction toward the frontalis muscle, which was almost vertical with respect to the anode), O (the direction toward the orbicularis oculi muscle), and F-O (in between F and O). In the upper branch stimulation, the LSR measurement was the highest when temporal branch stimulation was performed, and in the lower branch stimulation, the LSR measurement was the highest during the mandibular branch stimulation

Table 5 LSR, lateral spread response; F, direction which was almost vertical to stylomastoid foramen; F-O, in between F and O; O, direction toward orbicularis oculi muscle

n (%)	
Location of maximal LSR	
F	325 (66.9)
F-O	91 (18.7)
O	12 (2.5)
Preoperative LSR	428 (88.1)
Intraoperative LSR positive	419 (86.2)
Post-decompression disappeared	404 (96.4)
Post-decompression persistent	15 (3.6)

These results are from Lee et al. [56]

We do not know the exact meaning of the LSR, but I think it is very important to observe that the LSR is lost due to any manipulation during surgery.

So every time we measure the LSR, we give the electric stimulus to the lowest intensity at which the LSR is measured. For example, if the LSR starts to be measured from 5 mA, the electric stimulus intensity at which the amplitude of the LSR waveform becomes the maximum is found by increasing the electric stimulus intensity by 1 mA.

Then, in the reverse order, the stimulus intensity is gradually lowered and the test progresses. Then, the minimum intensity of the electric stimulation in which the waveform is formed in proximity to the maximum amplitude is searched for, and the stimulus intensity is continuously tested during the operation.

When the offending vessel is slightly removed from the facial nerve during surgery, the LSR disappears immediately. Once the vessel is in place again, the LSR is measured again. In this way, it is very advantageous to discriminate whether it is a true offending vessel because LSR changes are observed very sensitively with the minimum electric stimulus intensity.

If you do not do this and do your tests with strong electrical stimulation, after the decompression, the LSR is not lost or the amplitude of the waveform is slightly decreased. Therefore, additional procedures that are not necessarily required are performed, resulting in hearing loss or facial palsy [35, 44, 49].

We were able to obtain a very sensitive response by testing with minimal electrical stimulus intensity in the LSR test and when it is judged that the decompression is perfect even if the LSR is not lost, additional procedures are not performed [50].

Compare Conventional and New Method

We performed LSR monitoring using centrifugal and conventional, centripetal methods simultaneously in 62 patients and compared the outcomes of the methods.

The conventional LSR measurement was performed in 34 patients (61.8%) before decompression, after decompression, 16 (29.1%) LSRs were observed without loss. In the new method, LSR measurement was 54 (98.2%) before decompression, after decompression, 1 (1.8%) LSR was observed without disappearance. In other words, the LSR measurement was smoother and the LSR was lost even after decompression (Table 6).

Among 419 patients, in LSR patients, 404 patients (96.4%) lost LSR after decompression, 15 patients (3.6%) remained unresolved (Table 5).

New methods of testing than the conventional method had to give a stronger electrical stimulus intensity of 0.476 mA on average, the latency of

Table 6 Comparison of efficacy of LSR monitoring by using the conventional and new methods

	Conventional method (%)	New method (%)	p value
No LSR	9.1	0.0	<0.0001
Disappearance of LSR	61.8	98.2	0.0012
Persistence of LSR	29.1	1.8	0.0051

LSR lateral spread response
These results are from Lee et al. [56]

Table 7 Parameters of intraoperative LSR monitoring (mean ± SD)

	Previous	New	p-value
Latency (ms)	11.255 ± 2.145	11.980 ± 1.567	0.0017
Amplitude (mV)	46.659 ± 50.081	38.767 ± 45.163	0.0600
Stimulation intensity (mA)	11.633 ± 5.594	12.109 ± 5.051	0.4305

LSR lateral spread response, *SD* standard deviation
These results are from Lee et al. [56]

the waveform was measured an average of 0.725 ms later, the amplitude of the waveform was observed to be 7.892 μV small (Table 7).

Facial Nerve Innervation

The conventional LSR measurement method gives the stimulus to the stem direction in the zygomatic branch, in this case, it is impossible to exclude that the zygomatic branch and the buccal branch are connected to each other [69, 70].

Comprehensive analysis of the branching pattern of the facial nerve is classified into six types, and zygomatic branch and buccal branch are connected to each other, accounting for more than 60% of C, E, and F. therefore, The facial nerve branch used for LSR measurement is the most effective temporal branch and mandibular branch. If the zygomatic branch or buccal branch is selected as the stimulation site for LSR measurement, residual LSR may be present because the upper and lower portions of the facial nerve branch can be stimulated at the same time (Fig. 20).

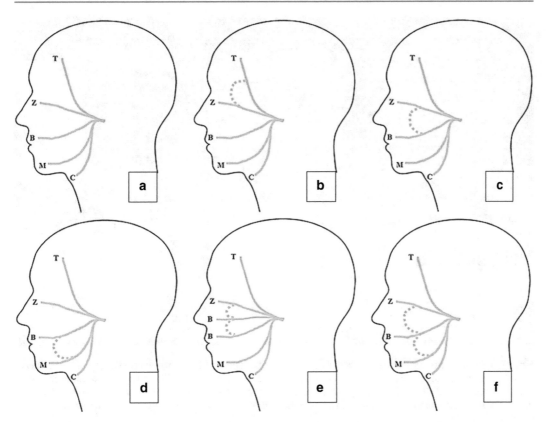

Fig. 20 The phenomenon in which the facial nerves are finely connected to each other is divided into six major categories. (**a**) is the basic shape of the branches of the facial nerve. (**b**) is a shape in which temporal branch and zygomatic branch are finely connected to each other. (**c**) is the shape that the zygomatic branch and the buccal branch are finely connected to each other. (**d**) is a shape in which the buccal branch and mandibular branch are finely connected to each other. (**e**) is the shape that the buccal branch is very developed and is finely connected to the zygomatic branch. (**f**) is a shape in which zygomatic branch, buccal branch, and mandibular branch are finely connected to each other. T, indicates temporal branch; Z, zygomatic branch; B, buccal branch; M, mandibular branch; and C, cervical branch

In general, the residual LSR measurement rate is observed in Thirumala 17.0% [54], Damaty 24.0% [51], Lee 25.6% [52].

However, if we stimulate the upper stimulation to the temporal branch in the opposite direction, the stimulation response of the frontalis and oculi was significantly increased, indicating that the upper branch was fully stimulated.

The LSR that occurs in this state, even if a patient with a Zygomatic branch and a Buccal branch are connected to each other, they can stimulate only the upper branch, which can help to eliminate residual LSR caused by insufficient stimulus.

According to the study, LSR loss rate during postoperative decompression was 96.4%, only 57 of 419 patients remained undisturbed and the measurements were very low at 3.6% [56].

In addition, the LSR measured by upper branch stimulation only observes changes measured in mentalis among the lower branch responses, and the LSR measured by lower branch stimulation only observes changes measured in frontalis among the upper branch responses. You can increase your sensitivity in evaluating whether you have (Fig. 21).

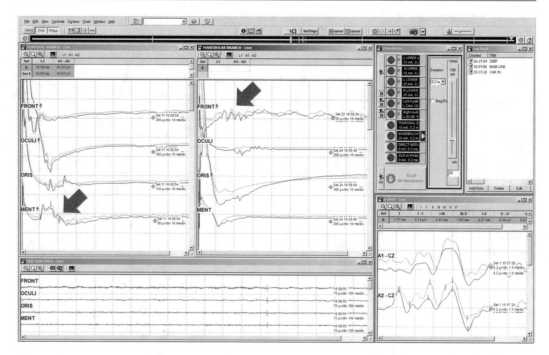

Fig. 21 LSR measurements observed changes in mentalis at upper branch stimulation and changes in frontalis at lower branch stimulation. It is advantageous to configure on the same screen in order to discriminate with the free-running EMG how much touching the facial nerve during surgery. The LSR should also be measured while observing the changes in the very sensitive BAEPs waveforms, so it is best to configure the same screen

References

1. Møller AR. Neural generators of auditory evoked potentials. In: Jacobson JT, editor. Principles and applications in auditory evoked potentials. Boston: Allyn & Bacon; 1994. p. 23–46.
2. Jewett DL, Williston JS. Auditory-evoked far fields averaged from the scalp of humans. Brain. 1971;94(4):681–96.
3. Møller A, Jannetta PJ. Neural generators of auditory evoked potentials. In: Jacobson JT, editor. The auditory brainstem response. SanDiego: College-Hill Press; 1985. p. 13–31.
4. Thirumala PD, Carnovale G, Habeych ME, Crammond DJ, Balzer JR. Diagnostic accuracy of brainstem auditory evoked potentials during microvascular decompression. Neurology. 2014;83(19):1747–52.
5. Amagasaki K, Watanabe S, Naemura K, Nakaguchi H. Microvascular decompression for hemifacial spasm: how can we protect auditory function? Br J Neurosurg. 2015;29:347–52.
6. El Damaty A, Rosenstengel C, Matthes M, Baldauf J, Dziemba O, Hosemann W, Schroeder HWS. A new score to predict the risk of hearing impairment after microvascular decompression for hemifacial spasm. Neurosurgery. 2017;81(5):834–43.
7. James ML, Husain AM. Brainstem auditory evoked potential monitoring: when is change in wave V significant? Neurology. 2005;65(10):1551–5.
8. Legatt AD. Mechanisms of intra operative brainstem auditory evoked potentialchanges. J Clin Neurophysiol. 2002;19:396–408.
9. Sindou M, Mercier P. Microvascular decompression for hemifacial spasm : surgical techniques and intraoperative monitoring. Neurochirurgie. 2018;64(2):133–43.
10. Park S-K, Joo B-E, Park K. Intraoperative neurophysiology during microvascular decompression for hemifacial spasm. Clin Neurophysiol. 2012;123:78.
11. Miller LE, Miller VM. Safety and effectiveness of microvascular decompression for treatment of hemifacial spasm: a systematic review. Br J Neurosurg. 2012;26:438–44.
12. Lee MH, Jee TK, Lee JA, Park K. Postoperative complications of microvascular decompression for hemifacial spasm: lessons from experience of 2040 cases. Neurosurg Rev. 2016;39(1):151–8.
13. Soriano-Baron H, Vales-Hidalgo O, Arvizu-Saldana E, Moreno-Jimenez S, Revuelta-Gutierrez R. Hemifacial spasm: 20-year surgical experience, lesson learned. Surg Neurol Int. 2015;6:83.
14. Amagasaki K, Watanabe S, Naemura K, Amagasaki HNK, Watanabe S, Naemura K, Nakaguchi H. Microvascular decompression for hemifacial

spasm: how can we protect auditory function? Br J Neurosurg. 2015;29:347–52.

15. Polo G, Fischer C, Sindou MP, Marneffe V. Brainstem auditory evoked potential monitoring during microvascular decompression for hemifacial spasm: intraoperative brainstem auditory evoked potential changes and warning values to prevent hearing losseprospective study in a consecutive series of 84 patients. Neurosurgery. 2004;54:97–104.

16. Pau HW, Just T, Bornitz M, Lasurashvilli N, Zahnert T. Noise exposure of the inner ear during drilling a cochleostomy for cochlear implantation. Laryngoscope. 2007;117:535–40.

17. Thirumala PD. Reduction in high-frequency hearing loss following technical modifications to microvascular decompression for hemifacial spasm. J Neurosurg. 2015;123(4):1059–64.

18. Polo G, Fischer C, Sindou MP, Marneffe V. Brain auditory evoked potentialmonitoring during microvascular decompression for hemifacial spasm: intra-operative brainstem auditory evoked potential changes and warning values to prevent hearing loss. Prospective study in a consecutive series of 84 patients. Neurosurgery. 2004;54:97–106.

19. Polo G, Fischer C. Intraoperative monitoring of brainstem evoked potentialsduring microvascular decompression of cranial nerves in cerebello-pontineangle. Neurochirurgie. 2009;55:152–7.

20. Grundy BL, Jannetta PJ, Procopio PT, Lina A, Boston JR, Doyle E. Intraoperative monitoring of brain-stem auditory evoked potentials. J Neurosurg. 1982;57(5):674–81.

21. Raudzens PA, Shetter AG. Intraopertive monitoring of BAEPs. J Neurosurg. 1982;57(3):341–8.

22. Watanabe E, Schramm J, Strauss C, Fahlbusch R. Neurophysiologic monitor-ing in posterior fossa surgery: part II-BAEP-waves I and V and preservation ofhearing. Acta Neurochir (Wien). 1989;98:118–28.

23. Martin WH, Stecker MM. ASNM position statement: intraoperative monitoring of auditory evoked potentials. J Clin Monit Comput. 2008;22(1):75–85.

24. American Electroencephalographic Society. Guideline eleven: guidelines for intraoperative monitoring of sensory evoked potentials. J Clin Neurophysiol. 1994;11(1):77–87.

25. Hatayama T, Møller AR. Correlation between latency and amplitude of peakV in th BAEPs: intraoperatve recordings in MVD operations. Acta Neurochir (Wien). 1998;140:681–7.

26. Jo K-W, Kim J-W, Kong D-S, Hong S-H, Park K. The patterns risk factors of hearing loss following MVD for HFS. Acta Neurochir (Wien). 2011;153:1023–30.

27. Jung NY, Lee SW, Park CK, Chang WS, Jung HH, Chang JW. Hearing outcome following microvascular decompression for hemifacial spasm: series of 1434 cases. World Neurosurg. 2017;108:566–71.

28. Park S-K, Joo B-E, Lee S, Lee J-A, Hwang J-H, Kong D-S, Seo D-W, Park K, Lee H-T. The critical warning sign of real-time brainstem auditory evoked potentials during microvascular decompression for hemifacial spasm. Clin Neurophysiol. 2018;129(5):1097–102.

29. Nielsen VK. Pathophysiology of hemifacial spasm: I. Ephaptic transmission and ectopic excitation. Neurology. 1984;34:418–26.

30. Møller AR, Jannetta PJ. On the origin of synkinesis in hemifacial spasm: results of intracranial recordings. J Neurosurg. 1984;61:569–76.

31. Thirumala PD, et al. Microvascular decompression for hemifacial spasm: evaluating outcome prognosticators including the value of intraoperative lateral spread response monitoring and clinical characteristics in 293 patients. J Clin Neurophysiol. 2011;28(1):56–66.

32. Møller AR. Vascular compression of cranial nerves II. Pathophysiology. Neurol Res. 1999;21:439–43.

33. Møller AR, Jannetta PJ. Microvascular decompression in hemifacial spasm: intraoperative electrophysiological observations. Neurosurgery. 1985;16:612–8.

34. Kiya N, Bannur U, Yamauchi A, Yoshida K, Kato Y, Kanno T. Monitoring of facial evoked EMG for hemifacial spasm: a critical analysis of its prognostic value. Acta Neurochir. 2001;143:365–8.

35. Mooij JJ, Mustafa MK, van Weerden TW. Hemifacial spasm: intraoperative electromyographic monitoring as a guide for microvascular decompression. Neurosurgery. 2001;49:1365–71.

36. Tobishima H, Hatayama T, Ohkuma H. Relation between the persistence of an abnormal muscle response and the long-term clinical course after microvascular decompression for hemifacial spasm. Neurol Med Chir (Tokyo). 2014;54:474–82.

37. von Eckardstein K, Harper C, Castner M, Link M. The significance of intraoperative electromyographic Blateral spread in predicting outcome of microvascular decompression for hemifacial spasm. J Neurol Surg B Skull Base. 2014;75:198–203.

38. Hyun S-J, Kong D-S, Park K. Microvascular decompression for treating hemifacial spasm: lesion learned from a prospective study of 1,174 operations. Neurosurg Rev. 2010;33:325–34.

39. Kondo A. Follow-up results of microvascular decompression in trigeminal neuralgia and hemifacial spasm. Neurosurgery. 1997;40:46–51.

40. Shimizu K, Matsumoto M, Wada A, Sugiyama T, Tanioka D, Okumura H, Fujishima H, Nakajo T, Nakayama S, Yabuzaki H, Mizutani T. Supine no-retractor method in microvascular decompression for hemifacial spasm: results of 100 consecutive operations. J Neurol Surg B Skull Base. 2015;76:202–7.

41. Yang M, Zheng X, Ying T, Zhu J, Zhang W, Yang X, Li S. Combined intraoperative monitoring of abnormal muscle response and Z-L response for hemifacial spasm with tandem compression type. Acta Neurochir (Wien). 2014;156:1161–6.

42. Kim C-H, Kong D-S, Lee JA, Park K. The potential value of the disappearance of the lateral spread response during microvascular decompression for predicting the clinical outcome of hemifacial spams: a prospective study. Neurosurgery. 2010;67:1581–8.

43. Wei Y, Yang W, Zhao W, Chunhua P, Li N, Yu C, Shang H. Microvascular decompression for hemifacial spasm: can intraoperative lateral spread response monitoring improve surgical efficacy? J Neurosurg. 2017;128(3):885–90.

44. Haines SJ, Torres F. Intraoperative monitoring of the facial nerve during decompressive surgery for hemifacial spasm. J Neurosurg. 1991;74(2):254–7.

45. Ishikawa M, Ohira T, Namiki J, Kobayashi M, Takase M, Kawase T, Toya S. Electrophysiological investigation of hemifacial spasm after microvascular decompression: F waves of the facial muscles, blink reflexes, and abnormal muscle responses. J Neurosurg. 1997;86(4):654–61.

46. Joo W-I, Lee K-J, Park H-K, Chough C-K, Rha H-K. Prognostic value of intraoperative lateral spread response monitoring during microvascular decompression in patients with hemifacial spasm. J Clin Neurosci. 2008;15(12):1335–9.

47. Hatem J, Sindou M, Vial C. Intraoperative monitoring of facial EMG responses during microvascular decompression for hemifacial spasm: prognostic value for long-term outcome: a study in a 33-patient series. Br J Neurosurg. 2001;15(6):496–9.

48. Isu T, Kamada K, Mabuchi S, Kitaoka A, Ito T, Koiwa M, Abe H. Intra-operative monitoring by facial electromyographic responses during microvascular decompressive surgery for hemifacial spasm. Acta Neurochir (Wien). 1996;138(1):19–23.

49. Huang B-R, Chang C-N, Hsu J-C. Intraoperative electrophysiological monitoring inmicrovascular decompression for hemifacial spasm. J Clin Neurosci. 2009;16(2):209–13.

50. von Eckardstein K, Harper C, Castner M, Link M. The Significance of Intraoperative Electromyographic "Lateral Spread" in Predicting Outcome of Microvascular Decompression for Hemifacial Spasm. J Neurol Surg B Skull Base. 2014;75(3):198–203.

51. El Damaty A, Rosenstengel C, Matthes M, Baldauf J, Schroeder HWS. The value of lateral spread response monitoring in predicting the clinical outcome after microvascular decompression in hemifacial spasm: a prospective study on 100 patients. Neurosurg Rev. 2016;39(3):455–66.

52. Lee SH, Park BJ, Shin HS, Park CK, Rhee BA, Lim YJ. Prognostic ability of intraoperative electromyographic monitoring during microvascular decompression for hemifacial spasm to predict lateral spread response outcome. J Neurosurg. 2017;126(2):391–6.

53. Park JS, Kong DS, Lee JA, Park K. Chronologic analysis of symptomatic change following microvascular decompression for hemifacial spasm: value for predicting midterm outcome. Neurosurg Rev. 2008;31:413–9.

54. Thirumala PD, Wang X, Shah A, Habeych M, Crammond D, Balzer JR, Sekula R. Clinical impact of residual lateral spread response after adequate microvascular decompression for hemifacial spasm: a retrospective analysis. Br J Neurosurg. 2015;15(6):496–9.

55. Hirono S, Yamakami I, Sato M, Kado K, Fukuda K, Nakamura T, Higuchi Y, Saeki N. Continuous intraoperative monitoring of abnormal muscle response in microvascular decompression for hemifacial spasm; a real-time navigator for complete relief. Neurosurg Rev. 2014;37(2):311–9.

56. Lee S, Park S-K, Lee J-A, Joo B-E, Kong D-S, Seo D-W, Park K. A new method for monitoring abnormal muscle response in hemifacial spasm: a prospective study. Clin Neurophysiol. 2018;129(7):1490–5.

57. Zhang X, Zhao H, Tang Y-D, Zhu J, Zhou P, Yuan Y, Li S-T. The effects of combined intraoperative monitoring of abnormal muscle response and ZL response for hemifacial spasm. World Neurosurgery. 2017;108:367–73.

58. Son B-C, Ko H-C, Choi J-G. Intraoperative monitoring of Z-L response (ZLR) and abnormal muscle response (AMR) during microvascular decompression for hemifacial spasm. Interpreting the role of ZLR. Acta Neurochirurgica. 2018;160(5):963–70.

59. Kircher ML, Kartush JM. Pitfalls in intraoperative nerve monitoring during vestibular schwannoma surgery. Neurosurg Focus. 2012;33(3):E5.

60. Kimura J. Current understanding of F-wave physiology in the clinical domain. Suppl Clin Neurophysiol. 2006;59:299–303.

61. Ishikawa M, Ohira T, Namiki J, Kobayashi M, Takase M, Kawase T, Toya S. Electrophysiological investigation of hemifacial spasm after microvascular decompression: F waves of the facial muscles, blink reflexes, and abnormal muscle responses. J Neurosurg. 1997;86:654–61.

62. Wilkinson MF, Kaufmann AM. Monitoring of facial muscle motor evoked potentials during microvascular decompression for hemifacial spasm: evidence of changes in motor neuron excitability. J Neurosurg. 2005;103:64–9.

63. Debruyne F. Latency differences between ipsilateral and contralateral auditory brainstem responses. Scand Audiol. 1983;12(3):223.

64. American Clinical Neurophysiology Society. Guideline 9C: guidelines on short-latency auditory evoked potentials. J Clin Neurophysiol. 2006;23(2):157–67.

65. Joo B-E, Park S-K, Cho K-R, Kong D-S, Seo D-W, Park K. Real-time intraoperative monitoring of brainstem auditory evoked potentials during microvascular decompression for hemifacial spasm. J Neurosurg. 2016;125(5):1061–7.

66. Lee MH, Lee S, Park S-K, Lee J-A, Park K. Delayed hearing loss after microvascular decompression for hemifacial spasm. Acta Neurochir (Wien). 2019;161(3):503–8.

67. Møller AR, Jannetta PJ. Monitoring facial EMG responses during microvascular decompression operations for hemifacial spasm. J Neurosurg. 1987;66(5):681–5.

68. Wilkinson MF, Kaufmann AM. Monitoring of facial muscle motor evoked potentials during microvascular decompression for hemifacial spasm: evidence of changes in motor neuron excitability. J Neurosurg. 2005;103(1):64–9.

69. Katz AD, Catalano P. The clinical significance of the various anastomotic branches of the facial nerve. Report of 100 patients. Arch Otolaryngol Head Neck Surg. 1987;113(9):959–62.

70. Myint K, Azian FA, Khairul K. The clinical significance of the branching pattern of the facial nerve in Malaysian subjects. Med J Malaysia. 1992;47(2):114–21.

Anesthetic Management of MVD

Jeong Jin Lee

Current anesthetic management during surgery for hemifacial spasm follows standardized anesthetic protocols. Most patients present with a good general condition and the surgery is often performed electively. Regardless, anesthesiologists should keep in mind potential problems that may arise during posterior fossa surgery, including possible injury to vital brainstem centers; pneumocephalus; and, with unusual patient positioning, C-spine injury and upper airway swelling or decubitus injury of the dependent portion [1–3]. Other considerations include the potential usage of intraoperative neurophysiologic monitoring (IONM). In this chapter, anesthetic management during posterior fossa surgery will be discussed, focusing on the completion of microvascular decompression (MVD) for the resolution of hemifacial spasm.

Preoperative Management

During the process of medical care, patient records and laboratory data are reviewed for the presence of hypertension, diabetes mellitus, cardiovascular disease, and other medical problems. If uncontrolled or active disease is found, elective surgery should be postponed to ensure a good physical state. Airway examination should be performed. In cases of high Mallampati scores, limited neck extension, or loose teeth, the anesthesiologist should anticipate the likelihood of difficult intubation and prepare rescue intubation modalities. Patients should fast for at least 8 h beforehand.

Intraoperative Management

When the patient arrives in the operating room, electrocardiography, oxygen saturation monitoring, and noninvasive blood pressure (BP) monitoring begin and a neuromuscular monitoring probe is applied to the ulnar nerve. The induction of anesthesia begins with an intravenous bolus injection of thiopental or propofol and opioid analgesic agents such as remifentanil, fentanyl, or sufentanil. After confirmation of an abolished eyelid reflex, baseline twitch height (T_0) measurements should be collected and the injection of a neuromuscular-blockade (NMB) agent (NMBA) such as rocuronium or vecuronium should be completed to establish an optimal intubation condition. Anesthesia is maintained with total intravenous anesthesia (TIVA) using propofol and remifentanil, or balanced anesthesia with volatile anesthetics and an opioid. Without IONM, there is no significant restriction on the choice of anesthetic agents.

Arterial catheterization can detect sudden changes in BP and identify blood gases and

J. J. Lee (✉)
Department of Anesthesiology and Pain Medicine,
Samsung Medical Center, Sungkyunkwan University
School of Medicine, Seoul, Republic of Korea
e-mail: jjeong.lee@samsung.com

© Springer Nature Singapore Pte Ltd. 2020
K. Park, J. S. Park (eds.), *Hemifacial Spasm*, https://doi.org/10.1007/978-981-15-5417-9_12

electrolytes as needed. A bispectral index (BIS) monitor should be attached to the forehead to measure the depth of anesthesia, and the propofol infusion rate should be titrated to maintain a target BIS value of 40–60. Importantly, the patient must not move during microsurgery; unanticipated bucking or straining may induce a catastrophic outcome. Such can be prevented with a cautious titration of appropriate anesthetic agents including NMBA and analgesics. Spontaneous activity as seen via facial nerve electromyographic (FNEMG) monitoring may precede physical movement. Volatile anesthetic agents had more hemodynamic variability and spontaneous activity on FNEMG, whereas TIVA was proven to be a more effective anesthetic for preventing patient physical movement when clinically titrated to produce stable operation conditions [4]. When spontaneous activity on FNEMG occurs, increasing the dose of NMBA or opioid can prevent the patient's movement. If NMBA is not used at all during surgery, deeper anesthesia is needed, but, with such, extreme hypotension and bradycardia can develop. This problem can be mitigated by infusing inotropic agents. Arterial BP should be controlled to within 30% of the preoperative value measured on the day before surgery.

Positioning

Most MVD surgeries can be performed using the park-bench position (three-fourths lateral prone decubitus) (Fig. 1), although the prone position or the sitting position has been preferred by some surgeons [1]. The assurance of careful positioning and padding can help to avoid injuries; for example, a gel roll should be placed under the axilla of the dependent side, while other pressure points such as the elbows, wrists, ischial spines, and heels should be protected. Excessive neck flexion may induce venous obstruction or cervical spine injury, which causes increased intracranial pressure (ICP), upper airway edema, or quadriplegia. Preexisting cervical spinal stenosis may predispose patients to cervical spine injury. Confirming that the airway pressure under controlled ventilation with fixed tidal volume has not risen during head fixation can avoid problems of excessive neck flection. Avoid compression or protrusion of tongue to prevent ischemia or venous congestion of the tongue. After positioning, patients should be checked carefully for signs of obstruction such as discoloration of the face, lips, or tongue, especially given that, during the operation, the patient's face can be more difficult to access because the operating table rotates 90 or 180 degrees. The corrugated tube and monitoring lines should be carefully secured before the start of the operation.

Anesthesia for Intraoperative Neuromonitoring

Various electrophysiologic monitoring techniques may be used during surgery for the central nervous

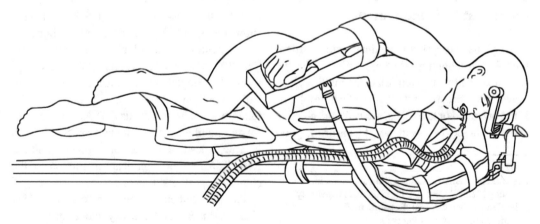

Fig. 1 Park-bench position

system. These include somatosensory evoked potentials (SSEPs), visual evoked potentials (VEPs), motor evoked potentials (MEPs), brainstem auditory evoked potentials (BAEPs), and electromyographic (EMG) monitoring of the cranial nerves (e.g., facial nerve, glossopharyngeal nerve). In general, during MVD surgery, SSEPs, BAEPs, and facial nerve EMG (FNEMG) are essential intraoperative monitoring. Most anesthetics affect IONM to varying degrees (Table 1).

Sensory evoked potential (SEP) monitoring affects the selection of anesthetics. In general, volatile anesthetics affect SEPs more than intravenous anesthetics. Evoked potentials (Eps) of cortical origin (i.e., the cortical portion of SSEPs and VEPs) are considered more prone to modification by anesthetics than brainstem potentials (i.e., BAEPs and subcortical portions of SSEPs). However, the neurologist can receive an interpretable SSEP wave under volatile anesthetics. In contrast, BAEPs are almost always entirely unaffected by anesthetics. It is essential to maintain constant anesthetic drug levels during the recording of Eps [5].

MEPs are affected by both the anesthetic agent and NMBA. Myogenic MEPs are easily suppressed by anesthesia, especially by volatile anesthetics [6]. Without NMBA, MEP waves can be monitored with volatile anesthetic agents but, with partial NMB, they are significantly abolished by a small dose of volatile anesthetics [7]. Intravenous anesthetics such as propofol and dexmedetomidine also affect myogenic MEPs, albeit to a lesser extent. Anesthesiologists prefer to use

Table 1 Influence of anesthetic agents on IONM

	SSEP	BAEP	MEP	EMG	Facial LSR
Sevoflurane	+	−	++	−	N/A
Desflurane	+	−	++	−	+
Remifentanil	+	−	±	−	N/A
Dexmedetomidine	±	−	+	−	N/A
Propofol	±	−	+	−	±
Muscle relaxant	−	−	++	++	++

SSEP somatosensory evoked potential, *BAEP* brainstem auditory evoked potential, *MEP* motor evoked potential, *EMG* electromyography, *Facial LSR* facial lateral spreading response

propofol in conjunction with an opioid during MEPs. NMBA does not affect D-waves but will affect myogenic MEPs. A complete withdraw of NMB can easily ensure a clear myogenic MEP wave, but the patient can move during transcranial electrical stimulation, which can interfere with the surgery and often puts the patient in a dangerous situation. Partial NMB may reduce but not abolish these movements and can complicate the interpretation of myogenic MEPs. Some centers generally omit NMB, whereas others tend to use partial NMB. If employed, partial NMB should be achieved with a continuous infusion of the NMBA with close titration under neuromuscular monitoring such as the assessment of the responses to train-of-four stimulation [8] to assess the degree of NMB. Bolus injections should be avoided because they induce too variable NMB levels [9]. Kim et al. reported that, under TIVA with propofol and remifentanil, partial NMB with a target T2/Tc of more than 50% can achieve acceptable MEP waves, but complete withdrawal of the NMBA can lead to more steady MEP waves [10].

Facial nerve-triggered EMG is essential monitoring during MVD. The lateral spreading response (LSR) is considered to be an effective diagnostic tool for complete decompression and also to be an important prognostic factor in MVD surgery [11]. Because EMG waves are influenced by NMBA, this requires that the patient not be paralyzed or been in a constant state of incomplete paralysis. In general, a TOF count of more than two has been recommended. It is well-known that facial muscles are more resistant to NMBA than the adductor pollicis muscle. However, Chung et al. reported that the maintenance of partial NMB with a target T1/Tc ratio of 50% rather than a TOF count of two resulted in a clinically acceptable success rate for LSR monitoring and surgical condition during MVD [12]. When NMB was completely removed, the success rate of LSR monitoring was not significantly different from when the target T1/Tc ratio was maintained at 50%, but the LSR response after decompression was not as easily defined in the patients without NMBA because of the delayed abolishment of LSR and spontaneous free-run

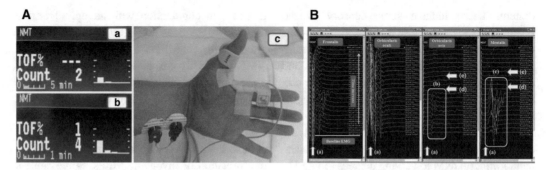

Fig. 2 (**A**) Images of neuromuscular blockade (NMB) monitoring using NMT module. The target of two counts of train-of-four (TOF) in TOF group (a), T1/Tc ratio of 50%inT1 group (b), and set up image (c). (**B**) Stack wave image of an intraoperative facial EMG upon stimulation of the zygomatic branch in a patient with partial NMBA maintained with a T1/Tc of 50%. Electrical stimulation of the zygomatic branch (a) produced an abnormal muscle response, called the LSR, in the orbicularis oris (b) and mentalis (c) muscles. This abnormal response disappeared (e) after decompression (d) [13]

EMG activity. Maintenance of partial NMB with a target T1/Tc ratio of 50% during LSR monitoring for MVD can therefore be recommended (Fig. 2) [13]. In general, facial nerve EMG is known to be affected by NMBA alone. However, a recent report suggests that desflurane (1 MAC) suppresses the LSR amplitude by 43% compared to TIVA alone, providing direct evidence that there is a central mechanism of action inherent in the origin of the LSR [14]. Most of all, close conversation between the neurosurgeon, neurologist, and anesthesiologist to optimize IONM, attain hemodynamic stability, and avoid unexpected movement of the patient is key.

Trigeminal Arrhythmia

Manipulation of the sensory branches of the trigeminal nerve is known to cause autonomic changes, such as bradycardia or asystole, known as the trigeminocardiac reflex (TCR). A risk for TCR should be considered in any craniomaxillofacial surgery, especially during surgery performed at or near the cerebellopontine angle (e.g., acoustic neuromas; microvascular decompression of the trigeminal, facial, and glossopharyngeal cranial nerves). One retrospective study showed that TCR occurred in 18% of microvascular trigeminal decompression patients: of note, their heart rate decreased by 46% and their mean

arterial blood pressure decreased by 57% during their operative procedures performed near the trigeminal nerve compared to the levels recorded immediately before the stimulus [15]. Knowledge of the applied anatomy will help in predicting this risk and choosing the appropriate prevention. If the risk of TCR is high, an intravenous anticholinergic agent like atropine or glycopyrrolate may be used to prevent such. Intraoperative predisposing factors are light plane of anesthesia, hypercarbia, hypoxia, and acidosis. If a TCR is suspected, anesthesiologists should check and correct those factors and inform the surgeon to stop stimulating and wait for the pulse to return to normal. Administration of intravenous anticholinergic agents after heart rate recovery can be used to prevent the recurrence of TCR. The surgeon then should continue the operation with gentle manipulation [16].

Cranial Nerve Dysfunction and Respiratory Center Injury

Operations in the posterior fossa can injure vital circulatory and respiratory brainstem centers and cranial nerves or their nuclei. Such injuries may occur as a result of direct surgical trauma or ischemia from retraction or other interruptions of the blood supply. Cranial nerve dysfunction, particularly of nerves IX, X, and XII, can result in a loss of control and patency of the upper airway. EMG of lower cranial nerves during surgery can help detecting the risk of cranial nerve injury ahead. Abnormal cardiovascular responses can result from irritation of the lower portion of the pons and upper medulla. These areas are most often stimulated during procedures on the floor of the fourth ventricle. Isolated damage to respiratory centers may rarely occur without preceding circulatory signs during surgery. Therefore, abrupt onset of bradycardia and hypotension, tachycardia and hypertension, or bradycardia and hypertension as well as ventricular dysrhythmias should alert the anesthesiologist to the possibility of such an injury [3, 17]. Meticulous attention to the electrocardiogram and a directly transduced arterial pressure during manipulation in this region are necessary to provide the surgeon with an immediate warning of the risk of damage to the adjacent cranial nerve nuclei and respiratory centers. The posterior fossa is a relatively small space, and its compensatory latitudes are even more limited than those of the supratentorial space. Relatively little swelling can result in disorders of consciousness, respiratory drive, and cardiomotor function. Abnormal respiration after MVD is more likely to occur when the brainstem injury is the result of vessel occlusion or hematoma (which can develop in a delayed manner) rather than direct mechanical damage caused by retraction of or dissection in the brainstem [18].

Macroglossia

An abnormal swelling of the tongue is a rare postoperative complication often associated with serious airway obstruction and prolonged intubation. Macroglossia has been most commonly reported after suboccipital and/or posterior fossa craniotomies and spine surgeries performed with the patient in the prone or park-bench positions. Risk factors for macroglossia include a surgery duration of more than 8 h, type of surgery (i.e., suboccipital and/or posterior fossa craniotomies and spine surgeries), and patient position (i.e., prone, park-bench, or sitting). The etiology of macroglossia is multifactorial and possible mechanisms include local mechanical tongue compression interfering with venous and/or lymphatic drainage, regional venous thrombosis, and/or local trauma. Prevention, awareness of the possibility, and early recognition are the best forms of treatment (see *Positioning*) [19, 20].

Pneumocephalus and Cerebrospinal Fluid Leakage

Although the sitting position notably increases the likelihood of pneumocephalus, this can also occur in the park-bench or prone positions [2]. Air enters the subarachnoid space, as cerebrospinal fluid (CSF) is lost during surgery. In patients with cerebral atrophy, drainage of CSF is marked; air can replace the CSF on the surface of the brain and in the lateral ventricles. Expansion of a pneumocephalus following dural closure can compress the brain. Postoperative pneumocephalus can cause delayed awakening and continued impairment of neurological function [21]. Because of these concerns, nitrous oxide is rarely used for sitting craniotomies [3]. Sometimes the surgeon asks for induction of a Valsalva maneuver (VM) to confirm a watertight dural closure. If the dural closure is not tight, CSF is expelled following a rise in ICP and air is sucked inside the cranium to equalize the ICP, opening up the possibility of pneumocephalus. In addition, forceful or prolonged VM can be associated with profound and complicated physiological responses and significant complications can occur. Therefore, clinicians should be well-acquainted with its applied physiology and use it judiciously to avoid the complications associated with it [22]. When we used high-dose volatile anesthetics, which were potent cerebral vasodilators, the brain could bulge out after dura opening and an infusion of mannitol or application of hyperventilation was mandatory. However, during TIVA anesthesia, infusing mannitol or applying hyperventilation in order to reduce the ICP during surgery should be done in consultation with the surgeon [23].

Venous Air Embolism

Venous air embolism is a very rare complication which can occur when the pressure within an open vein is subatmospheric. This scenario may occur in the context of any surgical position (and during any procedure) whenever the wound is above the level of the heart. The incidence of venous air embolism is greater during sitting craniotomy (20–40%) than in any other position. Air entry into large cerebral venous sinuses increases the risk. The most sensitive monitoring modalities are precordial Doppler and transesophageal echocardiography. However, these technologies are not necessary if the patient is not in a sitting position. An abrupt decrease in end-tidal CO_2 is another indicator of an air embolism. When an air embolism is detected, anesthesiologists should warn surgeons to stop surgery and apply saline or water-soaked gauze. Change the patient's position to head-down and right-side-up to prevent the air from traveling to the coronary or cerebral artery is another step [3, 17, 23].

Postoperative Management

Extubation and Airway Management

Most patients awake immediately after surgery and extubation is possible. However, in rare cases, respiratory center injury or macroglossia may occur. If there is evidence of macroglossia at the end of surgery, the patient should not be extubated. If there is suspicion of a circulatory center injury (i.e., severe hypertension, bradycardia, or arrhythmia during the operation) this should be taken into account during planning for extubation and postoperative care. If macroglossia develops after extubation, early airway management should be considered [24]. Airway management may be difficult and a surgical airway may be needed. Other supportive treatments may be required, such as placing the patient in a head-up position, preventing further tongue compression, steroids, keeping the tongue moist to prevent desiccation, and administering analgesics as needed [19, 20].

Nausea, Vomiting, and Headache

Patients, after MVD of the cranial nerves, frequently experience postoperative nausea and vomiting (PONV). MVD and acoustic neuroma resection were associated with an increased likelihood of PONV as compared with craniotomy performed for other tumor resections [25]. The risk was also higher among patients who underwent MVD of the trigeminal neuralgia than those treated for hemifacial spasm. Female sex was more susceptible to PONV and the use of a volatile agent increase the risk of PONV. Using intravenous anesthesia and prophylactic antiemetic agents like ondansetron, ramosetron, and metoclopramide can prevent PONV. However, despite the use of intraoperative prophylactic ondansetron, the overall incidence of PONV (nausea, emesis, or both) was reported to be 60% during the first 24 h after surgery. Application of a prophylactic transdermal scopolamine patch may also help to prevent PONV. It may be necessary to administer a combination of antiemetics to decrease the incidence of PONV after retromastoid craniotomy [26]. When vomiting occurs, the patient's head should be rotated to one side to avoid aspiration and symptomatic treatment should be provided.

Other Considerations

MVD surgery is known to be less invasive and safe. However, it is still challenging to perform and may result in a fatal outcome. In a retrospective analysis of 46 patients (0.66% of 6974 cases of MVD), who presented with a decline in consciousness after waking up from the anesthesia, 15 patients (0.2%) died. The authors reported that mortality is significantly higher in trigeminal cases with cerebellar hematoma and an immediate hematoma evacuation plus ventricular drainage could give the patient a greater chance of survival [27]. Most of the patients go to the intensive care unit after surgery. However, in case the patient is transferred to the postoperative care unit, continuous monitoring of blood pressure, pulse, respiration, and blood oxygen saturation is mandatory. The patient's consciousness and pupillary changes should also be observed closely. If the patient shows a sudden increase in BP accompanied by bradycardia and unconsciousness, slow and deep breathing or even arrest, decreased oxygen saturation, dilated pupils, and light reflex diminished or disappearing, the possibility of infarction, swelling, or bleeding in the cerebellum or brainstem should be considered [28]. Immediate communication with the neurosurgeon is required, an emergent head computed tomography scan should be considered, and other immediate actions as necessary should be taken.

References

1. Velho V, Naik H, Bhide A, Bhople L, Gade P. Lateral Semi-sitting Position: a Novel Method of Patient's Head Positioning in Suboccipital Retrosigmoid Approaches. Asian J Neurosurg. 2019;14:82–6.
2. Toung TJ, McPherson RW, Ahn H, Donham RT, Alano J, Long D. Pneumocephalus: effects of patient position on the incidence and location of aerocele after posterior fossa and upper cervical cord surgery. Anesth Analg. 1986;65:65–70.
3. Butterworth Iv JF, Mackey DC, Wasnick JD. Anesthesia for neurosurgery. Morgan & Mikhail's clinical anesthesiology. 6th ed. New York, NY: McGraw-Hill Education; 2018.
4. Jellish WS, Leonetti JP, Buoy CM, Sincacore JM, Sawicki KJ, Macken MP. Facial nerve electromyographic monitoring to predict movement in patients titrated to a standard anesthetic depth. Anesth Analg. 2009;109:551–8.
5. Nunes RR, Bersot CDA, Garritano JG. Intraoperative neurophysiological monitoring in neuroanesthesia. Curr Opin Anaesthesiol. 2018;31:532–8.
6. Sloan TB, Heyer EJ. Anesthesia for intraoperative neurophysiologic monitoring of the spinal cord. J Clin Neurophysiol. 2002;19:430–43.
7. Legatt AD, Emerson RG, Epstein CM, et al. ACNS guideline: transcranial electrical stimulation motor evoked potential monitoring. J Clin Neurophysiol. 2016;33:42–50.
8. Sloan TB, Janik D, Jameson L. Multimodality monitoring of the central nervous system using motor-evoked potentials. Curr Opin Anaesthesiol. 2008;21:560–4.
9. Adams DC, Emerson RG, Heyer EJ, et al. Monitoring of intraoperative motor-evoked potentials under conditions of controlled neuromuscular blockade. Anesth Analg. 1993;77:913–8.
10. Kim WH, Lee JJ, Lee SM, et al. Comparison of motor-evoked potentials monitoring in response to transcranial electrical stimulation in subjects undergoing neurosurgery with partial vs no neuromuscular block. Br J Anaesth. 2013;110:567–76.
11. Kong DS, Park K, Shin BG, Lee JA, Eum DO. Prognostic value of the lateral spread response for intraoperative electromyography monitoring of the facial musculature during microvascular decompression for hemifacial spasm. J Neurosurg. 2007;106:384–7.
12. Chung YH, Kim WH, Lee JJ, et al. Lateral spread response monitoring during microvascular decompression for hemifacial spasm. Comparison of two targets of partial neuromuscular blockade. Anaesthesist. 2014;63:122–8.
13. Chung YH, Kim WH, Chung IS, et al. Effects of partial neuromuscular blockade on lateral spread response monitoring during microvascular decompression surgery. Clin Neurophysiol. 2015;126:2233–40.
14. Wilkinson MF, Chowdhury T, Mutch WA, Kaufmann AM. Is hemifacial spasm a phenomenon of the central nervous system? The role of desflurane on the lateral spread response. Clin Neurophysiol. 2015;126:1354–9.
15. Schaller B. Trigemino-cardiac reflex during microvascular trigeminal decompression in cases of trigeminal neuralgia. J Neurosurg Anesthesiol. 2005;17:45–8.
16. Sandu N, Chowdhury T, Meuwly C, Schaller B. Trigeminocardiac reflex in cerebrovascular surgery: a review and an attempt of a predictive analysis. Expert Rev Cardiovasc Ther. 2017;15:203–9.
17. Butterworth JF, Mackey DC, Wasnick JD, Morgan GE, Mikhail MS, Morgan GE. Morgan & Mikhail's clinical anesthesiology. 6th ed. New York: McGraw-Hill; 2018.
18. Lee MH, Jee TK, Lee JA, Park K. Postoperative complications of microvascular decompression for hemifacial spasm: lessons from experience of 2040 cases. Neurosurg Rev. 2016;39:151–8.
19. El Hassani Y, Narata AP, Pereira VM, Schaller C. A reminder for a very rare entity: massive tongue swelling after posterior fossa surgery. J Neurol Surg A Cent Eur Neurosurg. 2012;73:171–4.
20. Brockerville M, Venkatraghavan L, Manninen P. Macroglossia in neurosurgery. J Neuroanesthesiol Crit Care. 2017;04:78–84.
21. He Z, Cheng H, Wu H, Sun G, Yuan J. Risk factors for postoperative delirium in patients undergoing microvascular decompression. PLoS One. 2019;14:e0215374.
22. Haldar R, Khandelwal A, Gupta D, Srivastava S, Rastogi A, Singh PK. Valsalva maneuver: its implications in clinical neurosurgery. Neurol India. 2016;64:1276–80.
23. Gracia I, Fabregas N. Craniotomy in sitting position: anesthesiology management. Curr Opin Anaesthesiol. 2014;27:474–83.

24. Toyama S, Hoya K, Matsuoka K, Numai T, Shimoyama M. Massive macroglossia developing fast and immediately after endotracheal extubation. Acta Anaesthesiol Scand. 2012;56:256–9.

25. Tan C, Ries CR, Mayson K, Gharapetian A, Griesdale DE. Indication for surgery and the risk of postoperative nausea and vomiting after craniotomy: a case-control study. J Neurosurg Anesthesiol. 2012;24:325–30.

26. Meng L, Quinlan JJ. Assessing risk factors for postoperative nausea and vomiting: a retrospective study in patients undergoing retromastoid craniectomy with microvascular decompression of cranial nerves. J Neurosurg Anesthesiol. 2006;18:235–9.

27. Xia L, Liu MX, Zhong J, et al. Fatal complications following microvascular decompression: could it be avoided and salvaged? Neurosurg Rev. 2017;40:389–96.

28. Li ST, Sun H. Surgical techniques of microvascular decompression for hemifacial spasm. In: Li ST, Zhong J, Sekula Jr R, editors. Microvascular decompression surgery. Dordrecht: Springer; 2016.

Botulinum Toxin Injection in Hemifacial Spasm

Jinyoung Youn, Wooyoung Jang,
and Jong Kyu Park

Hemifacial spasm (HFS) is a movement disorder characterized by involuntary, irregular, and recurring contractions of the muscles innervated by the facial nerve [1]. HFS is not a life-threatening disease, but can severely affect the quality of life [2–4]. Moreover, HFS could be associated with various psychologic symptoms, including low self-esteem, social embarrassment, social isolation, and depression [4, 5]. Although HFS could improve or even resolve without any treatment [6], HFS rarely resolves spontaneously because it mainly results from vascular compression of the facial nerve at the root exit zone [7]. Therefore, effective management is important for the quality of life, even tough HFS is not a fatal disease. Accordingly, the majority of HFS patients undergo treatment, such as symptomatic or curative options, and 1570 (88.5%) of 1775 HFS patients performed microvascular decompression or botulinum toxin injection therapy in previous study [6]. Additionally, the severity of HFS means severe indentation of facial nerve [8], and severe spasm is related with more impaired quality of life [2]. HFS patients with severe spasm might have even less chance for improvement or remission; thus, symptomatic or curative treatment is strongly recommended in severe HFS. Although the only curative option is surgical relief of the neurovascular compression in HFS, botulinum toxin injection is an evident, effective and safe treatment options to control the spasm, because HFS is a benign disease and the spasm itself could be well-controlled with injection therapy.

J. Youn (✉)
Department of Neurology, Samsung Medical Center,
Sungkyunkwan University School of Medicine,
Seoul, Republic of Korea

Neuroscience Center, Samsung Medical Center,
Seoul, Republic of Korea
e-mail: genian@skku.edu

W. Jang
Department of Neurology, Gangneung Asan Hospital,
University of Ulsan College of Medicine,
Gangneung, Republic of Korea
e-mail: neveu@gnah.co.kr

J. K. Park
Department of Neurology, Cheonan Soonchunhyang
Hospital, Soonchunhyang University College of
Medicine, Cheonan, Republic of Korea

Botulinum Toxin Injection as Management Option for Hemifacial Spasm

History of Botulinum Toxin Injection in Hemifacial Spasm

The initial trial with botulinum toxin was done in patients with strabismus [9], and botulinum toxin tested was unreliable, short-acting, or necrotizing at the early period [10]. Finally, with the purifying techniques and extensive animal experiments, botulinum toxin showed desired long-lasting, localized, dose-dependent muscle weakening without any previous side effects. By the early 1980s, botulinum toxin was injected for various diseases such as strabismus, blepharospasm, hemifacial spasm, cervical dystonia, and thigh adductor spasm. At Samsung Medical Center (Seoul, Korea), botulinum toxin injection clinic is run by the Neurology Department, and we usually perform more than 1500 injections annually. Among the patients with botulinum toxin treatment at Samsung Medical Center on 2017, 371 (50%) of total 744 patients were HFS patients. For all indications illustrated in Figure 1, HFS was the most commonly injected indication, and blepharospasm wa the second most common indications among various movement disorders for botulinum toxin injection therapy.

With various previous studies during two decades, botulinum toxin treatment has emerged as the first-choice treatment option for hemifacial spam, as well as microvascular decompression surgery [11–13]. Two randomized controlled trials and more than 30 open label studies, encompassing more than 2200 patients, have been already published, and botulinum toxin injection demonstrated excellent improvement in terms of symptom control [14, 15]. Additionally, evidence-based review by the Therapeutics and Technology Assessment Subcommittee of the American Academy of Neurology concluded that botulinum toxin injection is possibly effective with minimal side effects for the treatment of hemifacial spasm (one Class II and one Class III study) [16]. In

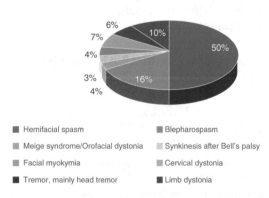

Fig. 1 The indications for botulinum toxin treatment at Samsung Medical Center for 1 year (2017). Total number of patients was 744, and almost half of the whole patients (*n* = 371) were injected because of HFS

addition, various botulinum toxin agents that are commercially available were also studied, and regarded to have similar effects for HFS [17].

For long-term efficacy, treatment with botulinum toxin appears to remain effective over long-term use of several years (from 4 to 20 years). In most cases, botulinum toxin treatmant will not require dosage increase, and even if required, the dosage increase usually occurs within the first 2 years of treatment [18].

Strategy to Decide Botulinum Toxin Injection in Patients with Hemifacial Spasm

Even though botulinum toxin treatment is an effective and safe treatment option in hemifacial spasm, botulinum toxin injection therapy is symptomatic management option unlike surgical relief of neurovascular compression. Considering the different characteristics of each treatment option, treatment modality should be decided with discussion based on three considerations [19].

- Patient-related factors: age at surgery, age at symptom onset, occupation, alcohol use, smoking history, family history, history of facial palsy, contralateral hearing loss and comorbidities, including, hypertension, diabetes, hyperlipidemia, and cardiovascular disease
- Disease-related factors: affected side, duration of symptoms, severity of spasm, associated symptoms (tinnitus and headache), preoperative facial palsy, and previous botulinum toxin treatment
- Surgery-related factors: offending vessel, number of offending vessels, compressive pattern, indentation on the facial nerve, discoloration of nerve, abnormal muscle responses, postoperative delayed facial palsy, and operation year

Among the patient-related factors, age should be considered. Although surgical decompression is effective in elderly patients as well as young patients, more complications were reported in elderly patients [20]. Higher surgical risk and more comorbidities in elderly patients has to be also considered the severity of spasm. Additionally, severe spasm is associated with severe indentation of facial nerve [8]; thus, there might be less chance for spontaneous improvement or remission. Although we reported improvement or remission without any treatments in half of the HFS patients [6], 86.0% of the enrolled subjects in this study had mild HFS (SMC grade 1–2), thus this result should be interpreted cautiously. Therefore, for the patients with severe HFS, treatment, such as macrovascular decompression or botulinum toxin injection, is strongly recommended, whereas we can follow up the patients with mild spasm without any treatment.

Interestingly, all these patient-related, disease-related, and surgery-related factors are not independent, but connected with each other. For example, severe spasm (disease-related factor) is marker for indentation of facial nerve (surgery-related factor), and associated with comorbidities, like hypertension or diabetes (patient-related factor), and disease duration (another disease-related factor). Additionally, associated factors with surgical outcome is also important during the decision of treatment option for HFS. For example, botulinum toxin treatment should be considered prior to surgical treatment in the people with the contralateral hearing loss, because ipsilateral hearing loss, possible complication from MVD, can make the patient deaf. The contents about the prognosis and complication for surgical decompression will be discussed in other chapters (Chapters "Possible complications of Microvascular Decompression" and "Prognosis of symptoms after microvascular decompression for hemifacial spasm").

Considerations Before Start Botulinum Toxin Injection in Hemifacial Spasm

In spite of the efficay and safety of botulinum toxin treatment in HFS, botulinum toxin injection is symptomatic management option, and not without side effects. Therefore, discussion with detailed information, such as the duration and latency for response, and possible side effects, should be done before to perform botulinum toxin treatment in HFS patients. Mostly, the effects and side effects happen based on the injected dose and target muscles. Possible side effects of botulinum toxin injections are erythema and ecchymosis of the injected site, dry eyes, mouth droop, ptosis, lid edema, and facial muscle weakness [15, 21]. Among the side effects, ptosis and facial muscle weakness tends to be transient and will resolve within 1–4 weeks. In terms of efficacy, the onset of effect occurs within 3 days to 2 weeks, generally with a peak effect at approximately 2 weeks. The beneficial effects of botulinum toxin injections are also transient with a mean duration of improvement of approximately from 2.8 to 3.1 months [13, 15]. At Samsung Medical Center, the mean duration of response was 3.46 months, and the mean frequency of injection was 2.15 per year for HFS patients. However, for the duration and onset of effect, there is a high variability of duration of the beneficial effect, thus all injections should be personalized for each patient.

The other consideration is the selection of botulinum toxin agent. Onabotulinumtoxin A (Botox, Allergan, Irvine, CA) is the most commonly used among the commercially available preparations. A large number of trials have validated the successful outcomes of botulinum toxin injection therapy with improvements in as many as 75–100% of individuals with hemifacial spasm [11, 22, 23].

Preparation for Botulinum Toxin Injection

Botulinum toxin injections in HFS are usually performed with the patient lying supine. We use specialized chair for botulinum injection on the face (Fig. 2), but it is not mandatory. If it is comfortable for both the patients and doctors for injection, any position is acceptable.

The toxin is diluted to minimum concentration of 10–50 Botox U/ml, 50–200 U/ml Dysport (Ipsen, Milford, MA) or 5000–10,000 U/ml Neurobloc/myobloc (Solstice Neurisciences, Malvern, PA) to minimize diffusion. At Samsung

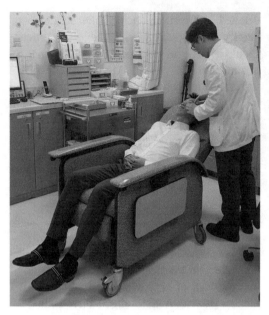

Fig. 2 The chair and position for botulinum toxin injection on HFS patients' face at Samsung Medical Center

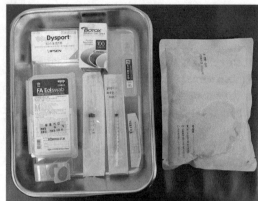

Fig. 3 The materials that are prepared before starting botulinum toxin injection at Samsung Medical Center. Botulinum toxins (Botox and Dysport), alcohol swab, syringes, needle, normal saline, and icepack

Medical Center, Botox 100 U is diluted in 2 ml of normal saline (50 Botox U/ml), and Dysport 500 U in 3.3cc (Dysport 151.5 U/ml). However, the amount of normal saline depends on the physician's experiences and the responses from each patient. For the muscles that need higher dose of botulinum toxin, like limb or neck muscles, botulinum toxin could be diluted in smaller amount of normal saline. On the contrary, if the patient easily showed weakness from botulinum toxin treatment or had already weakness even before injection, botulinum toxin could be mixed with a larger amount of normal saline.

Other materials for injection treatment should be prepared before starts treatment (Fig. 3). The injections are performed usually with 1-ml syringe with fine needle (30 gauge). Alcohol swab to sterilize injection site and gauge to compression in case of bleeding are also needed.

Other Specific Considerations

Usually, there is no need for management for injection pain, but if patients worry about pain during injection, the pain can be reduced either with skin cooling using ice or with EMLA Cream (lidocaine 2.5% and prilocaine 2.5%) [24, 25]. When we checked our patients at Samsung Medical Center, only 2.4% of HFS patients with botulinum toxin injection needed EMLA Cream for the pain during injections. Unlike in spasmodic torticollis, EMG recordings during injection are not necessary in HFS.

Botulinum Toxin Injection: Injection Site and Dose Selection

HFS usually begins in the periocular region and then progresses to involve the cheek and perioral muscles. The natural course is heterogeneous among the patients, and especially the time to spread to hemi-face and the time to visit clinic from the onset, is very different depending on the patients. The goal of botulinum toxin injection is a symptomatic management, not cure of HFS, thus the injection sites should be decided with careful observation of patients' symptoms (involved muscles) and history taking (main muscles that patients complain of). Sometimes, even though patients have spasm in hemi-face, they just complain of periocular spasm. In these cases, injections in perioral area are not necessary because this injection might cause side effects without any benefit. Additionally, the injection dose should be also personalized. With the same severity of spasm, some patients may experience severely impaired quality of life, but not for the others. Same for the side effects; it could be tolerable for some patients, but totally not tolerable for the others. Therefore, both injection site and dose should be decided based on full discussion with patients.

Facial Muscles and Injection Sites

For successful injection, the physicians should know the muscle anatomy and function in the face. The main muscles injected for HFS tend to be the orbicularis oculi, corrugator, frontalis, risorius, buccinator, and depressor anguli oris (Fig. 4).

The orbicularis oculi is composed of two parts: the pars palpebralis, which opens (with the help of the levator muscle) and closes the eyelid, and the pars orbitalis, which squeezes the eye shut. The pars palpebralis is composed of two parts: the preseptal and the pretarsal region. Typically, the injection sites in the orbicularis oculi at Samsung Medical Center are medial and lateral part of

upper lid, and middle and lateral part of lower lid (Fig. 5), and medial part of lower lid could be added based on the severity of symptoms. For the injection at pars palpebralis, pretarsal injection shows more effects with longer duration and less side effects compared to preseptal injection [26, 27]. Additionally, injection at pars orbicularis could be added. The most important rule is not to inject at the midline of upper eyelid to avoid ptosis. Ptosis could be due to local diffusion of the botulinum toxin affecting the levator palpebrae [28]; thus, too much dilution is not recommended especially for the patients with HFS.

The injection at orbicularis oris should be done carefully, because this injection could result in paralysis of the mouth producing further disability. Even though perioral spasm is controlled with botulinum toxin injection at perioral muscles, ipsilateral upper lip droop could be seen [29]. If patients complain of spasm more than the weakness of perioral muscles, bilateral injection to minimize asymmetry could be another option. However, considering that most HFS patients complain of periocular spasm rather than perioral spasm, the sites of injection should be fully

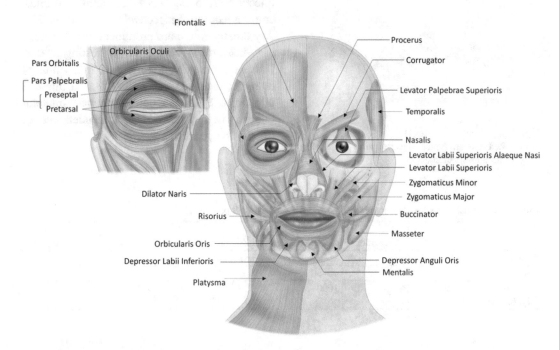

Fig. 4 Illustration of facial muscle anatomy for botulinum toxin injection. The injection at right muscle is the best way to maximize the efficacy and minimize the side effects

discussed with patients. Not all spasm needs to be controlled with botulinum toxin injection.

Injection Doses for Facial Muscles in HFS

The usual doses for commonly injected facial muscle are illustrated in Table 1 [30]. The average dose of botulinum toxin varies from 10 to 46

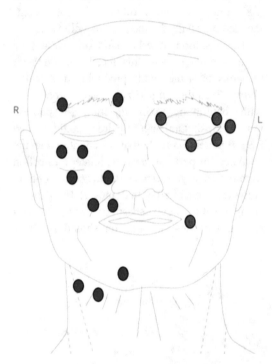

Fig. 5 Injections sites for HFS at Samsung Medical Center. Red dots are typical injection sites and blue dots are sites for additional injections

Botox U [12, 31], from 53 to 160 Dysport U [18, 32–34], and from 1250 to 9000 NeuroBloc/Myobloc U [35, 36]. Based on our experience at Samsung Medical Center, a small dose of botulinum toxin injection is recommended for the first injection, and the dose could be adjusted at follow-up injections based on the response at the previous injection. In particular, the dose for injection should be assessed by each muscle.

Unilateral Injection vs. Bilateral Injection

HFS is an unilaterally involved disorder, even though there are some cases with bilateral involvements. Therefore, botulinum toxin injection could be performed unilaterally. However, with unilateral injection, some patients could suffer from the asymmetry (sometimes subjective asymmetry) and bilateral injection could be helpful in these cases. When we compared bilateral injection ($n = 33$) with unilateral injection ($n = 45$), asymmetry in lower face was more common in unilateral injection than in bilateral injection (75.6 vs. 48.5%, unilateral and bilateral injection respectively, $p = 0.033$). Asymmetry was more prominent during voluntary movement compared to resting status. However, there was no difference in asymmetry of upper face between unilateral and bilateral injections. Based on our experience, bilateral injection in lower face is recommended, but not usually in upper face.

Table 1 The usual doses for commonly injected facial muscle in HFS patients [30]

	Frontalis	Corrugator	Orbicularis oculi	Zygomaticus major	Buccinator	Depressor angularis oris
Botox U	10	1	15–20	1	2	1
Dysport U	30	3	45–60	3	6	3
NeuroBloc/Myobloc U	500	50	1000	50	100	50

References

1. Wang A, Jankovic J. Hemifacial spasm: clinical findings and treatment. Muscle Nerve. 1998;21:1740–7.

2. Lee JA, Jo KW, Kong DS, Park K. Using the new clinical grading scale for quantification of the severity of hemifacial spasm: correlations with a quality of life scale. Stereotact Funct Neurosurg. 2012;90:16–9.

3. Reimer J, Gilg K, Karow A, Esser J, Franke GH. Health-related quality of life in blepharospasm or hemifacial spasm. Acta Neurol Scand. 2005;111:64–70.

4. Tan EK, Seah A. Health-related quality of life in blepharospasm and hemifacial spasm. Arch Ophthalmol. 2007;125:1141.

5. Heuser K, Kerty E, Eide PK, Cvancarova M, Dietrichs E. Microvascular decompression for hemifacial spasm: postoperative neurologic follow-up and evaluation of life quality. Eur J Neurol. 2007;14:335–40.

6. Lee JA, Kim KH, Park K. Natural history of untreated hemifacial spasm: a study of 104 consecutive patients over 5 years. Stereotact Funct Neurosurg. 2017;95:21–5.

7. Mauriello JA Jr, Leone T, Dhillon S, Pakeman B, Mostafavi R, Yepez MC. Treatment choices of 119 patients with hemifacial spasm over 11 years. Clin Neurol Neurosurg. 1996;98:213–6.

8. Na BS, Cho JW, Park K, et al. Severe hemifacial spasm is a predictor of severe indentation and facial palsy after microdecompression surgery. J Clin Neurol. 2018;14:303–9.

9. Scott AB. Botulinum toxin injection into extraocular muscles as an alternative to strabismus surgery. J Pediatr Ophthalmol Strabismus. 1980;17:21–5.

10. Scott AB, Rosenbaum A, Collins CC. Pharmacologic weakening of extraocular muscles. Investig Ophthalmol. 1973;12:924–7.

11. Jost WH, Kohl A. Botulinum toxin: evidence-based medicine criteria in blepharospasm and hemifacial spasm. J Neurol. 2001;248(Suppl 1):21–4.

12. Flanders M, Chin D, Boghen D. Botulinum toxin: preferred treatment for hemifacial spasm. Eur Neurol. 1993;33:316–9.

13. Batisti JP, Kleinfelder AD, Galli NB, Moro A, Munhoz RP, Teive HA. Treatment of hemifacial spasm with botulinum toxin type a: effective, long lasting and well tolerated. Arq Neuropsiquiatr. 2017;75:87–91.

14. Park YC, Lim JK, Lee DK, Yi SD. Botulinum a toxin treatment of hemifacial spasm and blepharospasm. J Korean Med Sci. 1993;8:334–40.

15. Yoshimura DM, Aminoff MJ, Tami TA, Scott AB. Treatment of hemifacial spasm with botulinum toxin. Muscle Nerve. 1992;15:1045–9.

16. Simpson DM, Blitzer A, Brashear A, et al. Assessment: botulinum neurotoxin for the treatment of movement disorders (an evidence-based review): report of the Therapeutics and Technology Assessment Subcommittee of the American Academy of Neurology. Neurology. 2008;70:1699–706.

17. Hallett M, Albanese A, Dressler D, et al. Evidence-based review and assessment of botulinum neurotoxin for the treatment of movement disorders. Toxicon. 2013;67:94–114.

18. Jitpimolmard S, Tiamkao S, Laopaiboon M. Long term results of botulinum toxin type A (Dysport) in the treatment of hemifacial spasm: a report of 175 cases. J Neurol Neurosurg Psychiatry. 1998;64:751–7.

19. Lee JA, Kim KH, Kong DS, Lee S, Park SK, Park K. Algorithm to predict the outcome of microvascular decompression for hemifacial spasm: a data-mining analysis using a decision tree. World Neurosurg. 2019;125:e797–806.

20. Youn J, Kwon S, Kim JS, Jeong H, Park K, Cho JW. Safety and effectiveness of microvascular decompression for the treatment of hemifacial spasm in the elderly. Eur Neurol. 2013;70:165–71.

21. Elston JS. Botulinum toxin treatment of hemifacial spasm. J Neurol Neurosurg Psychiatry. 1986;49:827–9.

22. Defazio G, Abbruzzese G, Girlanda P, et al. Botulinum toxin A treatment for primary hemifacial spasm: a 10-year multicenter study. Arch Neurol. 2002;59:418–20.

23. Costa J, Espirito-Santo C, Borges A, et al. Botulinum toxin type A therapy for hemifacial spasm. Cochrane Database Syst Rev. 2005;2005:CD004899.

24. Linder JS, Edmonson BC, Laquis SJ, Drewry RD Jr, Fleming JC. Skin cooling before periocular botulinum toxin A injection. Ophthalmic Plast Reconstr Surg. 2002;18:441–2.

25. Soylev MF, Kocak N, Kuvaki B, Ozkan SB, Kir E. Anesthesia with EMLA cream for botulinum A toxin injection into eyelids. Ophthalmologica. 2002;216:355–8.

26. Cakmur R, Ozturk V, Uzunel F, Donmez B, Idiman F. Comparison of preseptal and pretarsal injections of botulinum toxin in the treatment of blepharospasm and hemifacial spasm. J Neurol. 2002;249:64–8.

27. Lolekha P, Choolam A, Kulkantrakorn K. A comparative crossover study on the treatment of hemifacial spasm and blepharospasm: preseptal and pretarsal botulinum toxin injection techniques. Neurol Sci. 2017;38:2031–6.

28. Brin MF, Fahn S, Moskowitz C, et al. Localized injections of botulinum toxin for the treatment of focal dystonia and hemifacial spasm. Mov Disord. 1987;2:237–54.

29. Boghen DR, Lesser RL. Blepharospasm and Hemifacial Spasm. Curr Treat Options Neurol. 2000;2:393–400.

30. Frei K, Truong DD, Dressler D. Botulinum toxin therapy of hemifacial spasm: comparing different therapeutic preparations. Eur J Neurol. 2006;13(Suppl 1):30–5.

31. Jankovic J, Schwartz K, Donovan DT. Botulinum toxin treatment of cranial-cervical dystonia, spasmodic dysphonia, other focal dystonias and hemifacial spasm. J Neurol Neurosurg Psychiatry. 1990;53:633–9.

32. Elston JS. The management of blepharospasm and hemifacial spasm. J Neurol. 1992;239:5–8.

33. Van den Bergh P. Botulinum toxin treatment in clinical neurology. Acta Neurol Belg. 1995;95:70–9.

34. Yu YL, Fong KY, Chang CM. Treatment of idiopathic hemifacial spasm with botulinum toxin. Acta Neurol Scand. 1992;85:55–7.

35. Tousi B, Perumal JS, Ahuja K, Ahmed A, Subramanian T. Effects of botulinum toxin-B (BTX-B) injections for hemifacial spasm. Parkinsonism Relat Disord. 2004;10:455–6.

36. Wan XH, Vuong KD, Jankovic J. Clinical application of botulinum toxin type B in movement disorders and autonomic symptoms. Chin Med Sci J. 2005;20:44–7.

Medical Treatment of Hemifacial Spasm and Other Involuntary Facial Movement Disorders

Wooyoung Jang and Jinyoung Youn

Involuntary movement of facial musculature encompasses various etiology such as hemifacial spasm, blepharospasm, facial myokymia, oromandibular dystonia, tardive dyskinesia, and psychogenic origin [1]. Before establishing the treatment strategy, physicians should be cautious about what etiology could be involved. Therefore, careful history taking including medication, co-morbidity as well as delicate description for characteristics of involuntary movement could be essential [2]. For example, unilateral presentation is more likely organic etiology than psychogenic origin. While hemifacial spasm has usually a peripherally derived movement by ipsilateral facial nerve, involvement of masticatory muscle or tongue could designate central origin [3]. However, differentiation for each etiology for facial involuntary movement is beyond the scope of this chapter and we review what kinds of medical option are available for each facial involuntary movement disorders.

Hemifacial Spasm

Hemifacial spasm (HFS) can be defined as a movement disorder showing involuntary synchronous tonic or clonic muscle contraction of facial muscles that is innervated by ipsilateral facial nerve. HFS is usually unilateral, but bilateral HFS was reported and essential blepharospasm, Meige syndrome, Tardive dyskinesia, facial tic and myokymia should be differentiated [1, 4, 5].

Blepharospasm usually manifests with bilateral, symmetrical contractions of the eyelids [6]. In HFS, famous description by Joseph Babinski that internal part of frontalis muscle contract, when orbicularis oculi contracts, so called, the other babinski sign could help to differentiate HFS from blepharospasm [7]. Oromandibular dystonia could be differentiated from HFS in the point of involuntary muscle contractions that involve primarily the lower part and structures of the face, the mouth, the tongue, and the pharynx [8]. Facial tic presents with more bizarre and multifocal involvement patterns and alternating propagation from side to side of the face [9]. Compared to other involuntary movement

W. Jang (✉)
Department of Neurology, Gangneung Asan Hospital, University of Ulsan College of Medicine, Gangneung, Republic of Korea
e-mail: neveu@gnah.co.kr

J. Youn
Department of Neurology, Samsung Medical Center, Sungkyukwan University School of Medicine, Seoul, Republic of Korea

Neuroscience Center, Samsung Medical Center, Seoul, Republic of Korea
e-mail: genian.youn@samsung.com

© Springer Nature Singapore Pte Ltd. 2020
K. Park, J. S. Park (eds.), *Hemifacial Spasm*, https://doi.org/10.1007/978-981-15-5417-9_14

disorders, the tics can typically be suppressible and the patients feel urge to move.

As a treatment, four kinds of options could be suggested. First, if patient does not complain about symptoms and not particularly bothered by HFS, let HFS go by with no treatment at all. In this case, clinicians should exclude structural lesions such as mass (except compression or indentation of intracranial artery of facial nerve root entry or exit zone) and reassure their patients as it is a benign condition [4]. Lee et al. reported 5-year follow-up data of 104 untreated HFS revealing that 38.5% of patients remained stationary, 9.6% of patients showed partial improvement, and 41.3% experienced complete remission [10]. However, in another series, 93.4% of HFS showed spreading of muscle spasm [11]. Therefore, non-treatment option should be carefully considered. As second and tertiary options, local injection of botulinum toxin and microvascular surgery are the treatment of choice for HFS based on evidence with high success rate and low complication rate [2, 12, 13]. However, there could be HFS patients who do not want these types of treatment options or who are not tolerable of botulinum toxin. In the aspect of oral medication, there are several options available for exceptional patients who do not fit botulinum toxin or surgical intervention. A variety of oral medications have been suggested for treatment of HFS, but almost in non-controlled studies. Therefore, evidence data is hardly lacking. Wang and Jankovic reported that only 8% with oral medication treatment revealed meaningful benefit [14]. Most of employed medication for treatment of HFS are anti-epileptic drugs, which are regarded as decreasing nerve excitability and can reduce the spasm and provide symptomatic benefit.

Carbamazapine

In 1982, one case series including three patients reported efficacy of carbamazepine with a dose of 600–1200 mg for HFS [15]. In this series, authors suggest that 50% of 46 patients, including their patients, showed improvement.

Baclofen

There is a single case report showing a 58-year-old woman with baclofen (daily dose of 37.5 mg) experienced a dramatic effect and remained asymptomatic during 12 months [16]. No side effect was reported in this report. However, there were no controlled studies or further evidence.

Clonazepam

In 1985, one non-controlled series was reported. Standard dosage of clonazepam ranges from 0.5 to 4 mg for daily dose [17].

Zonisamide

Siniscalchi et al. reported a single case regarding complete remission of HFS in a 65-year-old woman by adding zonisamide (150 mg twice a day, 6 weeks) followed by failure of clonazepam treatment [18]. There were no further studies regarding the effectiveness of zonisamide for HFS.

Gabapentin

There are three case series that gabapentin was reported to be effective for HFS [19–21]. Each series included 1, 5, 23 patients and show its efficacy from 600 to 2400 mg per day. Caravaglios et al.'s series including 23 patients demonstrated that 69.6% of subjects showed a significant clinical reduction, defined as reduction above 70%. Gabapentin is usually well tolerated, but clinicians should be cautious about its side effects such as somnolence and dizziness avoiding non-compliance. Because gabapentin showed its efficacy with a relatively high dose, it should be introduced at the lowest dose possible and gradually escalated. In Caravaglios's series, transient weakness and marked reduction of anxiety were presented as a side effect.

Pizotifen

Pizotifen, which is a 5HT receptor antagonist, was reported as showing improvement of HFS with continual efficacy in two patients [22]. The dose of pizotifen was 1 and 1.5 mg per day for respective patients.

Levetiracetam

Relatively levetiracetam was recently reported as an option of HFS treatment. There are three case series showing efficacy of levetiracetam for HFS [23–25]. First, in 2004, Deleu described two patients with HFS showing improvement with levetiracetam (1500 mg per day). In this series, both patients also underwent treatment with botulinum toxin type A, but showed only short-term relief and disabling side effect. Second, in 2005, Biagio Carrieri also reported complete remission of HFS with 2 weeks' levetiracetam treatment (500 mg per day). Finally, in 2016, Kuroda et al. suggested mechanism of levetiracetam for HFS as anti-kindling effect with two case reports showing dramatic improvement after levetiracetam introduction without significant adverse effect (500 mg per day). Therefore, till now, levetiracetam could be good candidate of oral medication for HFS treatment, warranting well designed clinical trials.

Blepharospasm and Other Involuntary Facial Movement Disorders

Blepharospasm

Essential blepharospasm is regarded as focal dystonia, which is characterized by forced eyelid closure due to dystonia of orbicularis oculi and other periorbital muscles [6]. Besides medical treatment, chemodenervation using botulinum toxin is a treatment of choice for blepharospasm. The American Academy of Neurology suggests Level B recommendation for onabotulinumtoxin-A and incobotulinum toxin-A and level C recommendation for abobotulinum toxin-A for treatment of blepharospasm [26].

There has been increasing evidence that deep brain stimulation could be effective for blepharospasm, although this beneficial effect is reported from studies including patients with blepharospasm combined with other types of dystonia. Santos et al. reported successful treatment of isolated blepharospasm by pallidal stimulation, and there is also a single case report for successful symptomatic relief after bilateral pallidal deep brain stimulation in intractable blepharospasm [27–29]. Further studies of DBS as treatment options of blepharospasm are strongly warranted. On the other hand, repetitive transcranial magnetic simulation (rTMS) shows a beneficial effect on blepharospasm. One randomized control study including 15 patients with low-frequency rTMS provides Class II evidence for blepharospasm in the aspects of safety and efficacy [30]. Recently, rTMS combined botulinum toxin injection revealed enhanced efficacy and prolongation of the effect of botulinum toxin treatment on blepharospasm [31].

Oral medication as a treatment of blepharospasm encompasses anticholinergics, GABAergic, antidopaminergics, and mexiletine. Owing to unwanted detrimental side effects and lack of evidence in clinical trials, oral medication has many limitations for symptomatic relief of blepharospasm.

At first, there is only one double-blinded crossover study for investigating the efficacy of anticholinergics such as trihexyphenidyl and benztropine for blepharospasm [32]. In this study, anticholinergic side effect was presented with high incidence such as confusion, sedation, and dry mouth, and in eight patients, there was no significant difference among peripheral, central anticholinergics and placebo. Baclofen and clonazepam have been reported in single case reports showing efficacy, and mexiletine was reported in case series including three patients showing meaningful efficacy [33–35]. Tetrabenazine, which is a presynaptic monoamine-depleting drug, has also been reported as a treatment option for blepharospasm in one double-blinded study [36]. Finally, apraclonidine, an alpha-2 adrenergic agonist, was introduced revealing lid elevation through sympathetic activation of Muller muscle [37]. However, its therapeutic availability is known to be limited due to tachyphylaxis.

Other Facial Involuntary Movement Disorders

Meige's syndrome is a combination of two different forms of dystonia including blepharospasm and oromandibular dystonia [8]. The pathogenesis of Meige's syndrome is identical to other types of dystonia including abnormal plasticity and cortical disinhibition. Therefore, management of Meige's syndrome is also identical with treatment options of dystonia including blepharospasm. There are several clinical trials and case reports for botulinum toxin and pallidal deep brain stimulation as a treatment option for Meige's syndrome [38–40]. In the aspect of oral medication, anticholinergics, benzodiazepine, levodopa, haloperidol, tetrabenazine, and baclofen have been used for symptomatic relief. Marsden investigated many drugs in 39 subjects with Meige's syndrome, and found that anticholinergics and benzodiazepine were beneficial [41]. Taner et al. also reported beneficial effect of centrally acting anticholinergics in 13 patients with blepharospasm and oromandibular dystonia [42]. In another study, similar findings are reported and they also reported that lithium could be an alternative [43]. Jankovic and Ford suggested long-term effectiveness in 26% of tetrabenazine, 37% of trihexyphenidyl, and 26% of lithium [44]. Recently, levetiracetam and zolpidem monotherapy were reported being beneficial in some case reports [45, 46]. Facial myokymia is defined as involuntary contraction resulting in wavelike or vermicular propagation involving facial muscles, especially orbicularis oculi [47]. This phenomenon is characterized by electromyography findings, spontaneous muscle activity with different motor units and showing brief, repetitive discharges with rhythmic burst [47]. Usually, facial myokymia occurs in healthy subjects with no associated disease entity and most cases are transient and self-limiting [48, 49].

However, in some reports, facial myokymia with persistence and involvement of entire facial muscles could be suggested as ipsilateral pontine tegmentum lesion such as multiple sclerosis, timorous condition, and cysticercosis [50–53]. Therefore, atypical nature or long-standing facial myokymia should be regarded as potential structural lesion and cranial MRI could be considered.

Pharmacology of Botulinum Toxin

Botulinum neurotoxin is produced by Gram-positive anaerobic spore–forming bacteria called *Clostridium botulinum*. Botulinum toxin is divided into 7 serotype of neurotoxins from A to G. Since 1973, first application of treatment for strabismus, FDA approved its indication for hemifacial spasm or blepharospasm in 1989 and botulinum toxin is regarded as treatment of choice of hemifacial spasm [54, 55]. A detailed explanation of its application is beyond the scope of this chapter and the next chapter handles this issue. In this section, mechanism and basic pharmacology of botulism toxin will be discussed.

Mechanism

Botulinum toxin is produced by *Clostridium botulinum* in an anaerobic circumstance which is targeting intracellular substrate. The mechanism of botulinum toxin is divided into five major steps. (1) binding to membrane, (2) internalization or endocytosis, (3) low pH–driven membrane translocation, (4) secretion of the L chain in the cytosol, (5) cleavage of SNARE protein with following inhibition of neurotransmitter release and paralysis of targeted muscle [56, 57].

Synaptic vesicle is filled with neurotransmitters, including acetylcholine. A heavy chain of various types of botulinum toxin is binding to each receptors of ganglioside and translocated into cytosol. Subsequently, disulfide bond is cleaved between heavy chain and light chain and a light chain of each types of botulinum toxin binds each targeted proteins within SNARE complex, which result in inhibiting the release of acetylcholine. For example, a light chain of botulinum toxin A, E cleaves synaptosome-associated protein (SNAP25), which is necessary for fusion of vesicles, while botulinum toxin B,D,F,G target vesicle-associated membrane protein (VAMP, synaptobrevin).

Pharmacology

Botulinum toxin has some unique characteristics of pharmacological activity [56]. First, the mechanism of action of botulinum toxin is derived by modification of single target protein within neuron. Second, there is still no available information of pharmacokinetics of botulinum toxin at the doses for treatment. Finally, botulinum toxin cannot be reversed of its action once the toxin has arrived within the neuron [58].

There are many preparations for clinical use, and almost are based on A1 serotype botulinum toxin, while only one type is based on B1 serotype [59]. Three major brands are commercially available using A1 type toxin, including onabotulinumtoxinA known as Botox by Allergan (Irvine, CA), abobotulinumtoxinA as Dysport by Ipsen (Paris, France), and icobutulinumtoxinA as Xeomin by Merz Pharmaceuticals (Frankfurt, Germany). Three other manufacturers based in South Korea and China also produce further serotype A botulism toxin, which are Meditoxin by Medy-Tox (Korea), Botulax by Hugel Inc. (Korea), and Prosigne by Lanzhou Institute for Biologic product (China). One serotype B1 toxin, Myobloc, using rimabotulinumtoxinB, is available. Table 1 demonstrates the characteristics of each product.

Botulinum toxins are freeze-dried or vacuum-dried preparation. However, all toxin products show the similar range of diffusion and efficacy from the site of injection, which is dependent on the amount [60]. Usually, greater doses of serotype B1 toxin are necessary to acquire a comparable effect of serotype A1 toxin and the duration of action is shorter in B1 toxin in skeletal muscle [61]. Usually, the comparable dose ratio of A1 and B1 toxin is 1:25–30 [61]. The potency of botulinum toxin preparation is estimated by units. 1U in Botulinum Toxin A1 correspond to 1 lethal dose (LD) for 3 days in mice [62]. In case of male human with 70 kg, LD50 of botulinum toxin A is 2500~3000 U. Dose conversion ratio between Botox and Dysport are reported as 1:2.5–3, while Botox and Xeomin are reported as having near-equivalence in potency [63, 64]. Recently, Yun et al. suggested that a 1:2.5 conversion ratio between Botox and Dysport could be appropriate in treatment of cervical dystonia [65].

Owing that botulinum neurotoxin and related protein could act like antigen, injection into patient could elicit antibody formation, which could interfere its biological activity [66]. However, the frequency of developing neutralizing antibody for serotype A1 is reported to be very low (0–3%) [67]. Whereas, for botulinum toxin B1, the frequency of neutralizing immune response is reported to be relatively high (10–44%) [68]. Common factors influencing the immunogenicity of botulinum toxin are produced related factors such as protein load, presence of inactive toxins and treatment associated factors including dose, frequency of injections, and sites of injection [69].

Table 1 Comparison of various representative commercial botulinum toxin products

Product	Botox	Dysport	Xeomin	Myobloc	Prosigne	Meditoxin
Company	Allergan (USA)	Ipsen(UK)	Merz(Germany)	Solstice Neurosciences (USA)	Lanzhou Institute of Biological Products(China)	Medy-Tox (Korea)
Unit	100 or 200 units	300 or 500 units	50, 100, 200 units	2500 units per 0.5 mL, 5000 units per 1 mL, 10,000 units per 2 mL	50, 100 units	50, 100, 150, 200 units
Serotype	A	A	A	B	A	A
Molecular weight	900 kDa	<900 kDa	150 kDa		900 kDa	940 kDa
Pharmaceutical form	Powder	Powder	Powder	Solution	Powder	Powder
Storage after dilution	24 h/2–8 °C	4 h/2–8 °C	24 h/2–8 °C	4 h (if diluted)/2–8 °C	4 h/2–8 °C	4 h/2–8 °C
Storage before dilution	−20~−5 °C (Freeze), 2–8 °C	2~8 °C	2~8 °C		2~8 °C	−20~−5 °C (Freeze), 2~8 °C
pH	7.4	7.4	7.4	5.6		6.8
Generic	Onabotulinumtoxina (ONA)	Abobotulinumtoxina (ABO)	Icobutulinumtoxina (INCO)	Rimabotulinumtoxinb (RIMA)	CBTX-A	BONTA
Conversion ratio	ONA:INCO = 1:1	ABO:ONA = 3:1 or even lower		ONA:RIMA = 1:50 ~100		

References

1. Evidente VG, Adler CH. Hemifacial spasm and other craniofacial movement disorders. Mayo Clin Proc. 1998;73:67–71.
2. Rosenstengel C, Matthes M, Baldauf J, Fleck S, Schroeder H. Hemifacial spasm: conservative and surgical treatment options. Dtsch Arztebl Int. 2012;109:667–73.
3. Tan NC, Chan LL, Tan EK. Hemifacial spasm and involuntary facial movements. QJM. 2002;95:493–500.
4. Galvez-Jimenez N, Hanson MR, Desai M. Unusual causes of hemifacial spasm. Semin Neurol. 2001;21:75–83.
5. Tan EK, Jankovic J. Bilateral hemifacial spasm: a report of five cases and a literature review. Mov Disord. 1999;14:345–9.
6. Valls-Sole J, Defazio G. Blepharospasm: update on epidemiology, clinical aspects, and pathophysiology. Front Neurol. 2016;7:45.
7. Stamey W, Jankovic J. The other Babinski sign in hemifacial spasm. Neurology. 2007;69:402–4.
8. Pandey S, Sharma S. Meige's syndrome: history, epidemiology, clinical features, pathogenesis and treatment. J Neurol Sci. 2017;372:162–70.
9. Potgieser AR, van Dijk JM, Elting JW, de Koning-Tijssen MA. [Facial tics and spasms]. Ned Tijdschr Geneeskd 2014;158:A7615.
10. Lee JA, Kim KH, Park K. Natural history of untreated hemifacial spasm: a study of 104 consecutive patients over 5 years. Stereotact Funct Neurosurg. 2017;95:21–5.
11. Conte A, Falla M, Diana MC, et al. Spread of muscle spasms in hemifacial spasm. Mov Disord Clin Pract. 2014;2:53–5.
12. Barker FG 2nd, Jannetta PJ, Bissonette DJ, Shields PT, Larkins MV, Jho HD. Microvascular decompression for hemifacial spasm. J Neurosurg. 1995;82:201–10.
13. Costa J, Espirito-Santo C, Borges A, et al. Botulinum toxin type A therapy for hemifacial spasm. Cochrane Database Syst Rev. 2005;2005:Cd004899.
14. Wang A, Jankovic J. Hemifacial spasm: clinical findings and treatment. Muscle Nerve. 1998;21:1740–7.
15. Alexander GE, Moses H. Carbamazepine for hemifacial spasm. Neurology. 1982;32:286.
16. Sandyk R. Baclofen in hemifacial spasm. Eur Neurol. 1984;23:163–5.
17. Herzberg L. Management of hemifacial spasm with clonazepam. Neurology. 1985;35:1676–7.
18. Siniscalchi A, Gallelli L, Palleria C, De Sarro G. Idiopathic hemifacial spasm responsive to zonisamide: a case report. Clin Neuropharmacol. 2009;32:230–1.
19. Bandini F, Mazzella L. Gabapentin as treatment for hemifacial spasm. Eur Neurol. 1999;42:49–51.
20. Daniele O, Caravaglios G, Marchini C, Mucchiut L, Capus P, Natale E. Gabapentin in the treatment of hemifacial spasm. Acta Neurol Scand. 2001;104:110–2.
21. Patel J, Naritoku DK. Gabapentin for the treatment of hemifacial spasm. Clin Neuropharmacol. 1996;19:185–8.
22. Gross ML. Hemifacial spasm: treatment with pizotifen. J Neurol Neurosurg Psychiatry. 1996;61:118.
23. Biagio Carrieri P, Petracca M, Montella S. Efficacy of levetiracetam in hemifacial spasm: a case report. Clin Neuropharmacol. 2008;31:187–8.
24. Deleu D. Levetiracetam in the treatment of idiopathic hemifacial spasm. Neurology. 2004;62:2134–5.
25. Kuroda T, Saito Y, Fujita K, et al. Efficacy of levetiracetam in primary hemifacial spasm. J Clin Neurosci. 2016;34:213–5.
26. Simpson DM, Hallett M, Ashman EJ, et al. Practice guideline update summary: botulinum neurotoxin for the treatment of blepharospasm, cervical dystonia, adult spasticity, and headache: report of the Guideline Development Subcommittee of the American Academy of Neurology. Neurology. 2016;86:1818–26.
27. Luthra NS, Mitchell KT, Volz MM, Tamir I, Starr PA, Ostrem JL. Intractable blepharospasm treated with bilateral pallidal deep brain stimulation. Tremor Other Hyperkinet Mov (N Y). 2017;7:472.
28. Santos AF, Veiga A, Augusto L, Vaz R, Rosas MJ, Volkmann J. Successful treatment of blepharospasm by pallidal neurostimulation. Mov Disord Clin Pract. 2016;3:409–11.
29. Yamada K, Shinojima N, Hamasaki T, Kuratsu J-I. Pallidal stimulation for medically intractable blepharospasm. BMJ Case Rep. 2016;2016:bcr2015214241.
30. Kranz G, Shamim EA, Lin PT, Kranz GS, Hallett M. Transcranial magnetic brain stimulation modulates blepharospasm: a randomized controlled study. Neurology. 2010;75:1465–71.
31. Wagle Shukla A, Hu W, Legacy J, Deeb W, Hallett M. Combined effects of rTMS and botulinum toxin therapy in benign essential blepharospasm. Brain Stimul. 2018;11:645–7.
32. Nutt JG, Hammerstad JP, deGarmo P, Carter J. Cranial dystonia: double-blind crossover study of anticholinergics. Neurology. 1984;34:215–7.
33. Gollomp SM, Fahn S, Burke RE, Reches A, Ilson J. Therapeutic trials in Meige syndrome. Adv Neurol. 1983;37:207–13.
34. Merikangas JR, Reynolds CF 3rd. Blepharospasm: successful treatment with clonazepam. Ann Neurol. 1979;5:401–2.
35. Ohara S, Tsuyuzaki J, Hayashi R. Mexiletine in the treatment of blepharospasm: experience with the first three patients. Mov Disord. 1999;14:173–5.
36. Jankovic J. Treatment of hyperkinetic movement disorders with tetrabenazine: a double-blind crossover study. Ann Neurol. 1982;11:41–7.
37. Vijayakumar D, Wijemanne S, Jankovic J. Treatment of blepharospasm with apraclonidine. J Neurol Sci. 2017;372:57–9.

38. Hipola D, Mateo D, Giménez-Roldán S. Meige's syndrome: acute and chronic responses to clonazepan and anticholinergics. Eur Neurol. 1984;23:474–8.

39. Houser M, Waltz T. Meige syndrome and pallidal deep brain stimulation. Mov Disord. 2005;20:1203–5.

40. Inoue N, Nagahiro S, Kaji R, Goto S. Long-term suppression of Meige syndrome after pallidal stimulation: a 10-year follow-up study. Mov Disord. 2010;25:1756–8.

41. Marsden CD. Blepharospasm-oromandibular dystonia syndrome (Brueghel's syndrome). A variant of adult-onset torsion dystonia? J Neurol Neurosurg Psychiatry. 1976;39:1204–9.

42. Tanner CM, Glantz RH, Klawans HL. Meige disease: acute and chronic cholinergic effects. Neurology. 1982;32:783–5.

43. Cramer H, Otto K. Meige's syndrome: clinical findings and therapeutic results in 50 patients. NeuroOphthalmology. 1986;6:3–15.

44. LeDoux MS. Meige syndrome: what's in a name? Parkinsonism Relat Disord. 2009;15:483–9.

45. An JY, Kim JS, Kim YI, Lee KS. Successful treatment of the Meige syndrome with oral zolpidem monotherapy. Mov Disord. 2008;23:1619–21.

46. Yardimci N, Karatas M, Kilinc M, Benli S. Levetiracetam in Meige's syndrome. Acta Neurol Scand. 2006;114:63–6.

47. Gutmann L, Gutmann L. Myokymia and neuromyotonia 2004. J Neurol. 2004;251:138–42.

48. Banik R, Miller NR. Chronic myokymia limited to the eyelid is a benign condition. Journal of Neuroophthalmol. 2004;24:290–2.

49. Miller NR. Eyelid myokymia. Surv Ophthalmol. 2011;56:277–8. author reply 278

50. Bhatia R, Desai S, Garg A, Padma MV, Prasad K, Tripathi M. Isolated facial myokymia as a presenting feature of pontine neurocysticercosis. Movement Disord. 2008;23:135–7.

51. Öge AE, Boyaciyan A, Sarp A, Yazici J. Facial myokymia: segmental demyelination demonstrated by magnetic stimulation. Muscle Nerve. 1996;19:246–9.

52. Sedano MJ, Trejo JM, Macarron JL, Polo JM, Berciano J, Calleja J. Continuous facial myokymia in multiple sclerosis: treatment with botulinum toxin. Eur Neurol. 2000;43:137–40.

53. Sharma RR, Mathad NV, Joshi DN, Mazarelo TB, Vaidya MM. Persistent facial myokymia: a rare pathognomic physical sign of intrinsic brain-stem lesions: report of 2 cases and review of literature. J Postgrad Med. 1992;38:37–40, 40a–40b.

54. Lew MF. Review of the FDA-approved uses of botulinum toxins, including data suggesting efficacy in pain reduction. Clin J Pain. 2002;18:S142–6.

55. Scott AB. Botulinum toxin injection of eye muscles to correct strabismus. Trans Am Ophthalmol Soc. 1981;79:734–70.

56. Dressler D, Benecke R. Pharmacology of therapeutic botulinum toxin preparations. Disabil Rehabil. 2007;29:1761–8.

57. Dressler D, Saberi FA, Barbosa ER. Botulinum toxin: mechanisms of action. Arq Neuropsiquiatr. 2005;63:180–5.

58. Pirazzini M, Rossetto O. Challenges in searching for therapeutics against botulinum neurotoxins. Expert Opin Drug Discovery. 2017;12:497–510.

59. Samizadeh S, De Boulle K. Botulinum neurotoxin formulations: overcoming the confusion. Clin Cosmet Investig Dermatol. 2018;11:273–87.

60. Carli L, Montecucco C, Rossetto O. Assay of diffusion of different botulinum neurotoxin type a formulations injected in the mouse leg. Muscle Nerve. 2009;40:374–80.

61. Bentivoglio AR, Del Grande A, Petracca M, Ialongo T, Ricciardi L. Clinical differences between botulinum neurotoxin type A and B. Toxicon. 2015;107:77–84.

62. Sesardic T. Bioassays for evaluation of medical products derived from bacterial toxins. Curr Opin Microbiol. 2012;15:310–6.

63. Dashtipour K, Pedouim F. Botulinum toxin: preparations for clinical use, immunogenicity, side effects, and safety profile. Semin Neurol. 2016;36:29–33.

64. Frevert J. Pharmaceutical, biological, and clinical properties of botulinum neurotoxin type A products. Drugs R&D. 2015;15:1–9.

65. Yun JY, Kim JW, Kim H-T, et al. Dysport and Botox at a ratio of 2.5:1 units in cervical dystonia: a double-blind, randomized study. Mov Disord. 2015;30:206–13.

66. Dolimbek BZ, Aoki KR, Steward LE, Jankovic J, Atassi MZ. Mapping of the regions on the heavy chain of botulinum neurotoxin A (BoNT/A) recognized by antibodies of cervical dystonia patients with immunoresistance to BoNT/A. Mol Immunol. 2007;44:1029–41.

67. Dressler D, Rothwell JC. Electromyographic quantification of the paralysing effect of botulinum toxin in the sternocleidomastoid muscle. Eur Neurol. 2000;43:13–6.

68. Dressler D, Bigalke H. Botulinum toxin type B de novo therapy of cervical dystonia: frequency of antibody induced therapy failure. J Neurol. 2005;252:904–7.

69. Naumann M, Boo LM, Ackerman AH, Gallagher CJ. Immunogenicity of botulinum toxins. J Neural Transm (Vienna). 2013;120:275–90.

Possible Complications of Microvascular Decompression

Doo-Sik Kong

Background Microvascular decompression (MVD) is a safe and effective treatment modality for the treatment of trigeminal neuralgia and hemifacial spasm. Postoperative complications may be either general or specific to the MVD and should be managed with care. In this chapter, we aimed to address possible complications during or following MVD. The possible complications associated with MVD include associated cranial deficits such as facial palsy and hearing loss, hemorrhagic complications, etc. Considering a rare incidence of the events, it cannot be overemphasized that care should be taken to avoid and reduce the major morbidities associated with MVD.

Microvascular decompression (MVD) has been recommended as a major safe and definitive treatment option for trigeminal neuralgia and hemifacial spasm. But in general, brain surgery itself poses a potential risk associated with vascular manipulation, especially around the cranial nerve. Operative morbidity and mortality are often defined as events within 30 days of operation.

Common complications during or after MVD include the postoperative facial paralysis, newly developed hearing deficits, cerebrospinal fluid leaks, cranial nervous deficits, and cerebrovascular events. The frequency of operative and postoperative complications naturally increases with the complexity of the procedure. In most cases, incidence of postoperative complications is between 1 and 7 days after the operation, specific complications occur throughout the postoperative period and in the late postoperative period. Among them, cerebrovascular complications are relatively rare, but once they occur, they can be fatal and are very significant complications. These complications associated with cerebral hemorrhage can leave very serious side effects, requiring more advanced techniques and standardized surgical procedures to prevent them from occurring. This study examines factors that can predict vulnerabilities to complications with MVD. There is a slight difference in the incidence of post-MVD complications between trigeminal neuralgia and hemifacial spasm. In our series over 4000 patients undergoing MVD, middle ear effusion (MEE), hearing impairment, dysgeusia, etc., were common adverse events post-MVD for trigeminal neuralgia. In contrast, after MVD for hemifacial spasm, facial nerve palsy, MEE, and hearing deficits [1] were more common adverse events. Surgeons have to know that morbidities associated with MVD is transient form or permanent form. As transient complications, temporary facial palsy, transient hearing deficit, and cerebrospinal fluid leak can occur. Permanent complications such as permanent hearing loss, complete facial palsy, and hemorrhagic/thromboembolic events, despite a

D.-S. Kong (✉)
Department of Neurosurgery, Samsung Medical Center, Sungkyunkwan University School of Medicine, Seoul, Republic of Korea

K. Park, J. S. Park (eds.), *Hemifacial Spasm*, https://doi.org/10.1007/978-981-15-5417-9_15

rare incidence, should be carefully avoided [2, 3]. In the case of transient complications, it is important to keep a close eye on the changes in the symptoms, which can be overcome wisely by reassuring the patient sufficiently while realizing that the symptoms may last forever.

Facial Palsy

Facial palsy is relatively common complication of MVD for hemifacial spasm, of which incidence has been reported to be 14.3–18.6% [2, 3]. According to the mode of onset, early-onset facial palsy occurs immediately after surgery within 24 h, where late-onset facial palsy usually occurs over a 24-h to 7-day period after the operation [4]. The exact pathogenesis of newly developed facial palsy after MVD is not known yet. Early-onset facial palsy seems to be associated with the inserted Teflon between offending vessels and facial nerve. Direct facial nerve injury by the compression of Teflon felt, delayed facial nerve edema by long-standing manipulation, or disturbance of microcirculation can explain the occurrence of facial palsy [2, 5, 6]. Because most cases of early-onset facial palsy were common in hemifacial spasms rather than trigeminal neuralgia, direct impairment of the facial nerve may be closely associated with early-onset facial palsy. In terms of surgical technique, Teflon-interpositional technique may contribute the occurrence of early-onset facial palsy after MVD more than arterial transpositional technique. Main cause of late-onset facial palsy remains unclear. The theory of viral origin may have plausible hypothesis [5–9]. The incidence of late-onset facial palsy following MVD is known to be 2.8–8.3% [5, 6, 10] and most of them could be improved spontaneously.

Most facial weakness can be resolved without special treatment. However, extensive rehabilitation can be required for excellent outcomes in some cases [11, 12]. Careful attention to the preoperative MR images, complete decompression of the facial nerve, less manipulation of the facial nerve, and intraoperative monitoring are the best ways to avoid this complication.

Hearing Impairment

Hearing loss is a relatively rare but significant risk of MVD. Hearing impairment is generally defined as increasing more than 15 dB depending on the bone conduction or by more than 20% of speech discrimination associated with the reference hearing. In most cases, hearing impairment after MVD is caused by an excessive retraction of the cerebellum. Several reasons can explain the postoperative hearing loss after MVD [3, 13–23]. Most plausible cause of hearing loss is attributable to the stretch of the eighth nerve during retracting the cerebellum. As a result, spasm of the vasa nervorum or ischemic injury to the nerve can develop. Taking into account the flocculus overlying the VII–VIII nerve complex, it is challenging to access directly with the dorsal direction and to observe the root exit area of the facial nerve. Thus, during the search for the root exit zone of the facial nerve, an excessive cerebellar retraction with inappropriate direction may lead to the postoperative hearing deficits. Our series showed the correlation between cerebellar retraction and BAEP changes.

Another possible cause includes the ischemic injury from the block of the labyrinthine artery and/or the anteroinferior cerebellar artery, resulting in the cochlear ischemic damages. In addition, direct trauma by coagulation or instruments, over-compression with Teflon felt interposed between the offending vessel and 7–8th nerve complex [24, 25]. Rarely, there happens a sudden decrease of brainstem auditory evoked potentials (BAEP) during or immediately after craniectomy. It may be attributable to an edema in the intracanalicular tract of the nerve, sudden change of intracranial pressure [26], or drill-induced noise or transient loss of CSF during surgery. There are some differences in the amplitude of BAEP under individual situation including recording conditions or electrode impedance [27].

In some cases, hearing impairment can show a gradual improvement when it is caused by ischemic injury due to local vasospasm. However, if hearing impairment is caused by a direct retraction injury, postoperative hearing impairment remains persistent and constant [3, 13–23].

Intraoperative monitoring of BAEP during surgery is helpful to minimize the risk of hearing impairment in patients undergoing the cerebello-pontine angle (CPA) surgery including MVD [28, 29]. The traction or compression of the nerve is closely associated with BAEP deterioration. These BAEP changes are caused by direct mechanical damage to the brainstem, vascular ischemia, or infarction. Our experience supported the high probability of hearing loss if the amplitude reduction and delay time of wave V during MVD surgery are fast and the changes are not recovered with modification [30]. In the literature, although there is no consensus regarding the criteria for significant intraoperative BAEP change, a latency prolongation of as little as 0.5 ms of the wave V or a reduction of amplitude more than 50% in wave V was a strong indicator of hearing impairment [23, 31, 32]. During surgery, BAEP monitoring plays an essential role in alarms for hearing preservation. Thus, we need to understand the surrounding surgical anatomy to minimize the risk of hearing loss. In addition, the minimum retraction of the cerebellum is of great importance, and can be helpful if the endoscope can provide a second look to identify nerve root entry areas and identify the location of the Teflon felt [9, 33–35]. Whether the hearing loss or impairment is identified intraoperatively or post-operatively, the rehabilitation protocol should be modified appropriately.

Lower Cranial Nerve Deficits

Clinical manifestations of lower cranial nerve deficits include hoarseness, dysphagia, and swallowing difficulty. In most cases, lower cranial deficits have a self-limiting course and are not fatal to the quality of life in patients. Despite a rare incidence, they should be carefully monitored, because their natural prognosis is sometimes poor. Considering that lower cranial nerve deficits are more common complications in hemifacial spasm than trigeminal neuralgia, manipulation of lower cranial nerves during the

arachnoid dissection may be a major contributing factor. In review of our series, some patients can have various natures of dysgeusia (impairment of the sense of taste), continuous sour taste, and hypogeusia. In the literature, some authors suggested the hypothesis of this complication as injury to nervus intermedius or the trigeminal nerve can develop abnormal sensation of taste involving anterior two-thirds of the tongue and floor of the mouth and the palate [36, 37]. In general, the prognosis of dysgeusia seemed to be poor. Other cranial deficits include trochlear or abducens nerve palsy after MVD, vocal cord palsy occurs, resulting in asymmetric soft palate elevation or uvula deviation.

Hemorrhagic or Thromboembolic Complications

Hemorrhage or thromboembolic complications are relatively rare but serious complications of MVD. Hemorrhagic risk after MVD is similar to the risk inherent in the general cranial surgery. It must be the most serious and life-threatening complication. Cerebral infarction and dural arteriovenous fistula can be candidates of reasons of hemorrhagic risks. During operation, the sacrifice of the inferior petrosal vein does not always carry the risk of hemorrhagic events. Even though it must be rare, it is recommended that the inferior petrosal vein should be preserved. Managements for cerebrovascular events must be modified personally based on each situation. Sometimes, drainage of CSF during MVD may be associated with sudden slack-down of the cerebellum, leading to the tearing of bridging vein, resulting in subdural hematoma.

Conclusions

In conclusion, MVD is known to be a safe and effective treatment modality for hemifacial spasm and trigeminal neuralgia. Considering a rare incidence of the events, it should not be overemphasized that care should be taken to avoid and reduce the major morbidities associated with

MVD. It cannot be overemphasized that care should be taken to maintain the disciplined implementation of the surgical, such as proper direction and adequate retraction of cerebellum and preservation of arachnoid sheath over the CN VIII and CN VII complex, etc. Most of these complications can be prevented through standardized protocols to incorporate monitoring technologies and specific technical practices, team approach, and accumulating volume.

References

1. Lee CC, et al. Brainstem auditory evoked potential monitoring and neuro-endoscopy: two tools to ensure hearing preservation and surgical success during microvascular decompression. J Chin Med Assoc. 2014;77:308–16.
2. Huh R, Han IB, Moon JY, Chang JW, Chung SS. Microvascular decompression for hemifacial spasm: analyses of operative complications in 1582 consecutive patients. Surg Neurol. 2008;69:153–7.
3. Miller LE, Miller VM. Safety and effectiveness of microvascular decompression for treatment of hemifacial spasm: a systematic review. Br J Neurosurg. 2012;26:438–44.
4. Hongo K, Kobayashi S, Takemae T, Sugita K. [Posterior fossa microvascular decompression for hemifacial spasm and trigeminal neuralgia—some improvements on operative devices and technique]. No Shinkei Geka. 1985;13:1291–96.
5. Lovely TJ, Getch CC, Jannetta PJ. Delayed facial weakness after microvascular decompression of cranial nerve VII. Surg Neurol. 1998;50:449–52.
6. Rhee DJ, Kong DS, Park K, Lee JA. Frequency and prognosis of delayed facial palsy after microvascular decompression for hemifacial spasm. Acta Neurochir (Wien). 2006;148:839–43.
7. Furukawa K et al. [Delayed facial palsy after microvascular decompression for hemifacial spasm due to reactivation of varicella-zoster virus]. No Shinkei Geka. 2003;31:899–902.
8. Huang TS, Chang YC, Lee SH, Chen FW, Chopra IJ. Visual, brainstem auditory and somatosensory evoked potential abnormalities in thyroid disease. Thyroidology. 1989;1:137–42.
9. Badr-El-Dine M, El-Garem HF, Talaat AM, Magnan J. Endoscopically assisted minimally invasive microvascular decompression of hemifacial spasm. Otol Neurotol. 2002;23:122–8.
10. Han JS, Lee JA, Kong DS, Park K. Delayed cranial nerve palsy after microvascular decompression for hemifacial spasm. J Korean Neurosurg Soc. 2012;52:288–92.
11. Kondo A. Follow-up results of using microvascular decompression for treatment of glossopharyngeal neuralgia. J Neurosurg. 1998;88:221–5.
12. Samii M, et al. Microvascular decompression to treat hemifacial spasm: long-term results for a consecutive series of 143 patients. Neurosurgery. 2002;50:712–8.
13. Dannenbaum M, Lega BC, Suki D, Harper RL, Yoshor D. Microvascular decompression for hemifacial spasm: long-term results from 114 operations performed without neurophysiological monitoring. J Neurosurg. 2008;109:410–5.
14. Chung SS, Chang JH, Choi JY, Chang JW, Park YG. Microvascular decompression for hemifacial spasm: a long-term follow-up of 1,169 consecutive cases. Stereotact Funct Neurosurg. 2001;77:190–3.
15. Fritz W, Schafer J, Klein HJ. Hearing loss after microvascular decompression for trigeminal neuralgia. J Neurosurg. 1988;69:367–70.
16. Hyun SJ, Kong DS, Park K. Microvascular decompression for treating hemifacial spasm: lessons learned from a prospective study of 1,174 operations. Neurosurg Rev. 2010;33:325–34.
17. Jannetta PJ, Hirsch BE. Restoration of useful hearing following microvascular decompression of cochlear nerve. Am J Otol. 1993;14:627.
18. Jannetta PJ, Moller MB, Moller AR, Sekhar LN. Neurosurgical treatment of vertigo by microvascular decompression of the eighth cranial nerve. Clin Neurosurg. 1986;33:645–65.
19. Park K, et al. Patterns of hearing loss after microvascular decompression for hemifacial spasm. J Neurol Neurosurg Psychiatry. 2009;80:1165–7.
20. Rosseau GL, Jannetta PJ, Hirsch B, Moller MB, Moller AR. Restoration of useful hearing after microvascular decompression of the cochlear nerve. Am J Otol. 1993;14:392–7.
21. Vasama JP, Moller MB, Moller AR. Microvascular decompression of the cochlear nerve in patients with severe tinnitus. Preoperative findings and operative outcome in 22 patients. Neurol Res. 1998;20:242–8.
22. Shah A, et al. Hearing outcomes following microvascular decompression for hemifacial spasm. Clin Neurol Neurosurg. 2012;114:673–7.
23. Jo KW, Kim JW, Kong DS, Hong SH, Park K. The patterns and risk factors of hearing loss following microvascular decompression for hemifacial spasm. Acta Neurochir. 2011;153:1023–30.
24. Polo G, Fischer C, Sindou MP, Marneffe V. Brainstem auditory evoked potential monitoring during microvascular decompression for hemifacial spasm: intraoperative brainstem auditory evoked potential changes and warning values to prevent hearing loss—prospective study in a consecutive series of 84 patients. Neurosurgery. 2004;54:97–104.
25. Sindou MP. Microvascular decompression for primary hemifacial spasm. Importance of intraoperative neurophysiological monitoring. Acta Neurochir (Wien). 2005;147:1019–26.
26. Jo KW, Lee JA, Park K, Cho YS. A new possible mechanism of hearing loss after microvascular

decompression for hemifacial spasm. Otol Neurotol. 2013;34:1247–52.

27. Moller AR. Intra-operative neurophysiologic monitoring in neurosurgery: benefits, efficacy, and cost-effectiveness. Clin Neurosurg. 1995;42:171–9.

28. Lee SH, Song DG, Kim S, Lee JH, Kang DG. Results of auditory brainstem response monitoring of microvascular decompression: a prospective study of 22 patients with hemifacial spasm. Laryngoscope. 2009;119:1887–92.

29. Sindou M, Fobe JL, Ciriano D, Fischer C. Hearing prognosis and intraoperative guidance of brainstem auditory evoked potential in microvascular decompression. Laryngoscope. 1992;102:678–82.

30. Ying T, et al. Incidence of high-frequency hearing loss after microvascular decompression for hemifacial spasm. J Neurosurg. 2013;118:719–24.

31. Acevedo JC, Sindou M, Fischer C, Vial C. Microvascular decompression for the treatment of hemifacial spasm. Retrospective study of a consecutive series of 75 operated patients—electrophysiologic and anatomical surgical analysis. Stereotact Funct Neurosurg. 1997;68:260–5.

32. Ying T, et al. Empirical factors associated with brainstem auditory evoked potential monitoring during microvascular decompression for hemifacial spasm and its correlation to hearing loss. Acta Neurochir. 2014;156:571–5.

33. Bohman LE, Pierce J, Stephen JH, Sandhu S, Lee JY. Fully endoscopic microvascular decompression for trigeminal neuralgia: technique review and early outcomes. Neurosurg Focus. 2014;37:E18.

34. Cheng WY, Chao SC, Shen CC. Endoscopic microvascular decompression of the hemifacial spasm. Surg Neurol. 2008;70(S1):40–6.

35. Kabil MS, Eby JB, Shahinian HK. Endoscopic vascular decompression versus microvascular decompression of the trigeminal nerve. Minim Invasive Neurosurg. 2005;48:207–12.

36. Grant R, Ferguson MM, Strang R, Turner JW, Bone I. Evoked taste thresholds in a normal population and the application of electrogustometry to trigeminal nerve disease. J Neurol Neurosurg Psychiatry. 1987;50:12–21.

37. Natarajan M. Percutaneous trigeminal ganglion balloon compression: experience in 40 patients. Neurol India. 2000;48:330–2.

Prognosis of Symptoms After Microvascular Decompression for Hemifacial Spasm

Jeong-A Lee and Kwan Park

Patients with hemifacial spasm (HFS) are more likely to have obsessive-compulsive disorders, anxiety, and sexual dysfunction [1–3]. The more severe the spasms, the worse are the quality of life (QoL), headache, and depression [4–6]. Microvascular decompression (MVD) has been recognized as an effective and reliable treatment for HFS [7–8]. The post-surgery improvements of the HFS symptoms were associated with decreased social anxiety and improved QoL [9]. However, the postoperative course of MVD is variable for each patient. The QoL improvements were delayed in operated patients due to such postoperative course [10]. The endpoint to confirm whether the surgical outcomes are successful is still unclear. The decision for reoperation is difficult. Surgeons should be careful when deciding indications and timing of reoperation [11–13]. Predicting outcomes as accurately/soon as possible is needed to minimize patient discomfort and anxiety by providing specific and practical information and to determine the proper timing for reoperation and reduce unnecessary

reoperation. This chapter aims to investigate the predictors and the optimal prediction time of the surgical outcome through literature review.

Postoperative Course of Hemifacial Spasm

The postoperative course of MVD for HFS varies. In a systematic review for 12 articles with 2727 patients, the mean follow-up duration after surgery was 49 months (range 6.4–121.6 months). An average postoperative success rate was 85.1% with a range of 76.5–93.5%, but the immediate success rate after surgery is only 71.8% with a range of 59.5–84.0%. The mean rate of delayed improvement was 25.4% with a range of 18.8–37.1% [14]. Another review analyzed 82 publications counting more than 10,000 operating cases. The proportion of patients with total relief was between 85 and 90%. In many series, delayed relief was obtained in 33 ± 8% of patients and 12% of them were delayed for approximately 1 year. When the effect of MVD was considered to have been achieved, relief retained permanently, with the exception of 1–2% of the long-term followed patients [8].

The various postoperative courses of HFS were categorized into several patterns. This is an extreme representation of the postoperative course (Fig. 1). Patients were simply divided into two groups, with or without postoperative spasms [15]. Patients were divided into three groups

J.-A. Lee (✉)
Departments of Neurosurgery, Samsung Medical Center, 81, Irwon-ro Gangnam-gu, Seoul, Republic of Korea
e-mail: naja.lee@samsung.com

K. Park
Department of Neurosurgery, Konkuk University Medical Center, Seoul, Korea (Republic of)
e-mail: kwanpark@skku.edu

© Springer Nature Singapore Pte Ltd. 2020
K. Park, J. S. Park (eds.), *Hemifacial Spasm*, https://doi.org/10.1007/978-981-15-5417-9_16

according to the postoperative course: immediate cure, delayed cure, and failure [16]. Different outcomes were divided into four groups depending on the variable recovery period: immediate cure, spasms lasted with milder degree and slowly disappeared from 7 days to 2 years, spasms stopped immediately but relapsed 3 days later and ran the same course as in the second group, and failure [17]. The five groups included

the following: immediate cure without recurrence, temporary relapse followed by cure, slow but steady improvement that leads to cure after 1 month or more, recurrence with persistent symptoms, and no improvement or improvement to some extent that does not lead to cure [18]. On the other hand, disappearance of symptoms over time was classified into three groups: immediate disappearance of symptoms after surgery, delayed disappearance of symptoms, and unusual disappearance of symptoms [19].

This figure is a polarized representation of the remaining spasms after surgery compared to the preoperative spasms.

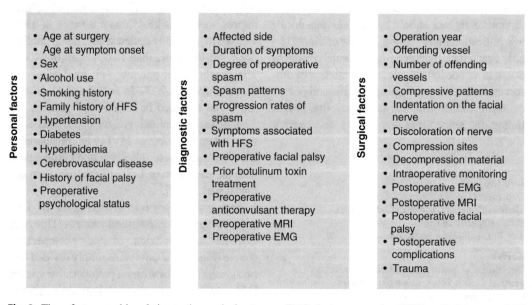

Fig. 1 Postoperative course of MVD for HFS. This figure is a polarized representation of the remaining spasms after surgery compared to the preoperative spasms

Predictors of Postoperative Outcome

English-language studies on MVD for HFS published since 2000 to the present have been retrieved from PubMed. Keywords of interest were prognosis, outcome, and result. Variables used in relation to the surgical outcome were arbitrarily classified into three factors: personal factors, diagnostic factors, and surgical factors (Fig. 2).

Personal factors
- Age at surgery
- Age at symptom onset
- Sex
- Alcohol use
- Smoking history
- Family history of HFS
- Hypertension
- Diabetes
- Hyperlipidemia
- Cerebrovascular disease
- History of facial palsy
- Preoperative psychological status

Diagnostic factors
- Affected side
- Duration of symptoms
- Degree of preoperative spasm
- Spasm patterns
- Progression rates of spasm
- Symptoms associated with HFS
- Preoperative facial palsy
- Prior botulinum toxin treatment
- Preoperative anticonvulsant therapy
- Preoperative MRI
- Preoperative EMG

Surgical factors
- Operation year
- Offending vessel
- Number of offending vessels
- Compressive patterns
- Indentation on the facial nerve
- Discoloration of nerve
- Compression sites
- Decompression material
- Intraoperative monitoring
- Postoperative EMG
- Postoperative MRI
- Postoperative facial palsy
- Postoperative complications
- Trauma

Fig. 2 Three factors used in relation to the surgical outcome. *EMG* electromyography, *HFS* hemifacial spasm, *MRI* magnetic resonance imaging

Personal Factors

Personal factors are potentially causative factors before diagnosis or surgery in patients with HFS, which are an individual-related trait that affects the postoperative course due to the general characteristics of the subject. These factors include the following: age at surgery, age at symptom onset, sex, alcohol use, smoking history, family history of HFS, hypertension, diabetes, hyperlipidemia, cerebrovascular disease, history of facial palsy, and preoperative psychological status. Although most variables were not related to the surgical outcome, there are a few studies that have described the association between age at surgery, sex, and the surgical outcome.

A few studies have shown that age at surgery considered being predictor, which significantly predicted the clinical outcome of patients following MVD [20–23]. The reason for this result is not clear. Engh et al. proposed that perhaps older patients who have failed MVD are more likely to have chronic, irreversible neuropathy of the facial nerve than younger patients [20]. Lv et al. presented that there is no doubt that (a) younger patients will have a relatively better operation condition for craniotomy, and (b) also indicate a relatively rapid recovery process than those of elderly patients. (c) With the extension of the duration of the disease and with aging, the incidence of microvascular complications may significantly increase accordingly, and hence associated with a relatively poor prognostic outcome [22].

Outcome of patients with HFS after MVD was predicted by sex [23–24]. Cheng et al. reported that female patients with HFS showed larger posterior fossa crowdedness and smaller posterior fossa cerebrospinal fluid (CSF) volume than male. These researchers suggested that posterior fossa crowdedness might increase the risk of developing HFS by causing congestion of the nervous and vascular structures [25]. Our study proposed that these results might be due to differences in hyperexcitability of facial motor nuclei in women and men [24].

Diagnostic Factors

Diagnostic factors are causative factors related to the disease or test result during the diagnosis of HFS, which are a disease-related trait that affects the postoperative course due to the characteristics of the disease. These factors include the following: affected side, duration of symptoms, characteristics of preoperative spasm (degree, patterns, and associated symptoms), preoperative facial palsy, prior botulinum toxin treatment, preoperative anticonvulsant therapy, preoperative magnetic resonance imaging (MRI), and preoperative electromyography (EMG). Although most variables were not related to the surgical outcome, there are a few studies that have described the association between duration of symptoms, prior botulinum toxin treatment, degree of preoperative spasm, and the surgical outcome.

Duration of symptom was the influential variable which may be useful for the prediction of prognosis in the patients who underwent MVD [22, 26]. Jin et al. proposed that delayed improvement is more likely to occur in patients with severe indentation of the facial nerve and/or longer duration of preoperative disease. The reasons they explained were as follows: (a) It could cause the pathological demyelinating changes in the root exit zone (REZ) or hyperexcitability of the facial motor nucleus; (b) MVD itself can lead to edema of the facial nerve, thus causing microinjury of the facial nerve and generating residual HFS; (c) Although the REZ is decompressed, it may still be compressed by pulsatile decompression materials and CSF, which may result in delayed resolution [26].

Regarding the relationship between prior botulinum toxin treatment and the surgical outcome, there have been different results. Ishikawa et al. suggested that the reason why this treatment produces good results is because it causes facial nerve paresis [15]. On the contrary, Zhao et al. explained that the possible reasons why this treatment produces bad results are as follows: (a) The neurotoxic effect of botulinum toxin may cause demyelination at different location in the same facial nerve; (b) Among patients, the incidence of hypertension was high, which can result in torturing and prolongation of blood vessels [27].

Some researchers have found that the degree of spasms was an important classifier that can predict patients' outcomes [23, 24]. However, different classification based on the tendency [16] or severity and extent [28] of spasms did not show a significant association with the outcomes. Such discrepancies in the results might be due to inconsistent criteria with regard to the degree of preoperative spasms. In our previous study, the degree of preoperative spasms based on the grading system had significant correlation with the duration of preoperative symptoms and QoL [28]. Furthermore, the preoperative duration of symptoms and psychological status have previously been reported to significantly affect the postoperative outcome [26]. These results imply a close relationship between the degree of preoperative spasms and the postoperative outcome.

Surgical Factors

Surgical factors are the factors related to the surgical findings or surgical outcome of MVD in the HFS, which are a surgery-related trait that affects the postoperative course due to the surgical findings or surgical outcome. These factors include the following: operation year, offending vessel, number of offending vessels, compressive patterns, indentation on the facial nerve, discoloration of nerve, compression sites, decompression material, intraoperative monitoring (IOM), postoperative EMG, postoperative MRI, postoperative facial palsy, postoperative complications, and trauma. Many studies have described the association between indentation on the facial nerve, offending vessel, postoperative facial palsy, and the surgical outcome.

Regarding intraoperative indentation on the facial nerve, most previous studies have proven that severe indentation on the REZ is closely related to postoperative improvement [22, 24, 25, 29–31]. The improved results observed in the case of indentation on the nerve might be attributed to the fact that the indentation is evidence of definitive compression by an offending vessel. Thus, it is necessary to identify indentation on the facial nerve during the surgery. Whereas some researchers described the reason for negative results may be because it produces ischemia of the nerve and gradually causes nerve degeneration [32]. Also, it is explained that severe indentation of the REZ could lead to the pathological demyelinating changes of the REZ or hyperexcitability of the facial motor nucleus [26].

Previous studies have demonstrated that venous compression may be related to a worse prognosis, despite thorough decompression in HFS [31, 33, 34]. Wang et al. mentioned that a vein can play an important role and can be the offending vessel in MVD for HFS, although HFS caused by a vein is rare [35]. Toda et al. guided that the surgical issues regarding venous compression are its resolution and the distinction of venous compression from venous contact [34]. Therefore, we suggest that it is necessary to carefully explore offending vessels to prevent surgical failure.

Previous studies have shown that significantly better results were observed in terms of overall disappearance of HFS in the patients with delayed facial palsy than in patients without delayed facial palsy [24, 36]. Lee et al. proposed that the occurrence of delayed facial palsy is due to manipulations or the gradual development of postoperative edema [36]. Although the causes of delayed facial palsy after surgery are unknown, 6.5–14.5% of patients who received MVD experienced it, and most patients have recovered completely [36, 37]. Therefore, it is important to inform patients of the possibility of delayed facial palsy postoperatively and to improve psychological stability by reassuring patients that it can be completely resolved and become an advantageous prognostic factor of the postoperative spasm pattern.

Numerous studies have been published on neurophysiological monitoring during surgery. However, the effectiveness of IOM on the surgical outcome was the most controversial. It is not yet conclusive, but it may be helpful if it is well utilized.

Models

The models were developed by analyzing various factors in a multidimensional manner. A risk assessment model was first created consisting of significant preoperative variables (Model 1) and intraoperative variables demonstrated little additive value (Model 2). This model demonstrated predictive value for persistent or recurrent spasm (Table 1) [23].

We established a prediction model for determining MVD outcome in HFS patients using decision tree analysis technique, a data-mining method. This prediction model was based on six categories including four items (postoperative delayed facial palsy, degree of preoperative spasm, indentation on the facial nerve, and sex) (Fig. 3). All patients of the first category con-

sisted of postoperative delayed facial palsy (yes) and the degree of preoperative spasm (grade I) showed improvement of spasm including delayed improvement. Meanwhile, 58.8% of the third category consisted of postoperative delayed facial palsy (no), indentation on the facial nerve (no or unknown), and the degree of preoperative spasm (grade IV) showed persistence of spasm [24].

Optimal Time for Outcome Prediction

It is necessary to find the earliest optimal time to determine the outcome of the surgery. The optimal time for the outcome prediction is divided as follows: POD 3 months, 6 months, 12 months, and 3 years.

Table 1 Risk scoring model of persistent HFS after MVD

Model		Variables			
		OR (95%CI)	p	AUC (95%CI)	p
Model 1		Age >50 + female gender + history of botulinum toxin use + platysma muscle involvement			
	Discharge	1.50 (1.03–2.18)	0.035	0.60 (0.50–0.70)	0.045
	Follow-up	3.01 (1.52–5.95)	0.002	0.75 (0.64–0.85)	0.001
Model 2		Age >50 + female gender + history of botulinum toxin use + platysma muscle involvement + LSR aberrancy after MVD + 2-stage surgery			
	Discharge	1.51 (1.16–1.97)	0.003	0.65 (0.56–0.75)	0.003
	Follow-up	1.61 (1.08–2.40)	0.019	0.67 (0.54–0.80)	0.023

LSR lateral spread response

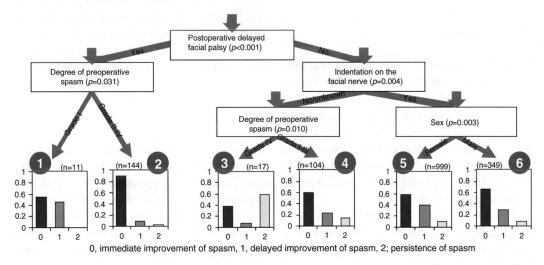

0, immediate improvement of spasm, 1, delayed improvement of spasm, 2; persistence of spasm

Fig. 3 Prediction model of outcome after MVD for HFS (N = 1624)

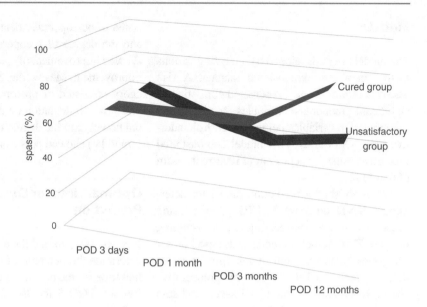

Fig. 4 Pattern of clinical improvement in cured and unsatisfactory group. Significant difference in symptom change between cured group and unsatisfactory group began to appear at 3 months postoperatively

POD 3 Months

In our previous studies, a chronological analysis of the changes in symptoms revealed that 3 months after surgery, symptoms of the cured group and the unsatisfactory group began to differentiate significantly (Fig. 4). This result implies that it might be possible to predict the outcome at least 3 months after surgery [18, 30].

Significant difference in symptom change between cured group and unsatisfactory group began to appear at 3 months postoperatively.

POD 6 Months

In our data, short-term and long-term outcomes of MVD in HFS patients were different, but the outcomes at 6 and 9 months were similar to those at 12 and more months (Fig. 5). Patients whose intraoperative offending vessel was not a vein, patients with intraoperative indentation on the facial nerve and patients with postoperative delayed facial palsy could predict good outcomes after 6 months of surgery [31].

The outcomes at 1, 3, 6, and 9 months were individually compared with the outcome after 12 months of surgery using the McNemar test with Bonferroni's correction.

POD 12 Months

Many authors commented that most cases were completely recovered within 1 year of observation and delayed resolution is uncommon after 1 year postoperatively. Therefore, even if MVD was expected to fail, it would be wise to wait 1 year before deciding on a reoperation [8, 11, 38]. Additionally, the patterns of postoperative outcome over time were divided into 10 groups. Of 267 patients in group 5–10, who showed persistence of spasm at the first month postoperatively (early nonresponders), 198 patients (74.2%) in group 5, 7, 8, and 9 showed improvement of spasm over 12 months (late responders) and the spasms improved at 3 months in 89 patients in group 5 (Table 2) [31]. This result can support the above opinions.

POD 3 Years

In our study, 12 out of 70 patients (17.1%) who experienced residual or recurrent spasms for more than 1 year postoperatively gradually improved after 1 year to even 3 years (Fig. 6). If the lateral spread response is resolved and severe indentation on the nerve is identified during surgery, reoperation may be delayed up to 3 years after MVD [29].

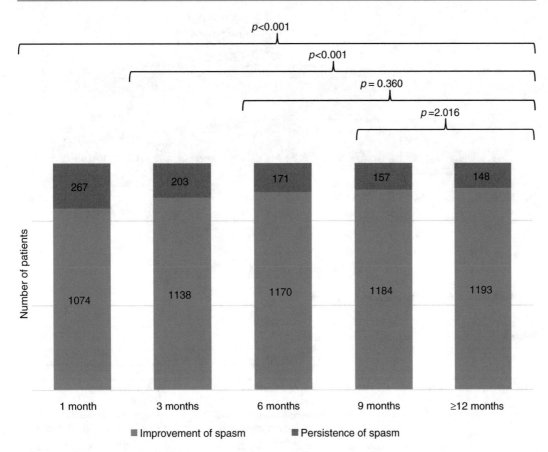

Fig. 5 Comparison of short-term and long-term outcomes ($N = 1341$). The outcomes at 1, 3, 6, and 9 months were individually compared with the outcome after 12 months of surgery using the McNemar test with Bonferroni's correction

Table 2 Patterns of postoperative outcome over time ($N = 1341$)

Group	Postoperative outcome over time					No. of patients
	1 month	3 months	6 months	9 months	≥12 months	
1	N	N	N	N	N	935
2	N	Y/N	Y/N	Y/N	N	60
3	N	N	N	N	Y	49
4	N	Y/N	Y/N	Y/N	Y	30
5	Y	N	N	N	N	89
6	Y	Y	Y	Y	Y	51
7	Y	Y	N	N	N	44
8	Y	Y	Y	Y	N	39
9	Y	Y/N	Y/N	Y/N	N	26
10	Y	Y/N	Y/N	Y/N	Y	18

N, improvement of spasm; Y, persistence of spasm; Y/N, improvement or persistence of spasm, with discontinuous patterns

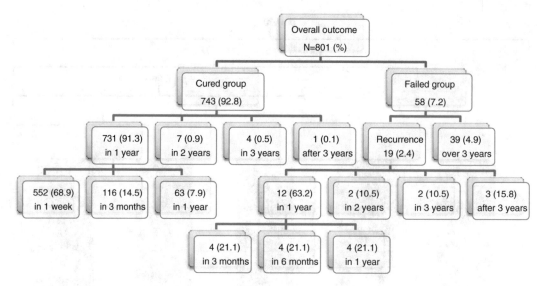

Fig. 6 Flowchart showing the postoperative course

Table 3 Summary of predictors and prediction time

			Prognosis	
			Better	Worse
Predictors	Weak	Pre-op	− Age at surgery Male − Symptom duration ± Botulinum toxin treatment − Spasm severity	+ Age at surgery Female + Symptom duration ± Botulinum toxin treatment + Spasm severity
	Moderate	Intra-op	Neurophysiologic findings	Neurophysiologic findings
	Strong	Intra-op	− Vein as offending vessel + Indentation on the nerve	+ Vein as offending vessel − Indentation on the nerve
		Post-op	+ Facial palsy	− Facial palsy
Prediction time	POD 3 months		− Spasm	+ Spasm
	POD 6 months		− Vein as offending vessel + Indentation on the nerve + Facial palsy	+ Vein as offending vessel − Indentation on the nerve − Facial palsy
	POD 12 months		− Spasm	+ Spasm

+/−, yes/no or increase/decrease

The above-mentioned points can be arbitrarily summarized in Table 3.

References

1. Mula M, Strigaro G, Marotta AE, Ruggerone S, Tribolo A, Monaco R, Cantello F. Obsessive-compulsive-spectrum symptoms in patients with focal dystonia, hemifacial spasm, and healthy subjects. J Neuropsychiatry Clin Neurosci. 2012;24(1):81–6.

2. Ozel-Kizil ET, Akbostanci MC, Ozguven HD, Atbasoglu EC. Secondary social anxiety in hyperkinesias. Mov Disord. 2008;23(5):641–5.

3. Perozzo P, Salatino A, Cerrato P, Ricci R. Sexual well-being in patients with blepharospasm, spasmodic torticollis, and hemifacial spasm: a pilot study. Front Psychol. 2016;7:1492.

4. Cheng J, Lei D, Hui X, Zhang H. Improvement of quality of life in patients with hemifacial spasm after microvascular decompression: a prospective study. World Neurosurg. 2017;107:549–53.

5. Peeraully T, Tan SF, Fook-Chong SM, Prakash KM, Tan EK. Headache in hemifacial spasm patients. Acta Neurol Scand. 2013;127(5):e24–7.

6. Tan EK, Lum SY, Fook-Chong S, Chan LL, Gabriel C, Lim L. Behind the facial twitch: depressive symptoms in hemifacial spasm. Parkinsonism Relat Disord. 2005;11(4):241–5.

7. Lee MH, Jee TK, Lee JA, Park K. Postoperative complications of microvascular decompression for hemifacial spasm: lessons from experience of 2,040 cases. Neurosurg Rev. 2016;39(1):151–8.

8. Sindou M, Mercier P. Microvascular decompression for hemifacial spasm: outcome on spasm and complications. A review. Neurochirurgie. 2018;64(2):106–16.

9. Kim YG, Jung NY, Kim M, Chang WS, Jung HH, Chang JW. Benefits of microvascular decompression on social anxiety disorder and health-related quality of life in patients with hemifacial spasm. Acta Neurochir. 2016;158(7):1397–404.

10. Shibahashi K, Morita A, Kimura T. Surgical results of microvascular decompression procedures and patient's postoperative quality of life: review of 139 cases. Neurol Med Chir. 2013;53(6):360–4.

11. Hatayama T, Kono T, Harada Y, Yamashita K, Utsunomiya T, Hayashi M, et al. Indications and timings of re-operation for residual or recurrent hemifacial spasm after microvascular decompression: personal experience and literature review. Neurol Med Chir. 2015;55(8):663–8.

12. Liu MX, Xia L, Zhong J, Li B, Dou NN, Li ST. What should we do for those hemifacial spasm patients without efficacy following microvascular decompression: expectation of delayed relief or early reoperation? World Neurosurg. 2018;110:e897–900. https://doi.org/10.1016/j.wneu.2017.11.118.

13. Wang X, Thirumala PD, Shah A, Gardner P, Habeych M, Crammond D, et al. Microvascular decompression for hemifacial spasm: focus on late reoperation. Neurosurg Rev. 2013;36(4):637–43.

14. Xia L, Zhong J, Zhu J, Dou NN, Liu MX, Li ST. Delayed relief of hemifacial spasm after microvascular decompression. J Craniofac Surg. 2015;26(2):408–10.

15. Ishikawa M, Nakanishi T, Takamiya Y, Namiki J. Delayed resolution of residual hemifacial spasm after microvascular decompression operations. Neurosurgery. 2001;49(4):847–56.

16. Terasaka S, Asaoka K, Yamaguchi S, Kobayashi H, Motegi H, Houkin K. A significant correlation between delayed cure after microvascular decompression and positive response to preoperative anticonvulsant therapy in patients with hemifacial spasm. Neurosurg Rev. 2016;39(4):607–13.

17. Li CS. Varied patterns of postoperative course of disappearance of hemifacial spasm after microvascular decompression. Acta Neurochir. 2005;147(6):617–20.

18. Park JS, Kong DS, Lee JA, Park K. Chronologic analysis of symptomatic change following microvascular decompression for hemifacial spasm: value for predicting midterm outcome. Neurosurg Rev. 2008;31(4):413–9.

19. Oh ET, Kim E, Hyun DK, Yoon SH, Park H, Park HC. Time course of symptom disappearance after microvascular decompression for hemifacial spasm. J Korean Neurosurg Soc. 2008;44(4):245–8.

20. Engh JA, Horowitz M, Burkhart L, Chang YF, Kassam A. Repeat microvascular decompression for hemifacial spasm. J Neurol Neurosurg Psychiatry. 2005;76(11):1574–80.

21. Heuser K, Kerty E, Eide PK, Cvancarova M, Dietrichs E. Microvascular decompression for hemifacial spasm: postoperative neurologic follow-up and evaluation of life quality. Eur J Neurol. 2007;14(3):335–40.

22. Lv MY, Deng SL, Long XF, Liu ZL. Long-term outcome of microvascular decompression for hemifacial spasm. Br J Neurosurg. 2017;13:1–5.

23. Shah A, Horowitz M. Persistent hemifacial spasm after microvascular decompression: a risk assessment model. Br J Neurosurg. 2016;1:1–9.

24. Lee JA, Kim KH, Kong DS, Lee S, Park SK, Park K. Algorithm to predict the outcome of microvascular decompression for hemifacial spasm: a data-mining analysis using a decision tree. World Neurosurg. 2019;125:e797–806. https://doi.org/10.1016/j.wneu.2019.01.172.

25. Cheng J, Fang Y, Zhang H, Lei D, Wu W, You C, et al. Quantitative study of posterior fossa crowdedness in hemifacial spasm. World Neurosurg. 2015;84(4):920–6.

26. Jin Y, Zhao C, Su S, Zhang X, Qiu Y, Jiang J. Residual hemifacial spasm after microvascular decompression: prognostic factors with emphasis on preoperative psychological state. Neurosurg Rev. 2015;38(3):567–72.

27. Zhao H, Li GF, Zhang X, Tang YD, Zhou P, Zhu J, Li ST. Long-term efficacy of initial microvascular decompression versus subsequent microvascular decompression for idiopathic hemifacial spasm. World Neurosurg. 2018;109:e778–82. https://doi.org/10.1016/j.wneu.2017.10.079.

28. Lee JA, Jo KW, Kong DS, Park K. Using the new clinical grading scale for quantification of the severity of hemifacial spasm: correlations with a quality of life scale. Stereotact Funct Neurosurg. 2012;90:16–9.

29. Jo KW, Kong DS, Park K. Microvascular decompression for hemifacial spasm: long-term outcome and prognostic factors, with emphasis on delayed cure. Neurosurg Rev. 2013;36:297–301.

30. Kim HR, Rhee DJ, Kong DS, Park K. Prognostic factors of hemifacial spasm after microvascular decompression. J Korean Neurosurg Soc. 2009;45(6):336–40.

31. Lee JA, Park K. Short-term versus long-term outcomes of microvascular decompression for hemifacial spasm. Acta Neurochir. 2019;161(10):2027–33.

32. Kim JP, Chung JC, Chang WS, Chung SS, Chang JW. Outcomes of surgical treatment for hemifacial spasm associated with the vertebral artery: severity of compression, indentation, and color change. Acta Neurochir. 2012;154(3):501–8.

33. Montava M, Rossi V, CurtoFais CL, Mancini J, Lavieille JP. Long-term surgical results in microvas-

cular decompression for hemifacial spasm: efficacy, morbidity and quality of life. Acta Otorhinolaryngol Ital. 2016;36:220–7.

34. Toda H, Iwasaki K, Yoshimoto N, Miki Y, Hashikata H, Goto M, Nishida N. Bridging veins and veins of the brainstem in microvascular decompression surgery for trigeminal neuralgia and hemifacial spasm. Neurosurg Focus. 2018;45:E2.

35. Wang X, Thirumala PD, Shah A, Gardner P, Habeych M, Crammond D, Balzer J, Burkhart L, Horowitz M. The role of vein in microvascular decompression for hemifacial spasm: a clinical analysis of 15 cases. Neurol Res. 2013;35:389–94.

36. Lee JM, Park HR, Choi YD, Kim SM, Jeon B, Kim HJ, Kim DG, Paek SH. Delayed facial palsy after microvascular decompression for hemifacial spasm: friend or foe? J Neurosurg. 2018;129:299–307.

37. Liu LX, Zhang CW, Ren PW, Xiang SW, Xu D, Xie XD, Zhang H. Prognosis research of delayed facial palsy after microvascular decompression for hemifacial spasm. Acta Neurochir. 2016;158:379–85.

38. Goto Y, Matsushima T, Natori Y, Inamura T, Tobimatsu S. Delayed effects of the microvascular decompression on hemifacial spasm: a retrospective study of 131 consecutive operated cases. Neurol Res. 2002;24(3):296–300.

Redo Surgery for Failed Microvascular Decompression for Hemifacial Spasm

Seunghoon Lee and Kwan Park

Failed Microvascular Decompression and Patient Selection for Redo Surgery

Sindou [1] reported successful results after microvascular decompression (MVD) in 87% of hemifacial spasm (HFS) patients. The recurrence rate was 2.4% and about 50% of patients who experience recurrent spasm proceeded to redo surgery from a meta-analysis comprising 5685 HFS patients from 22 reports [2]. What we can understand from these figures are (1) approximate 10% of the patients could not be spasm-free after MVD, (2) recurrence rate after MVD has been underreported so far, and the number of redo MVDs is much more limited. Therefore, we need to acknowledge that many patients after failed MVD are still suffering from spasm and a redo MVD can be considered for them.

Variable clinical responses are encountered after MVD and sometimes doctors as well as patients are confused about whether HFS recurs. For now, there is no established consensus on the observation period of variable responses after surgery or the timing of redo MVD. Sindou [1] and Hatayama [3] recommended at least 1 year of observation because no difference was observed in the MVD success rate between studies with a follow-up duration of less than 1 year and those with a follow-up duration of more than 1 year. Jiang [4] even advocated 2 years of observation before considering redo MVD. On the other hand, several other groups prefer early redo MVD. Engh [5] reported less favorable results in patients with late redo surgery and Zhong [6] showed a delayed spasm relief rate of <5% and recommended reoperation as early as possible. Liu [7] also compared results of early and late redo MVD and demonstrated the superiority of the early redo surgery. Our previously published data showed that 17.1% of patients with residual spasms at 1 year after MVD experienced gradual improvement, mostly within 3 years of MVD [8]. Therefore, the observation period required after MVD at our institution is at least 1 year and up to 3 years if the initial surgery was considered sufficient, with an indentation or discoloration of the facial nerve fully assessed, the offending vessel completely mobilized, and the facial nerve decompressed in the prior surgery. Referred patients, who had undergone surgery and usually spent several months postoperatively at other hospitals, were also followed up for at least 1 year and up to 3 years. However, earlier redo MVD was recommended in patients who did not undergo thorough exploration, had insufficient

S. Lee (✉)
Department of Neurosurgery, Samsung Medical Center, Sungkyunkwan University School of Medicine, Seoul, Republic of Korea
e-mail: shben.lee@samsung.com

K. Park
Department of Neurosurgery, Konkuk University Medical Center, Seoul, Korea (Republic of)
e-mail: kwanpark@skku.edu

© Springer Nature Singapore Pte Ltd. 2020
K. Park, J. S. Park (eds.), *Hemifacial Spasm*, https://doi.org/10.1007/978-981-15-5417-9_17

decompression during initial surgery, had no period of spasm relief or experienced worse spasm after the initial surgery, or had an evident neurovascular conflict revealed on MRI after the initial surgery.

However, there are many factors that make a decision to be reluctant in considering redo MVD. First of all, patient may not want to experience painful recovery time after surgery all over again. Many patients often complain of headache, dizziness, nausea, or vomiting at least couple of days after MVD. Neurosurgeons may be reluctant of any redo surgery, even more in surgery like MVD involving vulnerable cranial nerves (CNs) that are tangled with Teflon felt. Higher complication rates such as facial palsy or hearing loss can be easily anticipated. And magnetic resonance image (MRI) or electrophysiologic studies can hardly unveil the possible cause of failed MVD. Moreover, patient–doctor relationship has been somewhat compromised after the recurrence.

Intraoperative Considerations in Redo MVD: Surgical Guidance and Possible Causes of MVD Failure

Redo surgery is performed via the same route as the initial MVD. The cranial opening is usually larger than in the initial surgery, extending to the edge of the sigmoid sinus laterally and in the cephalic and caudal directions, too. Continuous intraoperative monitoring of brainstem auditory evoked potential (BAEP) and facial electromyography are performed to prevent CN dysfunction and to check abnormal muscle response (AMR). Due to postoperative adhesions, cautious arachnoid dissection should be guaranteed before cerebellar retraction to release the nerves and vessels from one another. An infrafloccular approach is used as usual to visualize the root exit zone (REZ) from the inferior aspects of the CN 7th–8th complex; however, if exploration is not available due to adhesions, another corridor is chosen from above the CN 7th–8th complex. The Teflon felt is withdrawn as much as needed to thoroughly inspect the neurovascular compression site, and we prefer not to remove all the Teflon felt. Because

we have not experienced Teflon granuloma, and complete removal of the Teflon felt may injure surrounding tissue. The neurovascular compression site is explored from REZ to internal auditory canal (IAC) until the offending vessel that triggered the spasm, even a small arteriole or vein, is detected. A 360° view of the facial nerve, including the medial and lateral sides, should be inspected. When exploration is performed near the IAC, the labyrinthine artery should be preserved; iatrogenic injury and vasospasm often occur. A 360° and whole-segment decompression of the facial nerve is performed during redo MVD. Disappearance of AMR is monitored during surgery to determine appropriate decompression of the culprit vessel. The change of BAEPs is also monitored to determine the degree of surgical manipulation, such as cerebellar retraction. If the BAEP changes remain abnormal or a vasospasm occurs at the exposed vessel, intravenous steroids or topical papaverine irrigation can be applied.

Regarding the causes of MVD failure in HFS patients, Zhong [6] postulated that insufficient decompression and a missed offending vessel located distal to the REZ can be the reason. Some reports have identified the growth of Teflon granuloma as the cause of recurrence [9, 10]. Furthermore, de novo artery compression may cause recurrent HFS [11]. We published our evaluation on causes of initial MVD failures based on redo MVD cases [12]. Possible causes can be classified into three categories. A neglected offending vessel refers to a causative vessel that was overlooked and not decompressed during the initial MVD, even though there are traces of attempted exploration of the REZ and decompression of other nearby vessels. Insufficient decompression results from surgical difficulties despite locating the offending vessel. Offending vessels with multiple perforators, encircling the facial nerve, or located deep in the brainstem were the major causes of surgical difficulty and resulted in insufficient decompression. An untouched neurovascular compression site is an area with no traces of Teflon felt or arachnoid adhesion after previous surgical attempts near the REZ or neurovascular compression site; instead, Teflon felt was found at a location far from the REZ. And noticeable findings were as follows; a

Table 1 Possible causes of initial failure after exploration during redo MVD

Cause	Description

Neglected offending vessel
  Medially located
  artery or vein
    at the REZ
    at the cisternal
    segment
	Traces of prior MVD at the REZ or
	Overlooked causative vessel at unusual location

Insufficient decompression	Traces of causative vessel manipulation Surgical difficulty
Untouched NVC site	No traces of prior attempts near the REZ or NVC sites

MVD microvascular decompression, *REZ* root exit zone, *NVC* neurovascular compression

vein was found to be the causative offending vessel, and cisternal segment or medial side of the facial nerve could be the location of compression. This is why a 360° and whole-segment inspection and decompression of the facial nerve are performed during redo MVD. Both a neglected offending vessel and insufficient decompression can cause MVD failure even in the hands of an experienced neurosurgeon, owing to the unusual location and surgical difficulty, whereas an untouched NVC site may be the result of inexperience (Table 1).

Clinical Outcomes of Redo MVD

In previous report by Jannetta [13], the spasm-free rate was inferior to the initial MVD at the postoperative 1-year follow-up. However, long-term spasm-free rate after redo MVD is similar to the rate after initial MVD, and complications such as hearing loss and immediate facial palsy occur more frequently after redo MVD than after initial MVD. Wang et al. [14] showed no significant difference in the relief rate (85.0 vs. 92.6%) and in the incidence of complications between repeat and first MVD. Bigder et al. [15] reported 91.7% of complete spasm resolution after reoperation for failed MVD without increasing complication rates. And most recently, three different Chinese groups reported their good outcomes after reoperations, ranging from 92.3 to 100%. However,

their complications including facial palsy or hearing loss were higher than initial MVD, ranging from 7.7 to 24.1% [4, 16, 17]. From our registry, spasm-free rates of redo MVD were 81.0% at immediate post-redo MVD, 81.0% at 1 month, and 80.5% at 1 year. At the last follow-up, 90.5% were spasm-free. Hearing loss and facial palsy were observed in 14.3% and 38.1% patients, respectively, and were persistent in 9.5% and 9.5% of these patients, respectively, at the last follow-up. Two patients with permanent hearing loss remained in unserviceable status. Two patients who presented with House–Brackmann (H–B) grade IV facial palsy after MVD still had H–B grade III palsy at the last follow-up. In one patient, cerebellar infarction due to an injury to the cortical branch of the posterior inferior cerebellar artery during dura opening was noticed on routine postoperative CT. One patient developed hoarseness postoperatively, and vocal cord palsy was noted but improved without any further management. There was no infection, hemorrhage, or mortality [12].

Feasibility of Redo MVD and Future Directions of MVD

From our experiences, spasm-free rates of redo MVD at the follow-ups earlier than postoperative 1 month showed higher than those of initial MVD due to the higher occurrence of postoperative facial palsy and masking residual or improving spasm. As facial palsy improves at around 1 year after redo surgery, the spasm-free rate is decreasing, whereas the spasm-free rate after the initial MVD is increasing. And thereby, at the last follow-up, spasm-free rates in redo MVD is comparable to those after the initial MVD. Therefore, it can be concluded that the long-term spasm-free rate after redo MVD is similar to that after initial MVD despite the different course of clinical symptoms.

In terms of complications, more frequent changes in BAEPs or permanent hearing loss and more facial palsy indicate an inherent higher risk in redo MVD. Although the frequent changes in BAEPs do not necessarily lead to permanent hearing loss, there is a high risk of nerve damage. Further manipulation of the facial nerve cannot be avoided and causes facial nerve damage in patients with an offending vessel located medial to or at the cisternal segment of the facial nerve. Moreover, overall hearing loss and immediate facial palsy occur more frequently after redo MVD than after initial MVD in our series as well as in other groups. Other severe complications, such as vascular complications or infection, can occur, although those do not develop significantly higher in redo MVD.

Based on prior studies including ours, redo MVD in HFS is a feasible treatment option for failed MVD in that spasm-free rates are comparable to those after the initial MVD. However, it is associated with higher risks of cranial nerve and vascular injuries. With the help of initial surgical records, images, and electrophysiologic data, patient selection as well as redo MVD surgery should be performed deliberately and carefully.

References

1. Sindou MP. Microvascular decompression for primary hemifacial spasm. Importance of intraoperative neurophysiological monitoring. Acta Neurochir (Wien). 2005;147:1019–26.
2. Miller LE, Miller VM. Safety and effectiveness of microvascular decompression for treatment of hemifacial spasm: a systematic review. Br J Neurosurg. 2012;26:438–44.
3. Hatayama T, Kono T, Harada Y, Yamashita K, Utsunomiya T, Hayashi M, et al. Indications and timings of re-operation for residual or recurrent hemifacial spasm after microvascular decompression: personal experience and literature review. Neurol Med Chir (Tokyo). 2015;55:663–8.
4. Jiang C, Xu W, Dai Y, Lu T, Jin W, Liang W. Failed microvascular decompression surgery for hemifacial spasm: a retrospective clinical study of reoperations. Acta Neurochir. 2017;159:259–63.
5. Engh JA, Horowitz M, Burkhart L, Chang YF, Kassam A. Repeat microvascular decompression for hemifacial spasm. J Neurol Neurosurg Psychiatry. 2005;76:1574–80.
6. Zhong J, Zhu J, Li ST, Li XY, Wang XH, Yang M, et al. An analysis of failed microvascular decompression in patients with hemifacial spasm: focused on the early reoperative findings. Acta Neurochir. 2010;152:2119–23.
7. Liu MX, Xia L, Zhong J, Li B, Dou NN, Li ST. What should we do for those hemifacial spasm patients without efficacy following microvascular decompression: expectation of delayed relief or early reoperation? World Neurosurg. 2018;110:e897–900.
8. Jo KW, Kong DS, Park K. Microvascular decompression for hemifacial spasm: long-term outcome and prognostic factors, with emphasis on delayed cure. Neurosurg Rev. 2013;36:297–301.
9. Deep NL, Graffeo CS, Copeland WR 3rd, Link MJ, Atkinson JL, Neff BA, et al. Teflon granulomas mimicking cerebellopontine angle tumors following microvascular decompression. Laryngoscope. 2017;127:715–9.
10. Dou NN, Zhong J, Liu MX, Xia L, Sun H, Li B, et al. Teflon might be a factor accounting for a failed microvascular decompression in hemifacial spasm: a technical note. Stereotact Funct Neurosurg. 2016;94:154–8.
11. Inoue H, Kondo A, Shimano H, Yasuda S, Murao K. Reappearance of cranial nerve dysfunction symptoms caused by new artery compression more than 20 years after initially successful microvascular decompression: report of two cases. Neurol Med Chir (Tokyo). 2016;56:77–80.
12. Lee S, Park SK, Lee JA, Joo BE, Park K. Missed culprits in failed microvascular decompression surgery for hemifacial spasm and clinical outcomes of redo surgery. World Neurosurg. 2019;129:e627–33.
13. McLaughlin MR, Jannetta PJ, Clyde BL, Subach BR, Comey CH, Resnick DK. Microvascular decompression of cranial nerves: lessons learned after 4400 operations. J Neurosurg. 1999;90:1–8.
14. Wang X, Thirumala PD, Shah A, Gardner P, Habeych M, Crammond D, et al. Microvascular decompression for hemifacial spasm: focus on late reoperation. Neurosurg Rev. 2013;36:637–43.
15. Bigder MG, Kaufmann AM. Failed microvascular decompression surgery for hemifacial spasm due to persistent neurovascular compression: an analysis of reoperations. J Neurosurg. 2016;124:90–5.
16. Xu XL, Zhen XK, Yuan Y, Liu HJ, Liu J, Xu J, et al. Long-term outcome of repeat microvascular decompression for hemifacial spasm. World Neurosurg. 2018;110:e989–97.
17. Zhao H, Zhang X, Tang YD, Zhu J, Ying TT, Yan Y, et al. Factors promoting a good outcome in a second microvascular decompression operation when hemifacial spasm is not relieved after the initial operation. World Neurosurg. 2017;98:872.e811–9.

Printed in the United States
by Baker & Taylor Publisher Services

Undergraduate Lecture Notes in Physics

Undergraduate Lecture Notes in Physics (ULNP) publishes authoritative texts covering topics throughout pure and applied physics. Each title in the series is suitable as a basis for undergraduate instruction, typically containing practice problems, worked examples, chapter summaries, and suggestions for further reading.

ULNP titles must provide at least one of the following:

- An exceptionally clear and concise treatment of a standard undergraduate subject.
- A solid undergraduate-level introduction to a graduate, advanced, or non-standard subject.
- A novel perspective or an unusual approach to teaching a subject.

ULNP especially encourages new, original, and idiosyncratic approaches to physics teaching at the undergraduate level.

The purpose of ULNP is to provide intriguing, absorbing books that will continue to be the reader's preferred reference throughout their academic career.

Series editors

Neil Ashby
University of Colorado, Boulder, CO, USA

William Brantley
Department of Physics, Furman University, Greenville, SC, USA

Matthew Deady
Physics Program, Bard College, Annandale-on-Hudson, NY, USA

Michael Fowler
Department of Physics, University of Virginia, Charlottesville, VA, USA

Morten Hjorth-Jensen
Department of Physics, University of Oslo, Oslo, Norway

Michael Inglis
SUNY Suffolk County Community College, Long Island, NY, USA

Heinz Klose
Humboldt University, Oldenburg, Niedersachsen, Germany

Helmy Sherif
Department of Physics, University of Alberta, Edmonton, AB, Canada

More information about this series at http://www.springer.com/series/8917

Amitabha Ghosh

Conceptual Evolution
of Newtonian and Relativistic
Mechanics

 Springer

Amitabha Ghosh
Department of Aerospace Engineering
    and Applied Mechanics
IIEST Shibpur
Howrah, West Bengal
India

ISSN 2192-4791   ISSN 2192-4805   (electronic)
Undergraduate Lecture Notes in Physics
ISBN 978-981-13-4840-2   ISBN 978-981-10-6253-7   (eBook)
https://doi.org/10.1007/978-981-10-6253-7

This Springer imprint is published by Springer Nature
The registered company is Springer Nature Singapore Pte Ltd.
The registered company address is: 152 Beach Road, #21-01/04 Gateway East, Singapore 189721, Singapore

To
dear
Meena

Preface

This book is the outcome of a science elective course offered at Indian Institute of Technology Kanpur by me along with another colleague of mine. The course used to cover both classical and quantum mechanics, but in this volume, only the classical part is being covered and the other part is expected to be penned down by my colleague. However, the motivation to design and offer a course on 'Conceptual Evolution of Mechanics' germinated from an interesting episode in the early life of the author of this volume. It may not be out of place to give a brief account of that here.

I was very fond of egg curry from my childhood, but my mother never gave me more than one egg at a time fearing it could cause stomach problems for me. As a young boy, I used to think that when I grow up and become independent, I would take as many eggs at a time as I pleased. Long after those childhood days, suddenly an opportunity came. I had a combined hand who used to do all cooking and other household chores for me during the last years of the 1960s when I was a young faculty at Bengal Engineering College, Shibpur, Howrah. In December 1970, when I decided to change over to IIT Kanpur, I had sent my combined hand to my native home 260 km away to take household belongings there. My wife was also there as my son was too young to undergo the problems of transferring residence to a faraway city. I was alone, and suddenly it comes to my mind the old desire of consuming as many eggs as I wanted. I had never learned cooking but prepared an egg curry with three eggs following whatever steps came to my mind. To my utter surprise, I could not eat the curry as it tasted horribly awesome. I realized that though I ate and digested (and enjoyed too) egg curry so much during the previous 25 years of my life, I did not know how to cook egg curry. This event gave me a realization that we teach our students only cooked science. As a result, it becomes difficult for them to create new science. From that time, I planned to design a course in which the students of mechanics can become familiar with the evolutionary process through which the science of motion developed and achieved maturity. Mechanics being a very basic subject and fundamental to many branches of physical science and engineering, I considered this subject to be the most suited for my experiment.

It should be noted that the 'conceptual evolution' is somewhat different from the 'history'. There are excellent books on history of mechanics. It is also not a textbook on mechanics. I have tried to emphasize the process through which the basic concepts evolved, transformed and led to the consolidation of the scientific principles involved. The course at IIT Kanpur was offered with a hope to give the students some taste of the process through which science is created. It was hoped that the effort would be somewhat useful in enabling the students to create new science when the occasion arises. As a secondary outcome, the course helped to remove many incorrect impressions about some major scientific discoveries in the field of mechanics.

Quite naturally, the major emphasis has been given on the development of Newtonian mechanics as that is considered as one of the starting points of modern science. The chapters on relativistic mechanics are much shorter as the evolutionary processes for the two theories were confined to a much smaller extent of 'space–time', to use a relativity terminology. The period of their development was only a couple of decades and involved a much smaller group, Einstein occupying the predominant place.

While leaving IIT Kanpur, I was requested by many of my colleagues to compile a book using the material used by me in the part of the course I developed for the use by younger faculty members desirous of offering similar courses. Realizing the desirability of their suggestion, I planned this book. It is only the students of the subject can decide if I have been successful (at least partially) in my original endeavour. If this book is used for offering similar courses, it will be my greatest satisfaction. It goes without saying that there are many short coming and mistakes in this book and I will remain perpetually grateful for suggestions and corrections.

It has taken a long time in writing this book, and I gratefully acknowledge the help and suggestions I received from my students and faculty colleagues of IIT Kanpur. Professors Ashok Kumar Mallik, Pinaki Guptabhaya, Raminder Singh, H.S. Mani and Late Himanshu Hatwal are the most prominent among them. I also gratefully acknowledge the kind help and encouragement received from Professor E.C.G. Sudarshan of University of Texas at Austin. I also acknowledge the help received from the Physics Department of UT Austin by giving me free access to the library. It would not have been possible to complete the book without the active support from my wife Meena who took the whole burden of running the household with negative help from my side. I also gratefully acknowledge the help from my sister-in-law Sabita Ghosh of Asansol for providing me the necessary refuge at her home that gave me free and undisturbed time required to finish the work. The financial support from Indian National Science Academy, New Delhi, and National Academy of Sciences, Allahabad, India, during the period when the book was being written is thankfully acknowledged. The enthusiasm shown by Ms. Swati Mehershi of Springer in publishing the manuscript and the careful typing of the handwritten manuscript by Mr. Sourav Kundu also deserve my sincere thanks.

The preparation of the manuscript had many interruptions, and it has taken almost nine years to complete the book. I will remain grateful to the readers for their suggestions for further improvement of the book.

Shibpur, Howrah, India
April 2017

Amitabha Ghosh
1 Boisakh, 1424 (Bengali New Year)

Contents

1 Evolution of Dynamics 1
  1.1 Early Concepts and Aristotelian Physics 1
      1.1.1 The Earth 2
      1.1.2 Earth–Moon Distance 2
      1.1.3 Measuring the Sun's Distance 2
      1.1.4 Size of the Universe 2
  1.2 Role of Astronomy 4
  1.3 Difficulties in Discovering the Laws of Motion 5
  1.4 Pre-copernican Astronomy 5
      1.4.1 Hipparchus 5
      1.4.2 The Epicycle–Deferent Model and Ptolemy 6
      1.4.3 Problems with Explaining the Observations
            with Ptolemaic Model 10
      1.4.4 Progress During the Period Between Ptolemy and Copernicus 11
  1.5 Copernican Model: Rediscovery of the Heliocentric Theory 12
  1.6 Tycho Brahe: Improvement in Accuracy for Naked-Eye Astronomy 14
  1.7 Kepler: Beginning of Modern Astronomy and Foundation
      of Science of Motion 14
      1.7.1 Discovery of the Laws of Planetary Motion 15
      1.7.2 Transition from Geometric to Physical Model 19
      1.7.3 Early Concept of Action-at-a-Distance and Gravitation,
            and the Concept of Force 19
  1.8 Galileo: Naked Eye to Telescopic Astronomy 20
      1.8.1 Observation of the Moon and Discarding the Concept
            of Fifth Element 20
      1.8.2 Discovery of Jupiter's Moons and its Implications 21
      1.8.3 Discovery of the Phases of Venus: A Further Proof
            of Heliocentric Model 21
  1.9 Galileo: Experimental Mechanics 22
      1.9.1 Early Works on Accelerated Change: Merton School 23
      1.9.2 Galileo's Work on Free Fall and Uniformly
            Accelerated Motion 23
      1.9.3 Discovery of the Law of Inertia of Motion in Its
            Primitive Form 25
      1.9.4 Laws of Compound Motion: Projectiles 25
      1.9.5 Galilean Relativity 27
  1.10 Collapse of the Old Science 27

1.11 Descartes: Beginning of Inertial Science   28
    1.11.1 Law of Inertia of Motion   28
    1.11.2 Collision Problems and Early Concept of Momentum
        Conservation.   29
    1.11.3 Descartes' Concept of Motion.   29
1.12 Huygens: Breakthrough in the Discovery of Dynamics   30
    1.12.1 Theory of Collision and Conservation of Momentum.   30
    1.12.2 Kinematics of Circular Motion and 'Centrifugal Force'   32
    1.12.3 Modern Concept of Force and the 'Force–Acceleration'
        Relation: Second Law of Motion in Primitive Form.   34
    1.12.4 Early Concept of the Principle of Equivalence   35
1.13 Halley, Wren and Hooke: Rudiments of Gravitation.   36
1.14 Newton and the Final Synthesis.   37
    1.14.1 Concepts of Mass, Momentum, Force and the Second Law
        of Motion.   37
    1.14.2 Collision Problem and the Discovery of the Third Law   38
    1.14.3 Law of Universal Gravitation and Planetary Motion.   39
    1.14.4 Universality of Gravitational Force   41
    1.14.5 Orbit for Inverse Square Law   42
1.15 Newtonian Dynamics in Matured State.   44
    1.15.1 Concept of Mass.   45
    1.15.2 Principia and Subsequent Development   45

2 **Some Basic Concepts in Newtonian Mechanics**   47
2.1 Nature of Motion and Space   47
    2.1.1 Newton's Concept of Absolute Space and Time   47
    2.1.2 Newton's Bucket Experiment   48
    2.1.3 Newton's Bucket Experiment Follow up   49
    2.1.4 Ernst Mach and Mach's Principle   50
    2.1.5 Quantification of Mach's Principle—Concept of Inertial
        Induction   50
    2.1.6 Origin of Inertia   52
2.2 Relative–Absolute Duality of Nature of Motion.   53
    2.2.1 The Nature of the Universe   53
    2.2.2 Absolute Motion in Terms of Relative Motion   53
2.3 Inertial and Gravitational Mass   55
    2.3.1 Inertial Mass.   56
    2.3.2 Gravitational Mass.   56
    2.3.3 Equivalence of Inertial and Gravitational Mass   57
2.4 Space–Time and Symmetry in Newtonian Mechanics.   58
2.5 Early Concept of Energy   59
2.6 The Principle of Relativity and Galilean Transformation   60
    2.6.1 The Principle of Relativity   61
    2.6.2 Symmetry and Relativity   62
    2.6.3 Form Invariance of Physical Laws.   63
    2.6.4 Energy and Energy Function.   64
    2.6.5 Energy Function   65
2.7 Laws of Motion and the Properties of Space and Time.   66
    2.7.1 The Second Law of Motion   67
    2.7.2 The Third Law of Motion   68

2.8   Action-at-a-Distance and Spatiotemporal Locality   70
      2.8.1   Early Work on Non-contact Forces   70
      2.8.2   Spatiotemporal Locality and Action-at-a-Distance   71
      2.8.3   The Concept of Field   71
      2.8.4   Field and Absolute Space   74

3   **Post 'Principia' Developments**   75
    3.1   Early Concepts and Aristotelian Physics   75
          3.1.1   Diffusion of Newton's Mechanics in Europe   75
          3.1.2   Multiplicity in the Concept of Force   76
          3.1.3   Degeometrization of Newtonian Mechanics   76
    3.2   Emergence of Analytical Mechanics   78
          3.2.1   New Principles for Dynamical Problems   79
          3.2.2   Principle of Virtual Velocity and Virtual Work   79
          3.2.3   D'Alembert's Principle   81
          3.2.4   Principle of Least Action   82
          3.2.5   Lagrangian Mechanics   83
    3.3   Dynamics of Rigid Bodies   84

4   **Special Theory of Relativity**   87
    4.1   Introduction   87
          4.1.1   Space–Time in Newtonian Mechanics   87
    4.2   Euler's Work on Relativity: Confrontation of Dynamics
          with Optics   88
          4.2.1   Principle of Relativity in Solving Rigid Body Dynamics
                  Problem   88
          4.2.2   Euler's Work on the Problem of Stellar Aberration   88
    4.3   Efforts to Detect Ether Speed: The Null Result of Michelson–Morley
          Experiment   91
          4.3.1   Early Attempts   91
          4.3.2   Michelson–Morley Experiment   92
    4.4   Electromagnetism: Challenge to the Principle of Relativity   93
    4.5   Einstein's Special Theory of Relativity   96
          4.5.1   Lorentz's Transformation from the Two Principles
                  of Relativity   97
          4.5.2   Special Relativity in Electromagnetic Phenomenon   98
          4.5.3   Need for a Relativistic Mechanics 100

5   **General Theory of Relativity and Extension of Mach's Principle** 103
    5.1   Introduction 103
    5.2   Transition to General Relativity 104
          5.2.1   Minkowski's Four-Dimensional Space–Time Continuum 104
          5.2.2   Principle of Equivalence 105
          5.2.3   Freely Falling Frames 106
          5.2.4   Uniformly Accelerating Frames and the 'Entwurf' Theory 109
          5.2.5   The Field Equation and Final Formulation 110
    5.3   Extension of Mach's Principle 111
          5.3.1   Velocity-Dependent Inertial Induction 112
          5.3.2   Some Features of Velocity-Dependent Inertial Induction 114
          5.3.3   Concluding Remarks 115

Bibliography 117

Index 119

About the Author

Amitabha Ghosh completed his bachelor's, master's and doctoral degrees in Mechanical Engineering at Calcutta University in 1962, 1964 and 1969, respectively. After serving as a lecturer in Mechanical Engineering at his alma mater, Bengal Engineering College, Shibpur (now an Institute of National Importance—Indian Institute of Engineering Science and Technology, Shibpur) from 1965 to 1970, Professor Ghosh joined the Indian Institute of Technology Kanpur in January 1971 as an Assistant Professor and served as Professor of Mechanical Engineering there from June 1975 till his retirement in 2006. From 1977 to 1978, Professor Ghosh was Senior Fellow of the Alexander von Humboldt Foundation at the RWTH Aachen and subsequently visited the university with a Humboldt fellowship many more times. He was director of the Indian Institute of Technology Kharagpur from 1997 to 2002. His primary areas of research are manufacturing science, robotics, kinematics and mechanism theory, and dynamics of mechanical systems. Professor Ghosh has written a number of textbooks, which are popular, both in India and abroad. He has guided numerous master's and doctoral students and published a large number of research papers. He received several academic awards, including a number of Calcutta University Gold Medals, D Sc (h.c.), Distinguished Teacher Award from IIT Kanpur and an award for excellence in research from the National Academy of Engineering. He is a fellow of all four national science and engineering academies in India.

Introduction

The Science of Motion

Once the mankind invented agriculture, it became relatively free from the continuous pursuit of food and the struggle for existence became relatively less severe. Agriculture also provided a concept of settlement, and man stopped being a constant wanderer. Once the food was ensured and shelters became permanent, man had some free time to think and observed the surroundings; in other words, man noticed the nature.

However, even with all these revolutionary changes in the history of civilization, most men were engaged in various services for the leaders and the kings, most important being the military activities for protecting the individual kingdoms and clans. Only a very few among them had the leisure to think of other matters; in Greek, 'schol' means 'leisure', these people who had the leisure to think were called 'scholars' and the places where they worked were termed as 'schools'. (Of course, hardly any school-going child today will agree to this definition.)

We should remember that 'science' is basically man's aim to understand natural phenomena. Earlier, the terminology 'natural philosophy' meant 'science'. Even the word 'physis', the origin of the word 'physics', means 'nature' in Greek. One of the major aims for understanding nature is to arrive at some general principles which can make reliable predictions about natural happenings. It must be remembered that such predictions are possible only because nature obeys well-structured general rules. In fact, one of the most important achievements of the Galilean revolution was the realization that 'The Book of Nature' is written in mathematical characters.[1] The key to such rules of Nature are the observational facts. Observations lead to hypotheses, a kind of tentative educated guess, and repeated experiments on a particular phenomenon lead to a theory. Observation is a process which does not manipulate the phenomenon, whereas in experiments, manipulation of certain aspects of nature may be essential.

However, quite a few Greek philosophers believed that one cannot deduce the true nature of the universe by simply observing it. In their opinion, sense may not reveal the true nature; instead, only the use of reason and the insights of human mind can lead to the correct understanding. A similar frame of mind existed even in the medieval Europe when people trusted more the teachings of the masters instead of conducting experiments to verify the truth. This, obviously, was a major reason for wrong theories to continue for a very long time causing a severe hindrance to the progress of science.

The first thing that attracted man's attention while observing his surrounding was change. Day changed into nights, the night sky changed with the progress of night, objects changed

---

[1]Galileo's one famous statement (in his treatise Il Saggiatore, 1624) is as follows: 'Philosophy is written in this very great book which always lies before our eyes but one cannot understand it unless one first learns to understand the language and recognize the characters in which it is written. It is written in mathematical language and the characters are triangles, circles and other geometrical figures; without these means it is humanly impossible to understand a word of it; without these there is only clueless scrabbling around a dark labyrinth'.

positions, water flows, birds flew and fruits fell from the tree. As a matter of fact, movement of objects was the most frequent happening all around. Thus, it is very natural that man wanted to understand the basic natural laws which explained motions, and the 'science of motion' began its long journey at the dawn of civilization. Nowadays, the students are introduced to the subject 'mechanics', i.e. the 'science of motion', at school level. They are trained to use the laws of motion as tools for solving mechanics problems; but, rarely do they get a chance to develop a thorough understanding of the basic essence of 'mechanics'. Since 'mechanics' forms the foundation of the 'physical sciences', a proper understanding of the subject is essential to develop a true scientific temperament. Furthermore, the habit of following the set laws without examining them is reminiscent of the situation prevailing in the Middle Ages. So, it is necessary for the progress of science to keep an open mind, and that requires a thorough and deep understanding of all these laws.

Furthermore, it is very important for a student of science to learn the thought processes, transitory concepts and the gradual conceptual evolution of scientific theories. As one cannot cook even after being fed a good dish repeatedly day after day, he or she can digest it well but will be totally clueless when asked to cook the same dish even if all the basic ingredients are supplied. So to create new science, it is essential to learn how science is created?

One may think that nothing more is there to learn about the laws of motion as the use of these laws has given correct and useful results for a long time. However, one may find it surprising to know that still there are unresolved issues and gaps in our understanding so far as the 'science of motion' is concerned. Furthermore, the 'laws of motion' which took 2000 years of hard work by giant intellectuals to be discovered are not as simple as they appear at a first glance. To develop a solid understanding, one must go through the same mental thought process which slowly led to the evolution of the subject.

Biological evolution is a very slow process and the raw human intelligence has not changed at all during the last 5000 years, though, of course, our knowledge base has expanded tremendously and man has improved his understanding of the nature significantly. So, the enormous time it has taken to unravel the basic aspects of the science of motion indicates how difficult the task must have been. One's understanding of the subject can be made very solid if he or she is taken through the same mental reasoning process which led the scientists and philosophers to arrive at the current level of understanding of the subject. It is essential for one to be familiar with the transitional concepts and the philosophical reasoning. Then only the understanding of the subject can be considered complete. Only then it is possible to make further progress and discover the uncharted territories of the subject.

Evolution of Dynamics

1.1 Early Concepts and Aristotelian Physics

The beginning of scientific thinking started in almost all the ancient civilizations in India, China, Babylon, Egypt and Greece. Though considerable progress was made in the preliminary ideas on natural processes particularly the movement of objects in India and other ancient civilizations, the most systematic and continued progress was made in Greece. The period 600–400 BC is considered to be the pre-Hellenistic period and the prominent philosophers of this period include Pythagoras (572–497 BC), Thales, Anaximander, Anaximenes (all in the sixth century BC), Anaxagoras (∼430 BC), Socrates (470–399 BC) and Plato (472–347 BC). Aristotle (384–322 BC) was Plato's disciple, and he shaped the thinking process in the subject of mechanics for the next one and half millennia. He systematized the subject (along with many other fields) based on the common experience.

It is needless to emphasize that by and large science develops from the attempt to understand the rules behind the observational facts. The bewildering variety of motions all around made it extremely difficult to figure out the fundamental ideas of the science of motion. Since the philosophers were unaware of the earth's gravitational pull, the presence of atmosphere and the buoyancy, friction, etc., the causes behind the motions of objects were considered to be the desire of those to reach their respective natural places. The synthesized scheme of the universe as systematized by Aristotle was inseparably intermingled with the science of motion and formed a complete science. That is one of the reasons why Aristotle's philosophy continued for such a long time. In Aristotle's scheme, there were four elements—earth, water, air and fire. For earth-like material, natural place was the earth's centre which was also considered to be the centre of the universe and stationary. Fire's natural place was in the heaven. Water and air were in between. Thus, a stone (consisting of earth-like matter) falls towards the earth's centre when dropped, and a flame (or smoke) goes up in an attempt to reach its natural place in the heaven. This type of motion was called 'natural motion'. However, it was also a common knowledge that a stone goes upwards when thrown up or an arrow moves as a projectile when shot. Aristotle termed these unnatural motions as 'violent motion'. Such motions could be induced by the act of an external agent like a hand or a bow. This led to the understanding why things move. However, the ancient men (and also the philosophers) were also aware of another type of motion—daily uniform circular motion of all the heavenly objects like the sun, the moon and the stars. Considering the luminosity of the heavenly objects and their unchanging nature for centuries (in contrast to all terrestrial objects which changed with time), Aristotle considered the heavenly bodies to be made of a different material—the 'fifth element'. The objects made of this 'fifth element' were considered to be unchanging, and their natural motion was 'uniform circular motion'. This scheme could successfully explain most of the observed motions at least qualitatively.

Figure 1.1 shows schematically the Aristotelian cosmology. The stable unmoving earth is at the centre of the universe. The outermost level consists of the stars and rotates once every day. The space in between is occupied by the various spheres containing the moon, sun and the planets, all rotating at different rates. It was conceived that these spheres were made of a transparent material. Beyond the sphere of the stars, there was absolutely nothing—not even space and time.

It will be interesting to consider some other scientific knowledge during the early civilizations. This will tell us about the wonderful reasoning power of the early scientists and will also give us a clue to the scientific thought process.

In the early period, the earth was considered to be stationary and at the centre of the universe. Heavenly objects were thought to go round the stationary earth, and the outermost dome was studded with the luminous stars.

© Springer Nature Singapore Pte Ltd. 2018
A. Ghosh, *Conceptual Evolution of Newtonian and Relativistic Mechanics*,
Undergraduate Lecture Notes in Physics, https://doi.org/10.1007/978-981-10-6253-7_1

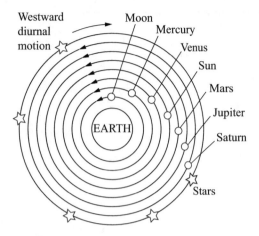

Fig. 1.1 Aristotelian two-sphere universe

1.1.1 The Earth

Although at the very dawn of civilization, the earth's spherical shape was unknown to man, but soon certain observational facts led the philosophers to realize that the earth was a sphere. It was a common knowledge that for approaching ships the first visible part was the mast. This happened irrespective of the directions (i.e. east, west, south or north) which indicated that the earth's surface slops down in all directions. This can happen only if the earth is spherical in shape. Astronomers and philosophers also noticed that when one travels towards the north the southerly stars gradually went below the horizon and new stars appear in the north. Aristotle also argued in favour of the spherical earth indicating the fact that the shape of the earth's shadow on the moon during eclipse is always circular. (It is interesting to note that even in Hellenistic astronomy the phenomenon of eclipse was well understood.) It is not only the shape; the size of the earth was estimated by ancient philosophers. Though there are references to a measurement of the earth's radius in Aristotle's writings, the first recorded details of a measurement is found to be by Eratosthenes, the librarian of the famous library in Alexandria, in the third century BC.

Figure 1.2 shows the procedure schematically. Eratosthenes noticed that when at a place called Seyne, 5000 stade due south of Alexandria, the noon sun did not cast any shadow of a vertical pole. However, at Alexandria, the angle subtended by the sunrays with a vertical pole was about 7.2°, i.e. about 1/50th of a full circle. Since all the sunrays can be considered to be parallel (because of the sun's vast distance), the angle $\angle AOS$ is also equal to about 1/50th of a full circle, implying the arc AS is about 1/50th of the earth's circumference. Thus, Eratosthenes concluded the earth's circumference to be about 250,000 stades, which is reasonably close to the actual value.

1.1.2 Earth–Moon Distance

It is really quite an astonishing fact that the astronomers during the Hellenistic period were aware of the phenomenon of parallax in astronomy. Parallax is the apparent shift in the position of an object in the background of a distant backdrop. Though it was very difficult to measure parallax due to the lack of instantaneous communication between two far off locations on the earth, this effect was used by Hipparchus to estimate the distance of the moon by observing moon at two different instants separated by some hours at the same night. When viewed from two diametrically opposite locations on the earth, the parallax is about 2°. The astronomers also knew that the parallax of the sun is much less than that of the moon and the sun is much further away than the moon (Fig. 1.3).

1.1.3 Measuring the Sun's Distance

Since the parallax in case of the sun is very small and beyond the capacity of the ancient astronomers to estimate it, a very novel scheme was suggested by Aristarchus in the second century BC. (He was among the ancient astronomers who believed in a heliocentric model of the universe).

Figure 1.4 shows the basic idea. He considered the fact that when the moon is seen from the earth exactly half full the angle $\angle EMS$ is 90°. At this instant, it is possible to measure the angular distance of the moon from the sun.[1] As the distance EM could be estimated using the parallax method, the hypotenuse ES of the right angle ΔEMS could be estimated. Aristarchus estimated $\angle MES$ as 87° which led to the result $ES = 19 \, EM$. In reality, the modern measurements show that $\angle MES = 89° \, 51'$ and ES is about 400 EM. Subsequently, Aristarchus developed another extremely innovative scheme using lunar eclipse for determining the sizes of the moon, sun and their distances.

1.1.4 Size of the Universe

The size of the universe was another important matter for the ancient astronomers. Using the astronomical observations and very ingenious logic, they demonstrated that the sphere of the stars was much bigger than the earth.

They noticed that when one equinoctial point sets in the west, the other equinoctial point rises in the east at the same moment. Figure 1.5a shows that if the size of the earth had been comparable with that of the sphere of the stars then the

---

[1] It is possible only on principle. Particularly, it is very difficult to have any accurate estimate of the angle $\angle MES$.

Fig. 1.2 Eratosthenes' scheme for measuring the earth's radius

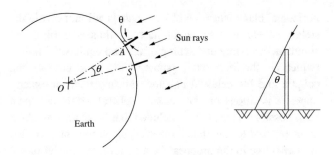

Fig. 1.3 Parallax of the moon

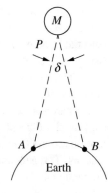

Fig. 1.4 Aristarchus' scheme to estimate the sun's distance

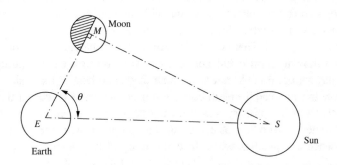

Fig. 1.5 Earth's size relative to the outermost sphere of the universe. **a** In a geocentric universe and **b** in a heliocentric universe

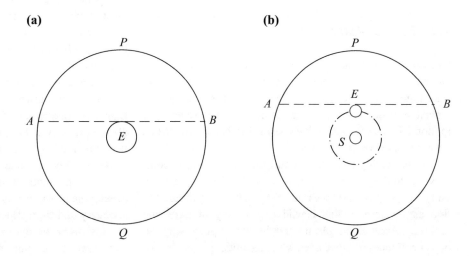

horizontal plane where an observe stands will not divide the stellar sphere is two equal halves. Thus, the observed simultaneous rising and setting of the two equinoctial points (which are the intersection points of two great circles—the ecliptic and the celestial equator—placing them in exactly opposite positions on the celestial sphere) would not have been possible. Thus, they concluded that the earth was like a point relative to the stellar sphere. This argument was also very effective in the proposal for a heliocentric model of the universe at a later time. Figure 1.5b shows the earth in its orbit around the sun. Since the orbit's size is much larger compared to that of the earth, the astronomers suggested that the horizontal plane of any observer cannot divide the celestial sphere in two equal halves. But the simultaneous rising and setting of the equinoctial points indicated otherwise. This reasoning required the celestial sphere to be so large that it was outright rejected.

Aristarchus reasoned against a geocentric model and suggested a diurnal spin rotation of the earth and an orbital motion around the sun. His logic was based on primarily two reasons. The geocentric model requires the stars to move at tremendously large speeds, and the sun is much bigger than the earth. Thus, philosophically, it was unacceptable that such a large object orbits around a much smaller body. He proposed that the observed motions were nothing but due to kinematic relativity. However, the heliocentric model was outright rejected. There was no idea of the law of inertia, and so a moving earth could not be conceived. When a stone is thrown up, it could not have come down and land at the same location had the earth been moving at such large speed —that was the primary reason for rejecting a moving earth hypothesis. Another objection to the proposal was the absence any parallax when the earth occupied two diametrically opposite positions in its orbit around the sun. It was not known then that the stars are immensely distant objects.

1.2 Role of Astronomy

One obvious point arises is that why the clue to the science behind the motion of objects on the earth was linked to astronomy? But the gradual development of astronomy played a major role in the evolution of the 'science of motion'. There are quite a few reasons for that which will be clear as this book progresses.

All the earthly motions get corrupted and become complex due to the presence of gravity, wind resistance, friction, buoyancy, atmospheric pressure, etc. So, it is very difficult to find any pattern in this bewildering variety of terrestrial motions. Aristotelian physics could not explain why an object continues to move even when the initial pushed leaves contact with the object as inertia of motion was not discovered. Aristotle unsuccessfully tried to explain that the

vacuum created by the displacement of a moving object is filled by air and that acts as a pusher. However, he himself knew about the unsatisfactory character of the theory and left the problem. His dissatisfaction was quite obvious as he used to be very argumentative on this subject. In the subsequent centuries, philosophers disproved this theory as they noticed that elongated objects with flattened and sharp tails moved almost identically when projected.

In this scenario, the motion of the heavenly bodies was a relief to the philosophers. These motions were regular, uniform and unchanging. In comparison, most of the earthly motions (like a dropping stone) were of very short duration and inadequate for any kind of scientific study, particularly in the absence of an accurate time measuring instrument. So, the celestial motions were very attractive to ancient philosophers; these motions were slow and regular.

Astronomy became important for another reason also. People thought that heaven is the abode of the gods and the heavenly bodies play a role in governing people and the earth. Such convictions became stronger as it was noticed that it becomes hot when the sun is in Cancer and it becomes cold when it is in Capricorn. The link between the tidal phenomenon with moon's phases and the matching of the menstrual cycle of the women world over with the lunar month consolidated the position of the subject 'astrology' in the society. People started thinking that stellar and planetary configurations can influence man's life, kingdoms' prosperity, results of the wars, etc. As astronomers were generally able to predict eclipses, it was felt that they, as astrologers, can predict future. Thus, astronomy became closely linked with a 'non-science' subject like astrology, and for many years, they formed a single profession. The astronomy which could predict the future stellar configurations was known as 'natural astrology' and that astronomy which predicts the future of men and nations from the stars and planets was known as 'judicial astronomy'. Actually, most of the famous astronomers of the antiquity like Ptolemy and astronomers like Tycho Brahe and Kepler received their bread and butter because they were considered to be good astrologers!

In fact, one of the major motivations behind the growth of astronomy was astrology. This may explain the decline of Indian astronomy during the Buddhist period as in Buddhism astrology did not have a good standing in the society. It was revived after the reintroduction of Hinduism in India.

As will be seen later in the text, prolonged and careful study of the motions of the celestial objects led to the development of certain branches of mathematics, viz. geometry and trigonometry which played important roles in the discovery of dynamics. Apart from this, the gradual development of improved observational techniques of naked-eye astronomy ultimately convinced Kepler that interaction of bodies is responsible for bodies to move.

1.3 Difficulties in Discovering the Laws of Motion

It may appear surprising to any school-going student today that it took almost two millennia to discover the laws of motion. The subject is taught to school students at a reasonably early stage, and in the way, the laws of motion are presented to them in a plate as well-cooked food it is inconceivable for all to recognize the enormous difficulties which blocked the vision and reasoning of the philosophers of the antiquity. This single fact should make the present-day students of science aware how difficult the problems of discovering these laws are? In fact, it is essential for one to grasp the foundational concepts behind these three laws of motion. Thus, a discussion on the difficulties in discovering the laws of motion and to understand why it took so long is desirable.

Most persons familiar with the science of motion have a tendency to feel that the first law of motion, i.e. uniform rectilinear inertial motion, should be obvious to all. On the contrary, this law is most difficult to unravel as there is no uniform rectilinear motion on the earth. All rectilinear motions on earth, like free fall on an object or horizontal motion of an object on a surface, are non-uniform. Projectile and most other violent motions are not rectilinear. Even the eternal uniform motions of the celestial bodies are circular. Thus, the most fundamental type of motions is non-existent. The ancient philosophy believed that a motion has to be either natural (which is a property of every material) or violent. If ice skating had been a popular sport (which was impossible in a Mediterranean country like Greece); perhaps, people could get some clue to this fundamental aspect of dynamics.

Another relatively more difficult aspect did not get unravelled till the time of Galileo. He was the first to recognize and study the composition of motion. Till the idea of decomposing motions into their more elemental components was developed, it was virtually impossible to make any scientific study of motion as all appeared too complex.

The third difficulty which is responsible for this delay in discovering the laws of dynamics was the absence of accurate time measuring clock. We know that 'acceleration' plays the central role in dynamics, but quantitative study of accelerated motion was very difficult. This is because most of terrestrial motions are of relatively short duration and without a good clock are impossible to study. Thus, for a very long time, a wrong notion persisted that impetus (old term that was used to represent something like force) leads to speed. Without the discovery of the role acceleration plays in dynamics, it was impossible to develop the science of motion.

The next difficulty that faced the philosophers of the antiquity (and also of the Middle Ages) was the absence of the concept of any non-contact interactive force between objects. Though electrostatic phenomenon was discovered by the ancient Greeks, it was considered more as a magical effect and had little to do with physical science, i.e. science of motion in those days. In fact, it was about 1800 years after those Hellenistic periods when Kepler realized the influence of sun on the motions of the planets and a primitive idea of gravitation emerged. He too was also enthused along this concept because of Gilberts' book on magnetism published in 1600.

Finally, perhaps most damaging influence was too much dependence on the teachings of the old masters. It was also considered blasphemous to conduct experiments for verifying the teachings of the masters, particularly in the Aristotelian schools of thought. This caused many completely erroneous concepts to continue for centuries. It was taught that heavier bodies fall faster, a projectile attains maximum speed in midflight, etc. It was also taught that when a stone drops from the top of the mast of a moving ship it does not fall at the foot of the mast. Such anti-scientific teachings did a great harm and delayed the process of unravelling the fundamental basic laws of an inertial physics.

In hindsight, it looks that, perhaps, the development of science of motion has taken the inevitable course of destiny. Most probably, one should not expect to have taken a different and faster path. In the subsequent chapters, it will be seen how complex the process of discovering the laws of motion had been. It took more than 2000 years for a large number of giant intellectuals to build this grand edifice of science. It also demonstrates the fact that science progresses in very small steps involving a large number of thinkers, although the limelight, in general, is enjoyed by the person who accumulates the knowledge and presents a synthesized version.

1.4 Pre-copernican Astronomy

It has already been emphasized that astronomy was the most important factor which led man to study motion (of course the kinematic aspect of to begin with). Thus, it is desirable to follow the chain of thoughts and transitional concepts which led to the ultimate unravelling of the laws of motion.

1.4.1 Hipparchus

One of the most important figures in Hellenistic astronomy is, undoubtedly, Hipparchus (~ 150 BC). His visit to Babylon made the prolonged observation-based Babylonian data available to Hipparchus, and as a theoretical astronomer, he could systematize this data into a physical scheme. He used the phenomenon of parallax for estimating the distance to the moon and its size. Apart from the detection of the phenomenon 'precision of the equinox', Hipparchus was the first to attempt developing astronomy as a geometro quantitative discipline of philosophy.

Because of the orbital motion of the earth around the sun, the sun appears to slowly drift across the stationary stellar background and completes one full round in one solar year. Figure 1.6 helps to make the point clear. This path, apparently traversed by the sun, is called the 'ecliptic' (because eclipse occurs whenever the moon is on the plane described by the ecliptic, as then only the sun, the earth and the moon can be in one straight line).

A second great circle can be imagined on the celestial dome by expanding the terrestrial equator; this great circle is called the 'celestial equator' (Fig. 1.6c). The path of the sun (used to be assumed as a perfect circle in Hipparchus time) has four special equispaced points; two of these points, obtained by the intersection of the ecliptic with the celestial equator are called 'equinoxes' (because when the sun is at an equinoctial point, the day and night, i.e. nox, are equal everywhere on the earth). The other two points represent the two extremely northerly and southernly positions of the sun during its journey along the ecliptic. These two points are called 'solstice' as the sun temporarily appears to be standing still at these two positions. The summer solstice coincides with the longest day in the Northern Hemisphere, and the day is shortest at winter solstice. These four special positions of the sun on the ecliptic are equispaced at 90° as shown in Fig. 1.7

According to well-established theory of uniform circular motion of all the heavenly bodies, the sun should take equal time to cross the four quadrants. But the Greek astronomers were surprised to note that the sun takes 94½ to travel from the vernal equinox to the summer solstice position and 92½ days from the summer solstice to autumn equinox. This was absolutely against the prevailing physics of the time as non-uniformity in the natural circular motions of the heavenly objects was beyond any compromise. Hipparchus was the first to forward a geometro kinetic solution to the problem starting the era of quantitative astronomy. His proposed solution is explained in Fig. 1.8.

According to this theory, the sun moves uniformly on the circle with centre O which is displaced from the centre of the earth. The angle $\angle VOS = 93.2°$ so that the sun takes 94½ days to cover arc VS. $\angle SOA = 91.2°$, making the sun's duration in the arc SA equal to 91½ days. The earth's position is such that from the earth the segments VS and SA subtend 90° angles at E. This way the theory of uniform circular motion of the heavenly bodies was rescued to the great relief of the contemporary philosophers and astronomers.

1.4.2 The Epicycle–Deferent Model and Ptolemy

The invention of the epicycle–deferent theory is not clearly known, but it is clear that even Apollonius, about a century before Hipparchus and about four centuries before Ptolemy, was aware of this hypothesis. The need for the development of such geometro kinetic schemes arose primarily to fit the irregularities of the planetary and solar motions which became increasingly prominent as more and more observational data started accumulating. It should be remembered that irregularity meant any discrepancy of the observation with predictions based on uniform circular motion theory. The non-uniformity of the sun's motion was resolved by

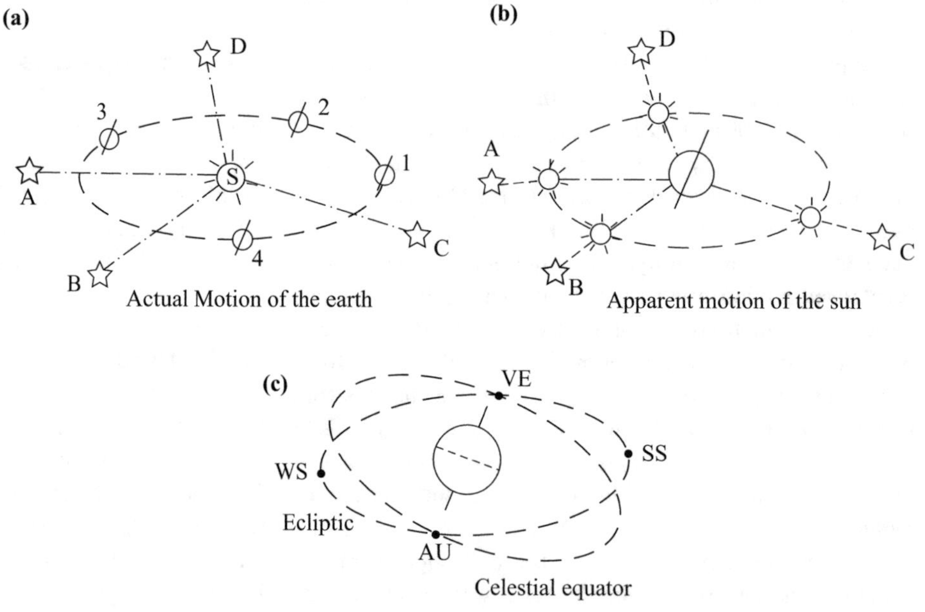

Fig. 1.6 Ecliptic and celestial equator

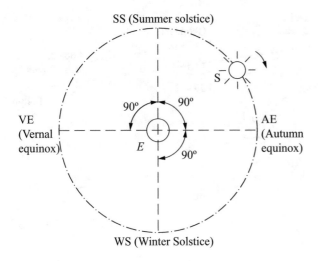

Fig. 1.7 Solstice and equinoctial positions of the sun

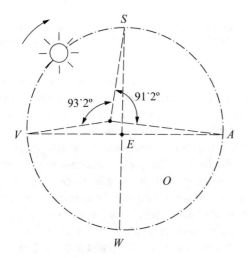

Fig. 1.8 Hipparchus's hypothesis of solar motion

Hipparchus as discussed in the previous section. However, more vexing problems started bothering the astronomers. It was observed that the brightness of the planets varied with time indicating the variation in their distances from the earth. Worst of all, it was found that except the sun and the moon all planets exhibited retrograde motion. A typical retrograde motion of mars is shown in Fig. 1.9.

This posed a severe challenge to the contemporary Aristotelian physics. The deferent–epicycle system was conceived to explain such complex motions using only uniform circular motions.

Figure 1.10 shows the basic idea of the theory. The circle with the earth as the centre is called the 'deferent'. Another smaller circle with its centre on the deferent is called the 'epicycle'. The heavenly body under consideration is P and rotates uniformly about the centre C. At the same time, the point C moves along the deferent at a uniform speed.

Suitable combinations of these two motions can generate complicated paths for the object P; one example is shown in Fig. 1.10b. It is clear from the figure that the object will appear to have a retrograde motion against the starry background from the earth when it is near the location A. It is further seen that at this location the object (a planet) is nearest to the earth and will appear to be brighter, which matches with actual observation. This mechanism is thus capable of explaining both the retrograde motion and varying brightness of the planets.

The deferent–epicycle model was also used as an alternative to Hipparchus's eccentric circle theory to explain the apparent non-uniformity of the sun's motion.

Figure 1.11 shows a deferent with the earth at its centre. The sun is placed on an epicycle with the desired eccentricity as its radius. Now, if the deferent point A moves uniformly and the epicycle does not rotate, then the sun's locus will be another circle with its centre shifted to O'. This can be a geometro kinetic model to explain the non-uniformity of sun's motion.

It is not difficult to demonstrate that observationally a deferent–epicycle system will be indistinguishable from a corresponding heliocentric system. Let P be a planet (for convenience, let it be an outer planet) orbiting the sun, S, along its circular path 1 (Fig. 1.12).

Let E be the position of the earth at the instant. Now, let it be assumed that the planet P is on an epicycle 2 of radius r (the radius of the earth's orbit) whose centre is at D on a deferent 3 with the earth as the centre and R (the radius of the planet's orbit around the sun). When the rotational speed of the planet in its epicycle is same as that of the earth's orbital motion and speed of point D along the deferent 3 around E be the same as the orbital speed of the planet around the sun, then observationally the result will be same to an astronomer on the earth. That is why the deferent–epicycle system provided a good computational instrument to the astronomers to both understand and explain the gross features of planetary motions. The position of P in the helioastral plane due to its rotation about the sun with an angular velocity will be identical with that if P is assumed to rotate in the epicycle with angular velocity ω the deferent point orbits the earth with angular velocity Ω and the earth rotates about the sun with an angular speed ω. The radius of the epicycle is equal to the radius of the earth's orbit, and the radius of the deferent is equal to the orbital radius of the planet P.

As more observational data accumulated, the astronomers needed to continuously adjust their Ptolemaic model. The most convenient method was to add epicycles over epicycles as shown in Fig. 1.13.

By adding more and more epicycles and adjusting the sizes and tinkering with the angular speeds, it was possible to fit any observational data and make reasonably good

Fig. 1.9 A typical retrograde motion of Mars

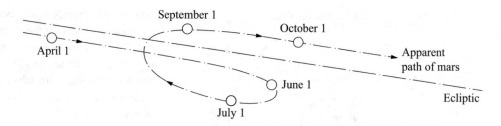

Fig. 1.10 Deferent–epicycle system

(a)   **(b)**

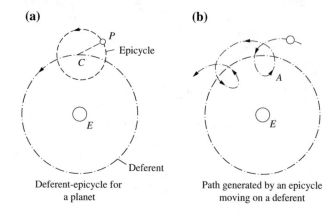

Deferent-epicycle for   Path generated by an epicycle
a planet   moving on a deferent

predictions. Astronomers after Ptolemy used multiple epicycles sometimes numbering even a dozen minor epicycle to fit the observation of a single planet taking care of all the irregularities in its motion.

Hipparchus used the concept of an eccentric to explain the irregularities of the motion of the sun (shown in Fig. 1.14a). Astronomers did not stop here. They sometimes used an eccentric whose geometric centre O (Fig. 1.14b) moves on a deferent with the earth as the centre. Sometimes, a scheme of using a second eccentric for the motion of the original eccentric O (Fig. 1.14c) was also proposed. Such combinations of epicycles and eccentrics lead to immensely complicated geometro kinetic models which could explain

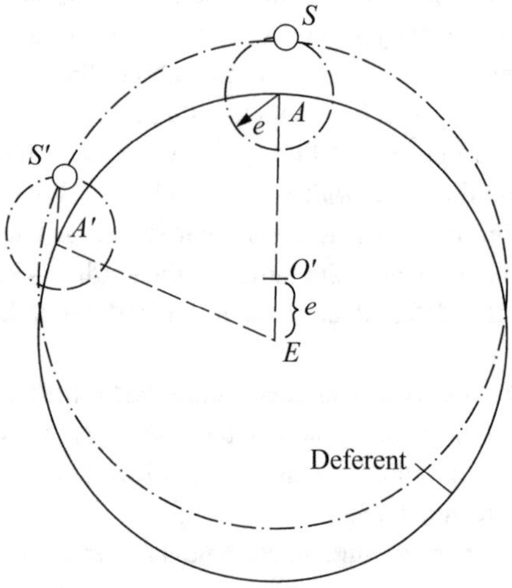

most of the irregularities in the motions of the heavenly objects. Ptolemy introduced a new concept in these approaches. He suggested that the irregular motion can be explained with the help of an 'equant'.

Figure 1.15a shows the circular orbit of a planet with the earth as the centre. The planet describes the circular path with a constant angular speed with respect to not its geometric centre where the earth is located but with respect to another point Q which was termed as the 'equant'. This new concept added another weapon to the arsenal of the theoretical astronomers. It is not difficult to realize why the equant concept yielded nice results.

Figure 1.15b shows the orbit of a planet around the sun. As was always assumed in the past, the orbit is circular and the sun is placed a little away from the centre O. Let SO be equal to e. The equant point E is also at a distance e from the centre O in the opposite direction on the line of apsides.

When θ is very small ($d\theta$), i.e. when the planet is near A, the following relation can be written

$$\frac{1}{2}(r+e) \times PA = \text{area swept by } PS \text{ in time d}t \text{ starting from } A$$
$$= dA$$

$$\therefore dA = \frac{1}{2}(r+e)(PA) \approx \frac{1}{2}(r+e) \times (r-e)d\theta$$

or

$$\frac{dA}{dt} = \frac{1}{2}(r^2 - e^2)\frac{d\theta}{dt} = \frac{r^2}{2} \times \omega \times \left(1 - \frac{e^2}{r^2}\right)$$

Fig. 1.11 Deferent–epicycle alternative model for sun's motion

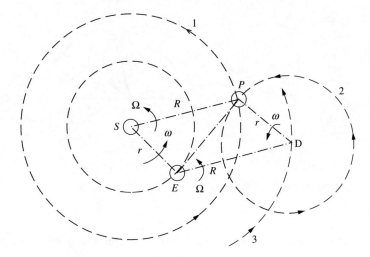

Fig. 1.12 Equivalence of heliocentric and geocentric deferent–epicycle model

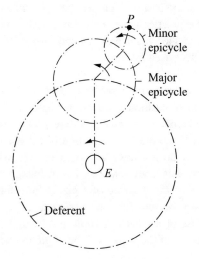

Fig. 1.13 Ptolemaic system with compounded epicycles

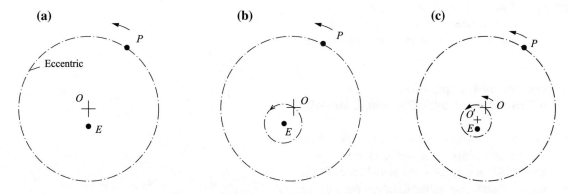

Fig. 1.14 **a** Eccentric, **b** eccentric on a deferent, **c** eccentric on an eccentric

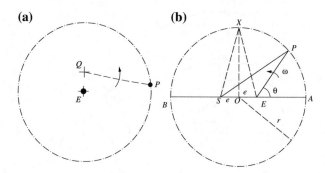

Fig. 1.15 **a** Concept of equant, **b** area law and concept of equant

Similarly at B (considering B to be the starting point and measuring θ from line OB)

$$dA = \frac{1}{2}(r - e) \times (r + e)d\theta$$

and

$$\frac{dA}{dt} = \frac{r^2}{2}\omega\left(1 - \frac{e^2}{r^2}\right)$$

near X it can be proved that

$$\frac{dA}{dt} = \frac{r^2}{2}\omega\left(1 + \frac{e^2}{r^2}\right)$$

Thus, when $\frac{e^2}{r^2} \ll 1$, the rate of area sweep remains constant if ω remains constant (the concept behind the equant).

The fixed idea in the minds of the people that heavenly objects are made of the fifth element whose natural motion was uniform circular motion as taught by the masters was the real culprit behind all these seemingly complex arrangements. It clearly shows how the Aristotelian mechanics shaped the astronomy in the ancient times. It is not difficult to imagine how happily an astronomer will retire after a lifetime's hard work resulting in the addition of one minor epicycle! Unfortunately, for them, there existed not many attractive prizes or awards for such accomplishments.

1.4.3 Problems with Explaining the Observations with Ptolemaic Model

As mentioned in the previous section, the Ptolemaic model could explain the observational results in a broad sense; using a number of minor epicycles, the model could fit into the increasingly complex observational data as the accuracy of naked-eye astronomy improved and the volume of accumulated observational data increased. However, the role of

these epicycles was nothing but introducing free adjustable parameters in a theory (as is seen to be not too uncommon a practice even in modern science).

However, even the ad hoc introduction of epicycles could not resolve some basic difficulties with the Ptolemaic model, and as is the practice even these days, such problems were pushed under the carpet and ignored. Some of these difficulties are discussed below:

(a) It was observed that a planet looked brightest whenever it was in opposition to the sun. Figure 1.16 shows the relative location of a planet with respect to the earth and the sun. Since the orbital motion of the sun and that of the planet do not have any rigid connection, such observation was unexplainable. It was recognized that the variation of the distance of a planet from the earth was the main reason for the variation of the apparent brightness of a planet. It could not be understood why the distance of the planet will be minimum when the sun

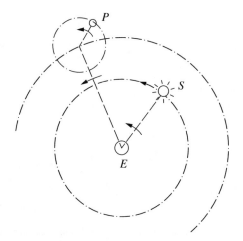

Fig. 1.16 Relative position of a planet with the earth and sun

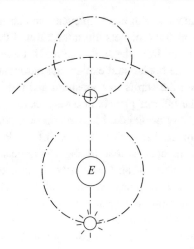

Fig. 1.17 A planet's nearest position to earth with the sun in opposition

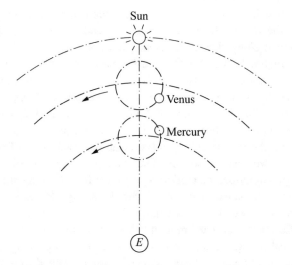

Fig. 1.18 Deferents of the sun, mercury and venus in the modified Ptolemaic model

was in opposition. So, why a configuration as shown in Fig. 1.17 is assured could not be understood with the help of a Ptolemaic model.

(b) It was also noticed that the planets mercury and venus always remain within a small angular distance from the sun. In the technical language of astronomy, the elongation of the inferior planets was limited. This difficulty was attempted to be resolved by another ad hoc assumption which tied the deferent of the sun, mercury and venus together as shown in Fig. 1.18. This additional constraint did put further strain on the credibility of the geometric model. The original attractive model proposed by Hipparchus lost most of its elegance by then.

(c) Another very uncomfortable feature of the Ptolemaic system was that all such gigantic bodies were moving around a void point which did not coincide with the earth's centre. It had no physical explanation.

1.4.4 Progress During the Period Between Ptolemy and Copernicus

Contrary to the common belief that no progress was made during the period between Ptolemy and Copernicus, some significant activities took place that had a reasonable degree of influence on the discovery of the inertial dynamics. The gradual decline of the Greek science and astronomy followed by the invasion from the Islamic countries in Asia led to the disappearance of the scientific activities in Europe. Some of the activities were forced to shift from the Mediterranean Europe to the northern parts of the continent. Though the old wisdom in the form of Aristotelian physics and Ptolemaic astronomy was lost form Europe, the Islamic civilization preserved a major part of that. During the later period of the Middle Ages, a resurgent Europe retrieved most of their old science and astronomy from the Arabs during the twelfth century. Medieval European scholars forged a number of concepts and sowed the seeds of many doubts about the Aristotelian science that helped the Copernican revolution and the emergence of modern science.

Jean Buridan (1295–1358) was one of most reputed scientist and was also the Rector of the Paris University who played a major role in introducing some new concepts in the science of motion. He developed a theory of impetus according to which if a body was given an impetus it could continue to move when not resisted by air or friction. It was against the Aristotelian mechanics which required a continued effort for a body to move. Buridan conceptualized the impetus of a body as the product of its speed and its quantity of matter. This may be considered to be the seed from which the matured concept of momentum was developed by Descartes and Huygens about three centuries later. Buridan also suggested that during free fall the weight of a body impresses equal increments of impetus upon the body in equal intervals of time. It was the forerunner of Galileo's theory that speed increases linearly with time during free fall. Buridan's student Nicole Oresme (1320–1382) argued in favour of a spinning earth that resulted in the apparent daily motion of the heavens.

Studies on motion were also conducted by Oresme in Paris, John of Holland and the scientists at Merton College, Oxford, more or less contemporarily. The concept of

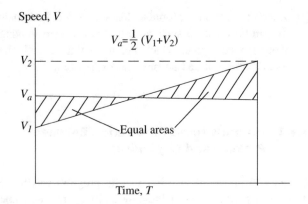

Fig. 1.19 Oresmes' geometrical proof of the Merton College rule

uniform motion in which an object traverses equal space in equal intervals of time was developed. Furthermore, we also notice that the rudiments of uniformly accelerated motion were also developed. It was shown that the distance traversed by a uniformly accelerated body during a period will be the same as that traversed by the mean speed (average of the initial and the final speeds). Oresme proved this theorem (first developed at Merton College) geometrically as shown in Fig. 1.19.

Oresme had the vision to assume the area under the $V - T$ curve represents the distance travelled. Thus, it is seen that the works done by Copernicus, Kepler and Galileo did not take place in vacuum. A reasonable degree of maturity and some early concepts of the science of motion did grow during the thirteenth and fourteenth century in northern Europe.

1.5 Copernican Model: Rediscovery of the Heliocentric Theory

Even in the Hellenistic period in Greece, quite a few eminent astronomers never felt very comfortable with the Aristotle's idea of a geocentric universe. Most notable was the hypothesis by Aristarchus (310–230 BC) that the earth spins around its own axis producing the relative diurnal motion of the heavens and a yearly orbital motion around the sun. It has been discussed earlier in this text why such a proposal was rejected by the philosophers and astronomers.

The popular belief that Copernicus proposed a simple heliocentric model of the universe that could explain the observations is, strictly speaking, not correct. Copernicus' book 'De Revolutionibus Orbium Celestium' published in 1543 had very strong resemblance with the writings of ancient and medieval astronomers. The model presented by Copernicus was, in fact, as complex as the modified

Ptolemaic models with many epicycles and eccentricities. In this model, also the planets did not go round the sun, but a void point away from the sun's centre. However, the model proposed in the book could explain the retrograde motion of planets when the sun is in opposition and the limited elongation of the inferior planets in a very natural way without introducing any artificial ad hoc conditions. Another major difference in the basic characteristics of the Ptolemaic and Copernican models is that in the Copernican model the motions and orbits of all the planets got linked. Even then the difficulties were enormous. In the absence of the concept of inertial motion (the first law of motion), people had extreme difficulty in understanding how a vertically thrown particle returns its original starting point. So, although De Revolutionibus was not a revolutionary book, it started the process. According to Kuhn, 'The significance of the De Revolutionibus lies less in what it says itself than what it caused others to say'.

It will be interesting to note first the natural occurrences of the retrograde motion of planets and the characteristic limited elongation of venus and mercury.

Figure 1.20 shows the sun, the earth and a superior planet, viz. mars. As the earth has a much faster motion than that of the mars, the corresponding positions of the two planets in their orbits will be somewhat as indicated. Examining the lines of sight from different positions of the earth, the path traced by mars in the backdrop of the fixed stars will be as shown. The natural occurrence of the apparent retrograde motion is very clear from the diagram. It should be noted at this point that the retrograde motion in Ptolemaic epicycle based model in real, whereas the retrograde nature of mars' motion in the heliocentric model is not a real retrograde motion of the planet mars. The retrograde motion of venus and mercury can be also explained with this model in a similar manner.

Though Ptolemaic model cannot explain the occurrence of retrograde motion only when the sun is in opposition in a heliocentric model, this observation is easily understood. A planet will also be at its brightest during retrograde motion as the distance from the earth will be the smallest as shown in Fig. 1.20.

The other major advantage of a heliocentric model is the observed limited elongation of venus and mercury. Figure 1.21 shows the sun and the orbits of the planets earth, venus, mercury.

It is clear from the diagram that from any position on the earth's orbit the maximum elongation of venus and mercury will be $\sin^{-1}\left(\frac{R_V}{R_E}\right)$ and $\sin^{-1}\left(\frac{R_M}{R_E}\right)$ where R_E, R_V and R_M are the radii of the orbits of earth, venus and mercury, respectively. Because of the limited elongation, venus is seen at

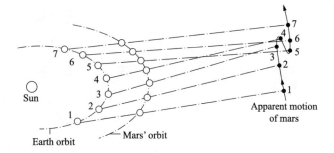

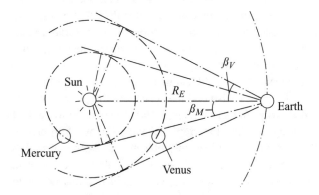

Fig. 1.20 Appearance of retrograde motion of Mars

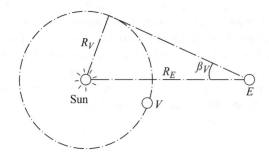

Fig. 1.21 Maximum elongation of venus and mercury

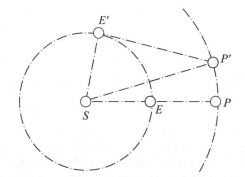

Fig. 1.22 Size of venus' orbit

Fig. 1.23 Determination of orbital radius for superior planets

heliacal rising and heliacal setting positions. In Ptolemaic model, one needs to artificially tie up the deferent of the sun, venus and mercury which mess up the aesthetic beauty of the model.

As mentioned earlier, in the Copernican system the sizes of the orbits of various planets were no longer independent and adjustable. All the orbits' radii could be expressed in terms of the radius of the earth's orbit. From Fig. 1.22, it is seen that if R_E and R_V be the radii of the earth and venus' orbit and β_V be the maximum elongation for the planet venus, then $R_V = R_E \sin \beta_V$.

Though a heliocentric model explains the observed characteristics of the planets' motion and their variations in brightness, the real grounds for Copernicus' belief in a spinning earth orbiting around the sun were different. He argued that as the earth is spherical its spinning motion is natural to it. He received psychological justification for an orbital motion also because of the earth's spherical shape. This demonstrated how ignorant the scientists had been about the science of motion (Fig. 1.23).

Figure 1.24 shows the fundamental difference between Ptolemaic and Copernican schemes. In both cases, the orbiting objects moved around some void points in space and not about material objects. So there was no possibility to guess that either the earth or the sun had anything to do with the motions of the orbiting bodies. Their motions had to be taken as natural motion. So the Aristotelian physics continued to play important roles even in the Copernican system. Apart from this, the Copernican model also looks as complicated and artificial as the Ptolemaic system. The motion of the earth, in Copernican model, was explained with the help of a scheme depicted in Fig. 1.25.

The earth moves in a circle with its centre at C_E which moves about a point C in a circle. The point C in turn rotates about the sun S in a circle. In case of other planets also equally complex mechanisms involving epicycles and deferents were devised. Therefore, one can call Copernicus as either the first modern astronomer or the last astronomer of the Aristotelian era.

Fig. 1.24 Fundamental difference between Ptolemaic and Copernican schemes

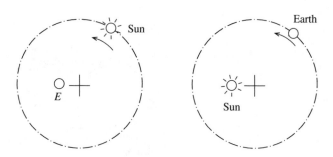

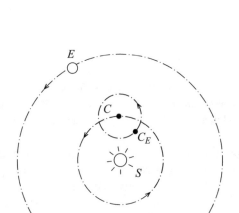

Fig. 1.25 Explanation of the earth's motion in Copernican model

1.6 Tycho Brahe: Improvement in Accuracy for Naked-Eye Astronomy

It is said that new concepts in science emerge when the accuracy of observation improves by an order of magnitude. The science of motion could not have been discovered without Tycho Brahe's phenomenal improvement in naked-eye astronomy. As the primary objective of the scientists is to explain observational phenomena with the help of theories, no motivation for either modifying or rejecting a scientific theory can be felt if the existing theories can explain the observations. Or, in other words, if the predictions by a theory agree with the observational or experimental results, there is no reason not to accept the theory as correct. In this sense, Tycho Brahe's contribution to the evolution of physical science and astronomy should be treated as very significant if not critical. Tycho Brahe (1546–1601), though played a revolutionary role in bringing many innovations to observational astronomy that improved the accuracy to about $1'$–$2'$ of arc (in traditional observation, the accuracy was rarely better than $10'$ of arc), was a lifelong opponent of the Copernican concept of a heliocentric model in which earth moved. Instead, he proposed his own Tychonic system in which the sun moved around the earth and the planets moved about the sun. Thus, Brahe's

tremendous energy and motivation led to the development of a huge bank of observational data with an accuracy that was almost an order of magnitude better than that of the traditional data. He was also very fortunate to observe a supernova in 1572 that appeared as a new star in the constellation of Cassiopeia. After systematic and careful observations over a year, he decided that there was no parallax of the new star placing it far beyond the spheres of the planets. When it faded away after a few years, it was the first time when doubt arose against the 'incorruptible' nature of the fifth element that constituted the heavenly objects. All these prepared the ground for emergence of new science and rejection of the old science. It was also a great coincidence that he invited Johannes Kepler (1571–1630) to work as his assistant and Kepler was, perhaps secretly, a devoted Copernican. It was also, perhaps, a still more important coincidence the Tycho Brahe died a very early, and untimely, death and all his excellent and accurate observational results fell in the hands of Kepler. What it resulted into will be seen in the next section. Though Tycho Brahe's cosmology and astronomical model had no importance in the development of new science, it is without doubt that without his invaluable accurate observations the so-called Copernican revolution would have not been possible.

1.7 Kepler: Beginning of Modern Astronomy and Foundation of Science of Motion

Normally, Johannes Kepler (1571–1630) is given the credit for discovering the three laws of planetary motion. But in reality, Kepler's contributions go far beyond transforming astronomy from a mystical medieval subject to an exact mathematical and physical science. He laid down the foundation for the modern science of motion by introducing a number of important concepts in the subject we call today 'dynamics'. The primary aim of this section is to present the evolutionary thought processes and some transitory concepts that later matured in the form of laws of motion. However, it will be desirable to start from his work on the motion of planets which played the pivotal role in exposing the hidden secrets of dynamics.

1.7.1   Discovery of the Laws of Planetary Motion

Typically, it is suggested that Kepler found the planets to describe elliptic orbits with the sun at a focus. In reality, the process of discovering this important fact of celestial mechanics was far more complex than popularly perceived. Though the role of determining mars' positions by triangulation was very important, the area law which was proposed by Kepler before finding the elliptic nature of the orbits played an equally if not more important role. It must be emphasized at the very beginning that all the laws could be discovered only because Kepler was unable to match only 8′ of arc in the martian orbit with theoretical predictions using only systems of perfect circles for geometric description of planetary motions. In fact, the difference between the predictions of the models and actual observations in old astronomy was most often below the detectable levels. Only with the drastically improved data, acquired by Tycho Brahe's superior astronomical observations, the discrepancy could be detected. Even then most astronomers would have, perhaps, ignored this mismatch of 8′ arc in the martian orbit. But for a strange reason, Kepler felt deep inside him that something far more fundamental reason exists for this error. Thus, it is also a matter of lucky coincidence that the planet mars has an eccentricity of its orbit that is small but significant.

Kepler was a Copernican believing in a heliostatic[2] system of the universe. But all the models were just geometric descriptions of the motions. There was no 'cause' for such movements. This disturbed Kepler deeply; he felt very strongly that there must be a reason for the orbital motions. He noticed that a huge object like the sun is not occupying the centre and the centre point was a void point in space! Psychologically, it was very disturbing to Kepler, and he felt that the sun must be the prime mover for all the planets' motions. In modern language, his arguments can be worded as follows:

- Why do all planets move around a void point which happened to be the centre of the earth's orbit?
- As the sun, a huge body, is always very near to the centre of the earth's orbit (called the mean sun) should not sun influence the planets to move as they do? Thus, should not these orbits be most simply described in a system of coordinates with the sun at the origin?

Kepler first established that the planets move on orbits which are fixed in helioastral space. Actually, Kepler was the

---

[2]The orbits were not really 'heliocentric' as the centres of the circles did not coincide with the sun. But as the sun was not an orbiting object, such models should be more appropriately termed as 'heliostatic'.

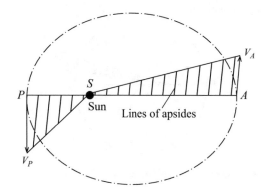

Fig. 1.26 Velocities of a planetal perihelion and aphelion positions

first to propose the concept of orbits. He found out from observations the nodal points of the orbits at which they crossed the plane of the ecliptic (i.e. the plane of the earth's orbit around the sun) and found that the planets return to their respective nodes at fixed periods. Kepler was also able to find out the inclinations made by the various orbital planes with the plane of the ecliptic. He noticed that all the orbital planes contain one common object—the sun. He also found that the apsidal lines of all the planets' orbits pass through the sun. However, Kepler continued to assume that all the planetary orbits were eccentric circles. Thus, to begin with, Kepler assumed all planets to move in perfect circles with their lines of apsides passing through the sun. He was of the impression that a perfect representation of the motion of all planets was possible by means of exact circles and Ptolemaic equants. Kepler realized that while going along the orbit a planet can sense only the distance from the sun. This directed him to describe the motion with the sun's distance from the planet to control the motion. Of course, he was still considering a circular orbit but with 'non-uniform' motion. His belief became a firm conviction when he noticed that the planets' speeds at the apsidal positions (Fig. 1.26) were inversely proportional to the distance from the sun.

Thus, he found that $\frac{V_A}{V_P} = \frac{SP}{SA}$, and he concluded that the planet's speed is inversely proportional to the distance from the sun. Such a law looked natural to him as he thought that larger distance will weaken the sun's driving influence, hence lower speed. Of course, we now know this to be wrong, but this was the forerunner for the famous area law which Kepler arrived at through a mixture of guesswork, intuition and serendipity. In an attempt to find out the time required for a planet to traverse a finite distance, through a complex argument, Kepler suggested that the rate at which area is described by the radius vector remained approximately constant. Noting the fact that a planet moves fastest when closest to the sun led Kepler to believe in a distance law according to which the speed of the planet to be inversely proportional to its instantaneous distance from the sun. To test this hypothesis, Kepler had to determine the

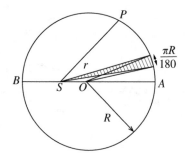

Fig. 1.27 Emergence of the area of law

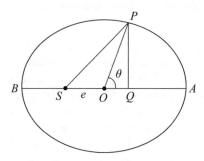

Fig. 1.28 Derivation of Kepler's equation from area law

time taken by a planet to reach a certain point in the orbit. Kepler divided the first half of the orbit from aphelion[3] A to perihelion B in 180 equal segments. Each segment was $\pi R/180$ as indicated in Fig. 1.27.

For each segment, Kepler found out the current distance r from the sun and assumed that the time taken by the planet to cover this distance, Δt, is proportional to r. The constant of proportionality was normalized to make the total time required to cover the whole distance from A to B, equal to half the orbital period which was well known. While going through this extremely laborious exercise, it occurred to him that all the instantaneous distances (which were in fact infinite in number) are contained in the area of the segment. Thus, he considered the area swept by a line joining the planet and the sun to be proportional to the time taken to traverse the corresponding segment. Of course, noticing the fact that the area law can be derived from the distance law (which according to Kepler was exact) only when the line joining the planet to the sun is at right angles to the instantaneous velocity of the planet (i.e. at A and B only), he treated his area law to be an approximate rule! Much later, he realized that the area law was exact. Kepler could show that the area law leads to predictions on planetary motion which were in closer agreement with the observations than those based upon the concept of an equant. Thus, the following scheme was developed by him (Fig. 1.28).

---

[3]The terms 'aphelion' and 'perihelion' were introduced by Kepler.

AB is the apsidal line of a planetary orbit (still considered to be a circle with centre O), and S is the position of the eccentrically located sun at a distance e from the centre. To determine the time t the planet takes to reach point P starting from A, it is necessary to find out the area of the segment ASP. Assuming the orbital period to be 2π and the orbit radius to be unity, the rate at which area is swept k, is given by

$$k = \frac{\pi \times 1^2}{2\pi} = \frac{1}{2}$$

Thus, time to reach P from A

$$t = \text{Area of } ASP/k = 2 \times \text{Area of } ASP$$

Now,

$$\text{Area of } ASP = \text{Area of } AOP + \text{Area of } OSP$$
$$= \frac{\theta}{2\pi} \times \pi \times 1^2 + \frac{1}{2} \times e \times PQ$$
$$= \frac{1}{2}\theta + \frac{1}{2} \times e \times PQ = \frac{1}{2}(\theta + e \sin \theta)$$

Hence, $t = \theta + e \sin\theta$

This equation is known as Kepler's equation. Kepler knew that this equation did not have a simple solution.

Only much later, Kepler discovered that in reality his 'area law' was exact and his favourite distance law was an approximation. He realized that the 'area law' was an empirically exact result, but, of course, he tried in vain to find out a physical significance through his favourite 'distance law'. This empirical rule eliminated the need incorporating the concept of Ptolemaic equant. Finally, Kepler concentrated his efforts to arrive at an exact description of the orbit of mars for which Tycho Brahe accumulated the most accurate and extensive observational data. One must recognize the tremendous difficulty of the problem. In Einstein's words 'If the planets moved uniformly in circles round the sun, it would have been comparatively easy to discover how their movements must look from the earth. Since, however, the phenomena to be dealt with were much more complicated than that, the task was a far harder one. To grasp how difficult a business it was even to find out the actual rotating movements, one has to realize the following. One can never see where a planet really is at any given moment, but only in what direction it can be seen just then from the earth, which is itself moving in an unknown manner round the sun. The difficulties thus seemed practically in surmountable. Kepler had to discover a way of bringing order into this chaos. To start with he saw that it was necessary to try and find out about the motion of the earth itself'.

Determination of the earth's orbit was a masterstroke of theoretical astronomy by Kepler. But Kepler had to follow a very torturous route. He struggled with the idea of a circular

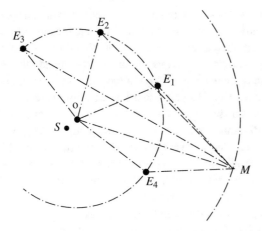

Fig. 1.29 Triangulation procedure for determining the earth's orbit

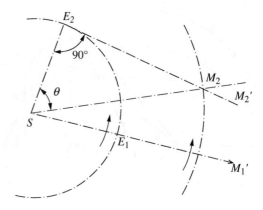

Fig. 1.30 Triangulation for determination of Mars's orbit

orbit and his area law. Unfortunately, the predicted positions of Mars still showed errors up to 8′ of arc. As he used a large number of acronychal[4] observations, it was possible for him to develop a theory of martian motion which yielded the martian longitudes almost correctly but not the positions on the orbits. Kepler called this theory as 'vicarious hypothesis' and used it for martian longitudes (as seen from the sun because during acronychal observations the sun, earth and mars are in one line). Kepler next determined the orbit of the earth in helioastral space. The orbital period of mars was known to be 687 days. Thus, Kepler assumed that mars must return to the same position in the helioastral space after every 687 days. So, observations of mars's position were recorded at an interval of 687 days; this makes the sun–mars line fixed in space. From the model of mars's motion, Kepler developed based on the assumption of a circular orbit, and using the acronychal observation,[1] very accurate longitudes could be estimated though the distances had errors. The observations gave the directions for both mars and the sun as seen from the earth. From the vicarious hypothesis, the sun–mars line *SM* is determined (Fig. 1.29).

Then, knowing the directions of the sun and the mars on three occasion (at the interval of 687 days), the positions of the earth with respect to the fixed sun–mars direction are found out as E_1, E_2 and E_3 by drawing the triangles with *SM* as the base. Assuming the earth's orbit to be circular, it can be drawn using a fourth point E_4 and it can be further verified if all the four points lie on the same circle. Luckily for Kepler, earth's orbit has very little eccentricity and he could establish the earth's orbit.[5]

Kepler's all attempts to apply the area law to find out mars's orbit (assuming it to be a circle) failed. He then attempted direct triangulation to determine the orbit. The method can be represented in Fig. 1.30.

The orbit of the earth is determined, and the position of earth at anytime can be located on its orbit. In the initial position, the sun, earth and mars are in one line as represented by the line SE_1 (mars's position is not known; only the direction is known). As the earth rotates at a faster rate around the sun than mars, it reaches a point E_2 so that $\angle SE_2M_2' = 90°$, M_2 being the corresponding new position of mars (location still unknown but the direction is known). If the time elapsed during the motion of the earth during this period is noted, $\angle E_1SE_2$ can be determined from the known motion of the earth. Similarly, $\angle M_1SM_2'$, through which mars rotates during the same period, can be also determined from the vicarious hypothesis Kepler developed. Now

$$\angle M_2'SE_2 = \angle M_1'SE_2 - \angle M_2'SM_1' = \theta$$

With this value of θ, the location of planet mars at the instant, M_2, is determined. Kepler found out three such positions and the circular orbit containing these three points. When he repeated the process a number of times, he found a new circle each time. With utter frustration, Kepler had to come to the conclusion that the orbit cannot be a circular one. This discovery led him to assume the orbit to be an ovoid.[6] Taking a non-circular geometry to be involved in planetary motion was unheard of, and this step by Kepler was truly revolutionary. But he found working with an ovoid orbit for applying his area law was very difficult. Then, by a stroke of luck and extreme case of serendipity, Kepler attempted to approximate the ovoid by an ellipse. He drew an auxiliary circle with line of apsides of mars as the diameter and inscribed an ellipse with the line of apsides as

---

[4]Acronychal observation means observation of a planet when it is in opposition; i.e., the planet, the earth and the sun are in one straight line. The acronychal longitudes are the longitude of a planet with the observer on either the earth or the sun.

[5]Kepler found that the sun was shifted from the centre of the earth's orbit by a very small amount. So it was only 1.8% of the orbit radius.

---

[6]It is surprising to note that 'ellipse' came to the picture at the end of Kepler's massive calculations and analyses.

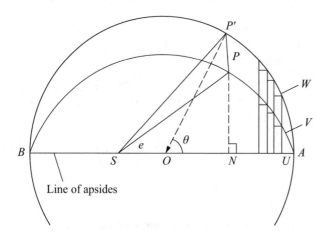

Fig. 1.31 Application of area law to an elliptic orbit

the major axis (Fig. 1.31). He could use his area law by applying a result obtained by Archimedes for elliptic curves.

Let P be a point on the ellipse and PN be normal to the major axis. When NP is extended, it cuts the auxiliary circle at point P'. Then, the property of an ellipse yields that the ratio of the lengths $P'P/PN$ to be constant. The area of the elliptic segment APN is equal to the sum of the elemental areas shown. Thus, the elemental areas constituting the circular segment $AP'N$ and the elliptic segment APN will be in proportion of the lengths of the elemental areas UW/UV, which remains constant for a particular ellipse. So,

$$\text{Area } AP'N / \text{Area } APN = \frac{P'N}{PN} = \lambda \text{ (a constant)}$$

Now, the areas of the two triangles $SP'N$ and SPN are also of the same proportion.

$$\text{Area } SP'N / \text{Area } SPN = \frac{P'N}{PN} = \lambda$$

Hence,

$$\begin{aligned} &\text{Area } SP'A / \text{Area } SPA \\ &= (\text{Area } SP'N + \text{Area } AP'N)/(\text{Area } SPN + \text{Area } APN) \\ &= \lambda \end{aligned}$$

Thus, Kepler could determine the point P after a time t (starting from position A) using his area law. It should be remembered that both the line of apsides for mars and the position of the sun on this line were well determined beforehand by Kepler. Now, for a given t, he could solve Kepler's equation.

$$t = \theta + e \sin \theta$$

to find out θ and through this the location of point P'. If t be the time for achronycal observation, then the martian

longitude as seen from the sun was known. Dropping a perpendicular onto the line of apsides, the location of P could be determined as the intersection of PN with the line SP with $\angle ASP$ known. With one point P determined, the whole ellipse is known and position of the planet can be found out at any time. Thus, Kepler's first law was discovered. Contrary to popular belief, the second law was discovered (in a serendipitous manner, of course) first and, actually, played a very vital role in discovering the first law, i.e. the planets move in elliptic paths. Kepler also soon discovered that the sun is at the foci of all the elliptic orbits of the planets. This, of course further convinced Kepler that the sun must be driving all the planets in their respective orbits.

Kepler discovered the third law that the orbital periods were proportional to the 3/2 power of the respective mean distances from the sun, about a decade after he completed his work on the first and the second laws. To begin with, he still continued with his erroneous concept that the driving force due to the sun decreases as $1/r$ where 'r' is the distance from the sun. He justified saying that as the purpose of this force is to drive the planets in their orbits the influence spreads only in the orbital plane and strength does not decrease as $1/r^2$. With a further proposal that orbital speed is directly proportional to the driving force, Kepler proposed the ratio of the periods of two planets 1 and 2 to be related as follows (in modern language):

$$\frac{T_2}{T_1} = \left(\frac{r_2}{r_1}\right)^2$$

However, Kepler worked on the basis of increment of period where one transfers from orbit 1 to orbit 2. One can get it from the above relation as follows (again in modern mathematical language):

The increase in time period for going from 1 to 2 is $\Delta T (= T_2 - T_1)$. Thus,

$$\begin{aligned} \frac{\Delta T}{T_1} &= \frac{T_2}{T_1} - 1 = \left(\frac{r_2}{r_1}\right)^2 - 1 = \left(\frac{r_1 + \Delta r}{r_1}\right)^2 - 1 \\ &= \frac{2\Delta r}{r_1} + \left(\frac{\Delta r}{r_1}\right)^2 \end{aligned}$$

where Δr is the increase in orbital radius for going from orbit 1 to 2. Kepler by mistake took

$$\frac{\Delta T}{T_1} = \frac{2\Delta r}{r_1}$$

The two errors, taking $\frac{T_2}{T_1} = \left(\frac{r_2}{r_1}\right)^2$ instead of $\frac{T_2}{T_1} = \left(\frac{r_2}{r_1}\right)^{\frac{3}{2}}$ and taking $\frac{\Delta T}{T_1} = \frac{2\Delta r}{r_1}$ instead of $\frac{\Delta T}{T_1} = \frac{2\Delta r}{r_1} + \left(\frac{\Delta r}{r_1}\right)^2$ had a mutual cancellation effect. Kepler was, therefore, initially encouraged by this chance agreement of observation with theory, though approximately. However, with more careful analysis of the observational results, Kepler finally arrived at the correct form of his third law.

1.7.2 Transition from Geometric to Physical Model

It has been demonstrated how Kepler transformed the very basic foundation of astronomy. Even till the time of Copernicus, the planetary motions were described in terms of the so-called natural motion of the heavenly objects. Thus, the models of the universe were based on the combination of uniform circular motions. All the motions were basically geometrical in nature, and there was no concept of cause and effect phenomenon. Of course, it is suspected that many astronomers of the antiquity considered the models as computing machines only for the purpose of describing the planetary motions.

Kepler was the first who introduced the concept that the planets are being driven by the sun. He also brought in the idea that the planets move in their respective paths fixed in the helioastral space—i.e. the concept of orbit was born. It is noticed that throughout his work on arriving at a correct description of the planetary motions Kepler constantly emphasized that the sun is behind the motions of the planets and these motions can be described with the help of a cause and effect relation. He established that all the orbital planes and the apsidal lines pass through the sun. Finally, once he was able to show that the planets move in elliptic orbits, it was shown that the sun is at the foci of all these elliptic orbits.

His distance law, that later changed to the area law, was based on a physical model in which the sun's distance played an important role. This concept ultimately gave birth to the third law. For the first time in the history of the astronomy, the whole solar system got bound by a common rule and became physically related.

1.7.3 Early Concept of Action-at-a-Distance and Gravitation, and the Concept of Force

Kepler's work not only transformed astronomy from its ancient structure to that belonging to the modern era, but he was the first since Aristotle who attempted to develop a new dynamical scheme. Though he was unsuccessful in his attempts, according to many researchers in the history of science, he came quite close to many basic aspects of Newtonian mechanics. He completely rejected any concept of natural motion and forwarded the suggestion that all motions are caused by some entity like force. At the same

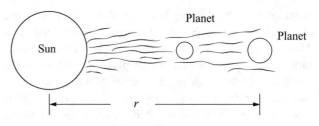

Fig. 1.32 Scheme of 'anima mortix'

time, he also suggested that all matter has the inherent tendency to remain at rest. Concept of inertia of rest was, thus, first suggested by Kepler, and the word 'inertia' was also first coined by him. Unfortunately, Kepler was unable to discover the subtle way in which force interacts with matter through 'acceleration'; he followed the mediaeval science in relating force to speed. Using modern methodology, his force law can be written as $F \propto v$.

He used this law to explain how a rotating sun can move the planets in their respective orbits. He considered forces to emanate from the body of the sun like spokes in a bicycle wheel which produces azimuthal force on a planet causing it to rotate around the sun. He gave a name of this entity coming out of the sun—'anima mortix'. He also assumed this force to expand radially in the plane containing the orbits of the planets. Thus, according to Kepler, this force was inversely proportional to the distance. Figure 1.32 shows his scheme of 'anima mortix'.

Kepler even estimated a possible rate at which the sun should rotate for the observed orbital motion of the planets. When, subsequently, Galileo observed the rotation of the sun and the rotational speed matched the prediction, Kepler was ecstatic and happy.

This phenomenon of 'action-at-a-distance', that causes the planetary motions under the influence of the sun, was considered as an occult phenomenon and rejected by most contemporary scientists including Galileo. Kepler's suggestion that the tidal phenomena were caused by the gravitational influence of the moon was also summarily rejected by Galileo. Kepler proposed his concept of this kind of force acting through void space after Gilbert published his work on magnetism. It was demonstrated that magnets can interact and generate force without physical contact. When Gilbert demonstrated earth to be a magnet, Kepler developed a scheme to explain the planetary motions considering the sun and the planets to be magnets.

While working on his ideas about how physically separate bodies can interact, Kepler came quite close to the concept of gravitation. Of course, without any concept of composition of accelerated motions with inertial uniform motions and dependence of acceleration on force rather than velocity, he could not make any useful contribution. Kepler made the following statement about gravity in his book

Astronomia Nova 'Gravity is a mutual propensity between like bodies to unite or come together... so that the earth draws to stone to it much rather than the stone seeks the earth... If two stones were to be placed anywhere in the world outside the range of influence of a third similar body, then each stone, like two magnetic bodies, would come together at an intermediate point, each stone travelling towards the other a distance proportional to the bulk of the other'.

It is quite clear from the passage that Kepler was very close to the truth. Perhaps, another point that strikes all is an anticipation of Newton's third law of motion. Kepler's all attempts to quantitatively explain his laws of planetary motion from the dynamics he developed failed. Like all astronomers before him, who spent lifetimes in observing near circular motions of the heavenly objects, he could hardly dream that the foundation of dynamics lies in the 'uniform rectilinear motion'. But Kepler sowed the seeds of many original and useful ideas which provided useful clues to the subsequent generation of scientists.

1.8 Galileo: Naked Eye to Telescopic Astronomy

Galileo Galilei (1564–1642) was born in 1564. His father Vincenzio Galilei was a cloth merchant but also an expert on music. Galileo spent his early student years in a monastery, and in 1581, his father got him admitted to the University of Pisa to study medicine. (In those years, even in academics medicine was considered a superior subject compared to mathematics and the salary of medicine professor was about six times that of a mathematics professor.) Galileo shifted to mathematics and physics finding little interest in medicine, particularly the way it used to be taught then.

Galileo made his scientific contribution to both the science of terrestrial motions and astronomy. Both brought revolutionary changes to physics. Furthermore, he was the first to bring the concept of scientific demonstration and experiment to compare the actual outcome with the mathematical prediction. In this sense, Galileo can be considered as the father of modern science.

Galileo started his work on the science of motion during his first posting at Pisa. He observed the motion of pendulums and considered the problem of free fall of bodies; however, he was Aristotelian in his outlook though quite critical of many of his teachings. He studied Copernicus's De Revolutionibus but was still not converted to Copernicanism.

His work with telescope and profound discoveries started much later—towards the end of 1609. Later, he again returned to the study of terrestrial motions and made specific revolutionary contributions that supplied the required ammunition to destroy the old science.

As this book has been dealing with astronomical concepts and ideas, this section of Chap. 2 is being developed to the astronomical discoveries made by Galileo.

1.8.1 Observation of the Moon and Discarding the Concept of Fifth Element

Johannes Kepler sent a copy of his first book 'Mysterium Cosmographicum' in 1597 to Galileo. This book showed that the orbits of planets in the Copernican model of the universe followed a harmony and showed that the corresponding spheres are circumscribed by the five regular solids. Galileo thanked Kepler and confessed that though he believed in Copernican theory he could not publish it for the fear of losing professional reputation and credibility. In the subsequent letter, Kepler requested Galileo to publicly support Copernicanism. In the same letter, he enquired whether Galileo had any astronomical equipment for measuring star positions with an accuracy of a thousand of a degree; with that, stellar parallax could be measured providing irrefutable proof for the orbital motion of the earth.

It has been found again and again that nothing goes waste in life. Galileo's father died in 1591 and Galileo became the head of an extended family. He had to support his mother and three siblings and had to arrange for substantial amounts for the dowry to get his two sisters married. Each time the dowry amounts were about twice his annual salary as a professor of mathematics. Over and above, he had his own family with three children and a mistress to take care. Therefore, Galileo had to run a parallel business of making and supplying precision instruments of various kinds, and business was good. This gave Galileo a reputation in precision instrument making. His capability in designing and making precision instruments was essential for his telescopic discoveries.

In 1592, Galileo had to leave the University of Pisa and joined a better post at the University of Padua near Venice. As Venetian society was much more liberal and tolerant, he found many like-minded scholars and an appreciating intellectual environment. Such an environment was ideal for continuing his investigations on science of motion. At Padua, he heard from his friend in 1609 that an unknown Dutch spectacle maker claimed to have invented an optical device that could make faraway objects nearby. The matter drew Galileo's attention when he was told by his friend that a foreign merchant had approached the Venetian Senate for selling a crude 'spyglass' at a substantial price. Such a device was considered very useful to Venice for military purpose. Knowing Galileo's ability to make precision instruments, the senate told him to make a better spyglass. Thinking that a considerably improved spyglass could make his business prospects much better, Galileo discovered the basic principle of spyglass using his knowledge in optics; after a considerable number of trial and error, finally Galileo succeeded in developing a spyglass with power '9'. He led the Venetian Senate to the top of the tower at San Marco for a demonstration. Galileo presented the spyglass to the senate as a gift and received a good rise in his salary. Later, he travelled to Tuscany and presented a similar spyglass to the ruler. Towards the end of 1609, Galileo could develop spyglass with power twenty and even thirty! These devices[7] he turned to the heavens to observe the moon and other objects in January 1610. With this astronomy entered a new era leaving a two-millennia legacy of naked-eye astronomy.

The results were astounding and soon demolished many old concepts. It also provided new evidence in favour of the Copernican hypothesis. It was quite natural that when Galileo turned his telescope to the heaven the first object he studied was the moon. He saw the craters, mountains and the valleys. Measuring the shadow of the hills, he did even estimate the heights of many mountains. He announced that moon was made of earth-like material and its illumination was due to sunlight. He, by such observation, announced that the moon (and other planets as well) was not composed with any 'fifth element' as was suggested in Aristotelian science; instead, they are made of ordinary materials found on the earth. They shone like stars not because they radiated their own light but because of the sunlight reflected by these bodies.[8] This was a major conceptual revolution that demolished the idea of uniform circular motion as the natural motion of the heavenly bodies.[9] This also brought the necessity of a cause of the motion of the planets and Kepler's suggestion of the sun being the driver became more plausible. This was one step forward in establishing the heliocentric theory proposed by Copernicus and perfected by Kepler. Figure 1.33a shows a copy of Galileo's sketch of the moon as seen through his telescope.

He also noticed that the Milky Way consists of a very large number of stars. Galileo's sketch of the sky near the constellation Orion (with many more stars compared to as seen with naked eye) is shown in Fig. 1.33b.

1.8.2   Discovery of Jupiter's Moons and its Implications

With the improved version of the telescope, Galileo could also observe four moons[10] of Jupiter. He was amazed to see this mini solar system, and its significance was enormous. One major objection of the scientists to a moving earth was that how moon could then not be left behind. Galileo demonstrated by actual observation that planet Jupiter could move (about the earth as a centre as proposed by the old geocentric theory) carrying not one but four moons along with it. So earth could move without leaving behind one moon. Thus, another major hurdle could be overcome towards establishing the sun-centred model of the universe.

Subsequently, Galileo also observed that the planet Saturn has two ear-like bulges, but he failed to understand that this was a ring around the planet.

1.8.3   Discovery of the Phases of Venus: A Further Proof of Heliocentric Model

With the naked eye, the planet venus looks like any other planet, the major difference being that it is always seen either in the evening western sky or morning eastern sky (implying that venus always remains within a definite angular distance from the sun), and it is much brighter than other planets. When Galileo observed venus with his telescope, he saw that like the moon venus also passes through various phases. The characteristics observed supported a heliocentric model proposed by Copernicus. This is explained with the help of Fig. 1.34.

In case of a geocentric model, the relative configuration of the earth, the sun and venus is shown in Fig. 1.34a. If sun and venus go round a stationary earth as shown in Fig. 1.34a, point A moves in a circle. Furthermore, E, A and S (the sun) must always remain on a line as the angular distance between the sun and venus is always within a limit. For four successive positions of the planet 1, 2, 3 and 4, the shape of the illuminated part of venus will appear as shown in the figure. On the other hand, if both the earth and venus go round a stationary sun (as shown in Fig. 1.34b), the shape of the illuminated part of venus will change in a manner as

---

[7]In 1611, Galileo visited Rome and the local academy arranged a banquet in his honour: on a hillside just outside Rome. During this banquet, one of the mathematician guests, Giovanni Demisiani, announced a name for Galileo's device. By combining two words 'tele' (meaning 'distant') and 'skopeo' (meaning 'to took'), he pronounced that Galileo's device be given the name 'Telescope'.

[8]It is very interesting to note that though this phenomenon got revealed to the European astronomers only about four centuries ago, astronomers of ancient India were aware of this fact.

[9]It is also very interesting that in ancient India it was known that the earth rotates about its axis and that causes the sun to rise in the east and sets in the west.

---

[10]Later, Galileo proposed these moons orbiting the planet Jupiter to be used as a clock from different far off places for solving the longitude determination problem. While observing eclipse of these moons by Jupiter for this purpose, Römer first found that light travels with a finite speed and estimated its approximate magnitude.

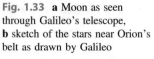

Fig. 1.33 a Moon as seen through Galileo's telescope, **b** sketch of the stars near Orion's belt as drawn by Galileo

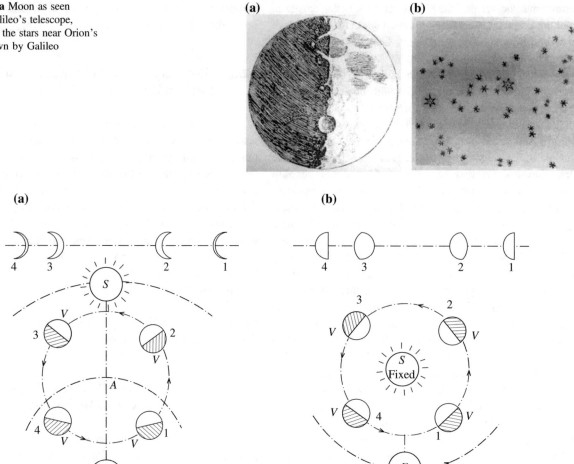

Fig. 1.34 Phases of venus and the model of the universe

shown in this figure. The observation supported the pattern with sun as fixed, and therefore, the universe is heliocentric.[11]

Of course, this could not completely demolish the geocentric model. The required clinching direct evidence was the detection of the parallax of stars at 6 months' interval (when the earth is at diametrically opposite positions in its orbit and at a distance of 2 AU). Unfortunately, even the nearest star is too far off for the detection of the extremely small parallax with Galileo's telescope.[12] Galileo argued the absence of any stellar parallax by suggesting the distance of the stars to be enormously large.

The astronomical observation by Galileo (from 1609 to 1612) with the help of his telescope could not completely demolish the geocentric model, but the new startling

discoveries he made brought his work and name to limelight. This helped him in gaining scientific respectability in spite of the wrath of the Catholic Church. His work on motion could not have gained popularity and quick acceptance without his reputation earned through the astronomical discoveries.

1.9 Galileo: Experimental Mechanics

According to most scientists, Galileo can be considered to be the first modern scientist. The four most important personalities of science in the medieval period can be easily classified into four different categories. If we consider Copernicus as a philosopher, Tycho Brahe as an observer, Kepler as a mathematician, then Galileo can be described as a physicist. He was the first to suggest that the nature follows strict mathematical rules and these rules can be discovered by observation and experimentation supported by logical thinking. Galileo had serious doubt about the Aristotelian scientific thoughts as many of the old and established

---

[11]It should be remembered that the model proposed by Tycho Brahe also satisfied the actual observation.

[12]The first stellar parallax was detected almost two centuries later by Bessel for 61 Cygnus 0' 31", by Henderson for α 0'.

theories (like heavier bodies fall faster in proportion to their weights) as he conducted experiments and the results did not agree with the old theory. He was definitely the major force in demolishing the old science and strove throughout his life to establish a new one. He was enormously successful in analysing kinematic problems but could not make any progress in the matter of creating a new dynamics. Perhaps, he still could not shake off some old concepts like 'natural motion', and the matter of force and motion relationship remained undisclosed to him. He, in fact, did not like Kepler's hypothesis of force causing motion of objects, and he totally rejected any idea of a force exerted by a body on another from a distance (like gravity).[13] So, he did brilliantly in unravelling the kinematics of motion and creating a solid foundation for one of the legs of 'mechanics' to rest. But the development of 'dynamics' had to wait for another half a century. As mentioned by Cohen very nicely 'Galileo was the last tenant of the old science. Like all tenants he complained and grumbled but never left the old premise'. The list of his major contributions in mechanics can be described in the following manner:

(i) Discovery of the law of inertia of motion and the Galilean relativity among all inertial frames.
(ii) Study of free fall and motion under constant acceleration.
(iii) Compound motion and projectiles.
(iv) Work on pendulum.

Apart from the above, Galileo did a considerable amount of pioneering work on hydrostatics, strength of materials and scaling laws. As the true secret of dynamics could be unravelled through 'acceleration' and not 'speed of motion', Galileo could not achieve the breakthrough, but his pioneering move to establish the mathematical rules for the behaviour of nature led to the development of 'dynamics'. According to Barbour, 'It was in this mathematization of empiricism that Galileo laid such secure foundation for dynamics'.

1.9.1 Early Works on Accelerated Change: Merton School

Normally, Galileo is given the full credit for studying accelerated motions. However, study of accelerated changes, i.e. not only accelerated motions (change of position) but any type of change, was started almost three centuries before

Galileo at the Merton College, Oxford, and little later by Nicole Oresme at Paris. At Merton College, work on studying the matter of accelerated change was carried out during the period 1328–1350. It was not restricted to only the 'change of position' (or motion) but any change. The 'instantaneous' rate of change used to be termed as the 'intensity' of the rate of change, and the total change used to be termed as the 'quantity' of change. When applied to the subject kinematics, the results relate to the case of uniformly accelerated motions. The important and fundamental kinematic theorem (that became known as the 'Merton Rule') can be described as follows (see also Sect. 1.4.4).

If a particle starts from rest and is uniformly accelerated at the rate 'a' for a time T, after a time $T/2$ the speed of the particle will be given by the following rule

$$V\left(\frac{T}{2}\right) = \frac{aT}{2}$$

The Merton Rule suggests that the total displacement of the particle in time T will be equal to the distance travelled by a particle in time T when moving with a constant speed equal to $V\left(\frac{T}{2}\right)$.[14]

1.9.2 Galileo's Work on Free Fall and Uniformly Accelerated Motion

One of the major difficulties in studying the kinematics of free fall and uniformly accelerated motion was the absence of an accurate time measurement device. Galileo overcame this hurdle by an extremely ingenious way. He started his study using objects rolling down inclined planes. How he came to this idea has been suggested by Stillman Drake in the following statement in his book 'Galileo' published by Oxford University Press in 1980. 'It was probably the long and heavy pendulum that Galileo used in 1602 which called his attention to the importance of acceleration in downward motion, and to the continuation of motion once acquired, things that soon led him to an entirely new basis for his science of motion which replaced his earlier causal reasoning'. This substantial reduction of acceleration due to gravity and his use of musical notes for timekeeping made it possible for him to discover the law of free fall and its associated phenomena.

In 1604, Galileo constructed a long incline of about 2 m length. The inclination to the horizontal was only 1.7°. A heavy bronze ball was made to roll down this inclined plane. Then, he placed small frets on the incline so that

---

[13]It is rather surprising that Galileo disbelieved any action-at-a-distance concept though by then magnetic attraction was an experimentally verified fact.

[14]As can be easily seen that this rule is the forerunner of the modern results used universally, $v = at$ and $s = \frac{1}{2} at^2$.

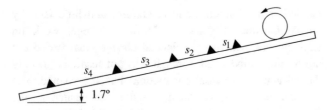

Fig. 1.35 Galileo experiment with incline plane

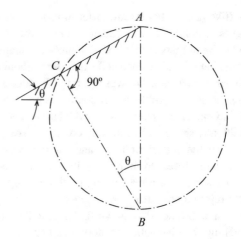

Fig. 1.36 Derivation of Galileo's theorem

whenever the ball passes over a fret a small sound can be heard. The arrangement is schematically shown in Fig. 1.35.

Galileo arranged the spacing of the frets so that the ticking sounds were at equal interval of time (managed by singing a musical tune). He then measured the increasing distances s_1, s_2, s_3 … as shown in Fig. 1.35. Like many before him, Galileo was also used to think that under constant acceleration the speed increases in proportion to the distance travelled.[15] When he measured the distances of the frets taking the distance covered in the first interval as unity

$$s_1 = 1$$
$$s_2 = 3$$
$$s_3 = 5$$
$$s_4 = 7$$

From these, he found that the total distance travelled increased as follows

$$s(1) = 1 = 1^2$$
$$s(2) = (s_2 + s_1) = 4 = 2^2$$
$$s(3) = (s_3 + s_2 + s_1) = 9 = 3^2$$
$$s(4) = (s_4 + s_3 + s_2 + s_1) = 16 = 4^2$$

Hence, the distance travelled is proportional to the square of the time. So if a be the constant acceleration, using Merton Rule

$$s(t) = 1/2\, a\, t^2$$

Thus, Galileo proved through experiment that motion in an incline plane is with constant acceleration.

He conducted further tests with free fall of objects by measuring the time it takes to hit the ground when dropped from a given height. For measuring time using a kind of water clock, he showed that free-falling objects also follow a constant acceleration, i.e. the acceleration due to gravity. Galileo also developed a theorem connecting the above two phenomena.

Figure 1.36 shows a particle at location A, and it is dropped from rest. It falls freely and reaches point B after time t. Next, a circle is constructed with AB as its diameter. Now, if the particle is allowed to slide from A freely along an inclined plane making an angle θ with horizontal, then Galileo's theorem states that the time taken by the particle to slide down along the incline to point C, the point of intersection of the incline with the circle, is the same as that taken by the particle to fall freely to point B.[16] We all know that Galileo at his young age noticed a swinging Chandelier in the Pisan Cathedral and discovered the time period to be independent of the amplitude of oscillation. He used his above theorem to prove the observational results as explained below. If the pendulum OC be long and the amplitude of swing (represented by the arc CB) be small, then the arc CB may be almost indistinguishable from the chord CB. Thus, the motion of the bob may be considered to be same as that of a sliding particle along an incline CB. Comparing Fig. 1.37 with an inverted version of Fig. 1.36 and applying Galileo's theorem, it can be seen that the time for the bob to reach point B is same as the time taken by a particle falling freely from A to B which is a constant quantity and remains independent of the location of point C (near B).

He extended his study of fall along an incline with the following postulate:

> I assume that the degree of speed acquired by the same moveable over different inclinations of planes are equal whenever the heights of those planes are equal.

---

[15]In fact it was a serious question before Galileo demonstrated that the dependence was on time (i.e. $v = at$). Logical thinking that leads to this conclusion is that if $v \propto s$ then it cannot start from rest when $s = 0$.

[16]To any modern school student, this is obvious. The acceleration along the incline is $g \sin \theta$, and the time t to cover the distance AC is $\sqrt{2 \times AC/g \sin \theta}$. But $AC/g \sin \theta = AB$. Hence, $t = \sqrt{2 \times AB/g}$ that is equal to the time for the particle to fall freely to B.

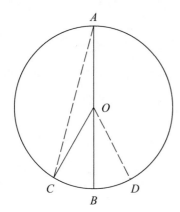

Fig. 1.37 Constancy of time period of long pendulums

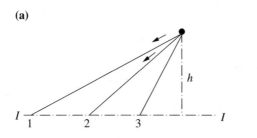 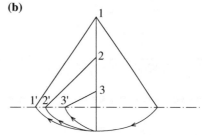

Fig. 1.38 **a** Independence of speed acquired for sliding along different inclines with same height, **b** Pendulums rise to the same original level from where they start

Figure 1.38a shows the situation. For three different inclines, the speed a particle acquires on reaching the level I-I remains same as the height '*h*' remains unchanged. Galileo also proposed that the bob of a swinging pendulum always reaches the original level from where is started as shown in Fig. 1.38b. (In the modern language, this demonstrates the conversion of gravitational potential energy to kinetic energy and vice versa.)

1.9.3 Discovery of the Law of Inertia of Motion in Its Primitive Form

While studying the pendulum motion, Galileo did have the glimpse of the tendency of the bob to continue in its motion after reaching the lowermost position. This did strike Galileo's thinking, and the persistent nature of motion was revealed to him. Subsequently, Galileo arrived at a crude form of the first law of motion—law of inertia of motion. The excerpt from Galileo's book given below demonstrates that he had arrived at a rudimentary idea of the law of inertia. He describes what 'happens to a heavy and perfectly round moveable placed on a very smooth plane'. In his own words 'If the plane were not inclined, but horizontal, then this round solid placed on it would do whatever we wish; that is,

if we place it at rest, it will remain at rest, and given an impetus in any direction, it will move in that direction, maintaining always the same speed that it shall have received from our hand and having no action to increase or diminish this, there being neither rise nor drop in that plane'.

However, in Galileo's scheme, the whole proposition is with respect to the earth and the continuity of motion was implied to be parallel to the earth's surface, i.e. along a circle. Thus, the true spirit of the most fundamental of the laws of motion is absent in his thought process. He considered his all laws of mechanics strictly with reference to the terrestrial surface. Galileo mentioned quite emphatically that the scheme of a horizontal plane is valid only locally. In reality, the eternal motion will be along a circle around the earth.

1.9.4 Laws of Compound Motion: Projectiles

One of the most significant contributions to the development of the science of motion made by Galileo was the development of the method to describe compound motion. A compound motion consists of two independent and simultaneous primordial motions. Galileo employed the technique used by the Greeks. It was known that if a point

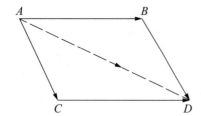

Fig. 1.39 Composition of two uniform motions

(a)   **(b)**

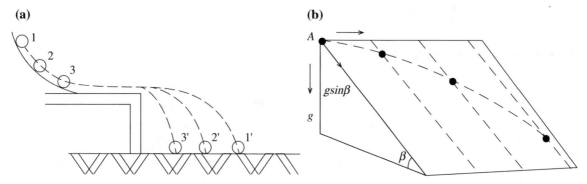

Fig. 1.40 **a** Experiment on law of inertia, **b** projectile experiment on an inclined plane

Fig. 1.41 Orbital motion as a compound motion

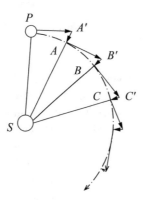

moves uniformly along a straight road *AB* (from *A* to *B*) and the rod is moved uniformly in a direction different from the line *AB* (Fig. 1.39), say along *AC*, then the point's resultant motion will be along the diagonal *AD*.

According to Drake, the discovery of the theory behind projectile motion was a by-product of Galileo's experiment to verify the law of inertia of motion.

Galileo allowed a smooth heavy ball to slide down an incline of known height acquiring a definite speed (depending on the height of decent). The ball is then allowed to drop to the floor below (Fig. 1.40a). Galileo considered the motion of the ball (after it leaves the table surface) to be composed of a uniform horizontal (inertial) motion and a vertical free fall motion. The distances of the point the balls

hit the floor should be proportional to the horizontal velocity acquired depending on the height of the decent. Galileo also used his inclined plane technique for studying projectile motion. It is indicated in Fig. 1.40b.

The reason for better accuracy was, again, the reduced component of the acceleration due to gravity along the plane.

On the fourth day, Galileo states in Proposition 1, Theorem 1 'when a projectile is carried in motion compounded from equable horizontal and from naturally accelerated downward, it describes a semi-parabolic line in its movement'.

This work on compound motion by Galileo was extremely important as all planetary motions were later described as a compound motion consisting of an inertial motion and a continuous free fall towards the centre of attraction.

Figure 1.41 explains the situation. An orbiting object at *P* moves to *A'* due to inertial motion but due to attraction towards the centre of attraction *S* falls to position *A*. Then, it continues due to inertia to reach *B'* but free fall towards *S* brings it to position *B* and the process continues.

Scientists and philosophers before Galileo also treated projectile problems for study. Unfortunately, their studies were purely qualitative in nature. Though they knew that projectiles follow curved paths they failed to notice any general pattern in those. Decomposition of motions into their basic elements which follow specific rules independently was a major breakthrough for the development of the science of motion.

1.9.5 Galilean Relativity

One of the major objectives Galileo pursued was to establish Copernican theory on firm scientific ground. One of the very natural and obvious objections to any suggestion of the daily rotation and the orbital motion of the earth was the complete absence of any sensory feeling and terrestrial observational[17] effects. The usual common sense told people that if the earth is a moving body then a stone dropped from the top of a tower should not fall at the base of the tower but a distance away opposite to the direction of motion of the earth at that location. Since the earth rotates once in every 24 h and the earth's equatorial circumference is about 24,000 miles, the speed of the earth's surface in the tropical region (due to its daily rotation only) is about 1000 miles/h! It is a large speed, and the effect should be very noticeable.

Galileo was the first to give explanation for the absence of any effect of the earth's motions on the objects moving on or near the earth's surface. In the process, he discovered one of the most profound scientific principles what Einstein termed as 'Galilean Relativity'. Galileo correctly mentioned that all objects on the earth share the same 'natural motion' of the earth and therefore cannot be noticed. He gave an extremely interesting example to clarify the statement. He suggests the readers to consider an artist sitting on the deck of a ship as it sails from Venice to Aleppo and drawing a picture. Next, he asks the reader to imagine the true path of the tip of the artist's pen as observed by an observer in one of the two cities. He suggests that it will be a long line from Venice to Aleppo with imperceptible deviations. But the picture drawn does not reveal any trace of the actual motion of the vessel. Thus, Galileo observes 'Motion, in so far as it is and acts as motion, to that extent exists relatively to things that lack it; and among things which all share equally in any motion; it does not act, and is as if it did not exist'.

Galileo gives the example of a cannon ball dropped from the tip of a mast of a ship sailing in the sea.[18] He mentions clearly that when a ball is dropped from the mast top the ball already has the motion of the ship. So when it makes its downward journey, it continues to move with the ship in the forward direction while accelerating down due to gravity. Thus, to an observer on the ship, who is also moving along with the ship, there is no sign of any deviation of the ball from its vertical straight path. He finds that the ball drops to a point at the base of the mast, the amount of forward motion of the ball being the same as that of the mast base during the period of fall. However, to an observer standing on the coast, the ball will appear to take the path of a projectile and describe a semi-parabolic path. At this point it is interesting to note that more than a thousand years before Galileo, Aryabhata I suggested the same principle giving the example of motions of objects on a moving boat in his famous book Aryabhatyam. He did it to support his proposition for a spinning earth.

Figure 1.42 explains the phenomena. The ship is moving towards right with a constant speed V, and the observer B is standing on the deck. There is another observer A standing on a fixed jetty. If the cannon ball takes a time t to reach the deck of the ship, the ship (and everything else attached to it) moves a distance Vt towards right (Fig. 1.42a) and the path of the ball is a semi-parabolic. This is because of the reason that when the ball started its decent it had a horizontal speed V (that it acquired at the start of the journey being a part of ship moving with a speed V). Observer A will notice this curved path, but to the observer B at every instant of time the horizontal displacement of the ball will be equal to that of observer B. Hence, B cannot see any relative movement of the ball in the horizontal direction. This study also reveals the significance of the concept of a frame of reference. Galileo described very correctly that if one shuts himself in a completely closed cabin in a ship he cannot decide whether the ship is sailing (in a calm sea) or not by observing all kinds of phenomena happening inside the cabin. It will be seen later how the application of Galilean relativity can help in discovering new principles of mechanics.

1.10 Collapse of the Old Science

Galileo's telescopic observations and his pioneering role in bringing experimental observation and verification of theory brought destruction to the most antique concepts in science. Kepler's mathematical genius and work leading to the establishing of the idea that an object can influence another body's motion from a distance without any physical contact paved the way for Newton to carry out the final grand synthesis leading to a matured science 'mechanics'. The most important of these was the elimination of the age-old idea of 'natural motion' and 'violent motion'. Kepler brought the idea of 'force' in a rudimentary form. He also brought the strict mathematical regularity obeyed by the solar system, and how the geometric and temporal aspects of the planetary system could link together was first unravelled by Kepler. Even today, one is awed to recognize the incomprehensible combination of utmost dedication and unparalleled drudgery Kepler's scientific life was composed of! The seed of future success in discovering the mystery of

---

[17]Foucolt's pendulum was too far in the future.

[18]In the seventeenth century, it was a common belief that such a cannon ball does not chop at the root of the mast but drops at a point somewhat behind the base at distance opposite to the direction of motion of the ship. This belief was so much firmly ingrained in the minds of the people that nobody ever thought of verifying it by conducting an actual observational test. In fact, the absence of any tendency of objects being left behind a moving earth's surface (when an object leaves physical contact with ground) was used by the people against a moving earth theory as a clinching evidence in favour of a geocentric model.

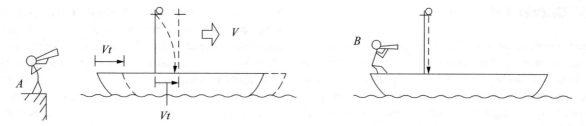

Fig. 1.42 Principle of Galilean relativity

the solar system was hidden in Kepler's concept of inter-action between the sun and planet, and the theory of compound motion as proposed by Galileo as the clue to the success was hidden in the understanding that all planetary motions are composed of inertial motion and a free fall to the sun.

However, the deepest mystery of dynamics is hidden in the fact that it is acceleration which plays the central role in the 'force and motion' type of causal relationship. It was beyond the capacity of both Kepler and Galileo to untangle this mystery. But Galileo helped to divert the attention of the scientists from the heaven to the motions occurring on the terrestrial surface. It was important. The terrestrial experiments have the advantage that a scientist can repeat such experiments with different initial and boundary conditions. On the contrary, the heavenly motions are unique and eternal. They can be observed but cannot be changed by altering the conditions. This shift of focus from the heaven to the terrestrial motions helped the scientists and the philosophers to uncover the deeper layers of the science of motion using better time measuring techniques and breaking free from the tradition of 'received wisdom'. Galileo also sowed the way how sometimes thought experiments help to deduce valuable hypotheses through 'hypothetico-deductive' method.

1.11 Descartes: Beginning of Inertial Science

It has been mentioned earlier that Galileo was the last link with the old science. He, undoubtedly, laid the stepping stones towards the new science but failed to take the necessary steps out of the old premise. The required necessary impetus to break away from the Aristotelian physics came from Rene Descartes (1596–1650). Though his contribution in mathematics, introduction of coordinate (or analytical) geometry, is well recognized, his contribution to the evolutionary process of the science of motion has remained somewhat obscured. It is, perhaps, due to the fact that his important work on dynamics was much entangled with his 'physics of the universe' that is unimpressive to most modern readers of science and obviously very wrong. The knowledge about the universe was so meagre in the early

seventeenth century that any attempts to explain the universal phenomena were bound to be trivial and wrong.

1.11.1 Law of Inertia of Motion

The fundamental difference between the approaches followed by Galileo and Descartes was Galileo wanted to study and analyse individual phenomenon related to motion, whereas Descartes took a holistic approach. Unlike Galileo, he proposed to establish the basic physics of the whole universe. He believed that the same laws of physics are valid for the whole universe. This was a revolutionary idea at that time which can hardly be appreciated nowadays. The idea of this universality of the laws of science came to him as he was building up a physics of the universe to explain all the observed phenomena. He considered that the two basic properties of all matter were 'extension' and 'motion'. According to his physics, God provided the 'primordial' motion to all material at the time of creation that could be only transferred from one body to another through interactions and collisions but could not be annihilated. This, perhaps, was the first concept of a conservation law which later got developed into very important fundamental principle of mechanics. This concept of 'indestructibility' of motion gave rise to the preliminary idea of 'inertia of motion'. The long-standing puzzling question of old science—'what keeps a thrown body to continue in its motion when the contact with the thrower is no more there?'—was reversed by Descartes. He suggested—'we should ask simply, why the body should not continue to move forever'. He also investigated the sling extensively. He observed that 'when one whirls a stone in a sling, not only does it go straight out as soon as it leaves the sling, it presses in the middle of the sling and causes the cord to stretch. It clearly shows thereby that it always has an inclination to go in a straight line and that it goes around only under constraint'. The language is very clear, and the principle is almost indistinguishable from the modern version of the first law of motion. One should notice the fundamental difference between the understanding of the law of inertia by Galileo and that by Descartes, whereas Galileo's whole premise was the surface of the earth

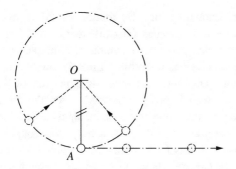

Fig. 1.43 Law of inertia from a sling

where a gravitational field exists; Descartes' law is irrespective of any particular situation and is in the form of a true universal law. Figure 1.43 shows the motion of a stone when the string of a sling in motion breaks.

That Descartes' concept of inertia was much matured can be judged by the three laws of nature he proposed in his book 'The World'. He proposed in his first rule the concept of inertia of rest as shown below[19]:

> The first is that each individual part of matter always continues to remain in the same state unless collision with others forces it to change that state.

His second rule was as follows:

> I suppose as a second rule that, when one of these bodies purchase another, it cannot give the other any motion except by losing as much of its own at the same time; nor can it take away from the other body's motion unless its own is increased by as much. This rule, joined to the preceding, agrees quite well with all experiences in which we see one body begin or ease to move because it is pushed or stopped by other.

In this rule, one can easily identify the rudimentary form of the conservation principle. Descartes' third rule is given in the following form:

> I will add as a third rule that, when a body is moving, even if its motion most often takes place along a curved line and can never take place along any line that is not in some way circular, nevertheless each of its individual parts tends always to continue its motion along a straight line. And thus their action, i.e. the inclination they have to move, is different from their motion.

So, the most fundamental principle in dynamics is expressed with no ambiguity, and the foundation of the inertial science of motion is laid down.[20] Nowadays, it is, almost, incomprehensible to all students of science how difficult the task was. No one before could imagine that the mystery of

dynamics was hidden in the uniform rectilinear motion as such motion was never observed either on the earth or in the heaven.

1.11.2 Collision Problems and Early Concept of Momentum Conservation

Descartes proposed that the primordial motions imparted by the God got transferred from one body to another and transformed from one type to another through collisions. Thus, it became absolutely essential for him to propose laws that are followed by such events. The emergence of the conservation principle and the concept of an entity like 'quantity of motion' were also inevitable in the process.

Descartes assumed the quantity of motion to be a product of the speed and volume of an object. He did not have any concept of 'mass' of an object. He assigned initial speeds of the colliding bodies 1 and 2 as u_1 and u_2 and the volumes of the bodies being β_1 and β_2. Given these quantities, the problem was how to predict the speeds after the impact, v_1 and v_2. As his basic fundamental principle was based on the conjecture that motion cannot be destroyed, he could write one relation as shown below:

$$\beta_1 u_1 + \beta_2 u_2 = \beta_1 v_1 + \beta_2 v_2$$

Though the above relation is correct when the directions of all the motions are same in a direct straight-line collision and the material of all objects is the same, it is not valid in a general sense. But the other necessary condition to determine the post-collision speeds could not be established by Descartes. Instead, he proposed six more ad hoc rules for finding the speeds. The solutions from these rules lead to confusing and contradictory results.

1.11.3 Descartes' Concept of Motion

In the old science, all motions were found to be observed with reference to the earth that was at a fixed centre of the universe. Thus, philosophically, the concept of motion posed no serious problem. After the dethronement of the earth from a fixed position at the centre of the universe, 'motion' became a problematic idea. Descartes was actually aware of the serious difficulty as one can find from his following statement:

> Furthermore, if we think that the earth moves, and travels from West towards the East exactly as far as the ship progress from the East towards the West; we shall once again say that the person seated in the stern does not change his place; because of course we shall determine his place by certain supposedly motionless points in the heavens. Finally, if we think that no truly motionless points of this kind are found in the universe; as will later be shown to be probable; then, from that, we shall

[19]Descartes', Rene 'Le Monde' (1664); 'The World: Le Monde' Translation by M. S. Mahoney, Abaris Books, New York (1979).
[20]Though Descartes' book 'Principles of Philosophy' was very influential at that time, a correct formulation of the law of inertia was published just before him by Pierre Gassendi. Unfortunately, he did not receive the kind of recognition that he deserved.

conclude that nothing has an enduring place, except in so far as its place is determined in our minds.

Thus, it is seen that Descartes had the preliminary idea about the relative nature of motion. The main contribution, therefore, in the development of dynamics was to delink it completely from the old science. Descartes was also the first to conceive that the laws of nature were valid everywhere in the universe. This was a revolutionary idea that time. In fact, this helped him to dissociate the inertial property of matter from its location and presented it in the form of a universal law and not just a particular type of behaviour of objects being experimented with on the surface of the earth. His concept of quantity of motion as the product of the volume of an object and its speed was obviously erroneous, but it should be remembered that for a new concept in fundamental science to gain maturity, thinking and experimenting by generations of great scientists are essential. In that sense, Descartes' work on collisions led to the beginning of the concept of momentum and the corresponding conservation principle. Unfortunately, for him, his attempt to develop a physics of the universe was too early and led to erroneous results, but in the process gave birth to the very important concepts in the science of motion.

1.12 Huygens: Breakthrough in the Discovery of Dynamics

Christiaan Huygens (1629–1695) was one of the brilliant scientists of the seventeenth century, and his contributions to the development of the science of motion were of prime importance. Looking at most books on mechanics and physics, it is amazing to note how his contributions are disastrously underestimated. He was a close friend and follower of Descartes. He also believed that all phenomena of mechanics could be understood by a process of interaction based upon physical contact. His attitude, inherited from Descartes (and also Galileo), prevented him from accepting the idea of an attractive gravitational force. As a result, he could not accomplish the synthesis of the science of celestial and terrestrial mechanics that was finally done by Newton. Huygens had the scientific talent to bring together all the elements of dynamics for the grand synthesis. His results were published properly after his death, and by then, Newton had already published his magnum opus—'The Principia'.

Besides making revolutionary breakthroughs in the science of motion, Huygens was a brilliant inventor and his invention of pendulum clocks made the progress in science possible. In a way, his work on the invention of clocks can be compared to that of Galileo for the telescope.

The seeds of the theory of relativity, that brought revolution in scientific thinking after more than two centuries, were first shown by him. He framed the Principle of Relativity of motion observed from different inertial frames, and using these along with some symmetry principles, he arrived at many conclusions resembling Einstein's work. According to Barbor, 'De Motu Corporum is a pre-run of the 1905 paper that created special relativity'. Even the principle of equivalence, that states the indistinguishability between inertial and gravitational forces locally, was conjectured by Huygens. It is this basic principle on which Einstein constructed his General Theory of Relativity. Though somewhat unrelated, his work on optics was also of phenomenal importance and he formulated the wavefront principle. It looks really strange that the students of mechanics do hardly get a chance to appreciate Huygens and all the limelight is taken away by Newton. As Barbor mentions, '... many of the formulas and results that students of dynamics must learn when they begin the subject were discovered by Galileo and Huygens before Newton published the Principle in 1687'.

1.12.1 Theory of Collision and Conservation of Momentum

To begin with Huygens reaffirmed Descartes' law of inertia of motion in a more succinct form as given below:

When once a body has been set in motion, it will, if nothing opposes it, continue the motion with the same speed in a straight line.

However, for tackling the problem of collision of two bodies, his second proposition was another important concept which first germinated in the mind of Galileo—the principle of Galilean relativity, but Huygens framed it in a very modern language. He mentions[21] very clearly that '... when two bodies collide but have in addition a further common uniform motion, they impart to each other impulses that, viewed by one that also partakes in the common uniform motion, are exactly the same as if the uniform motion common to all were not present'. Huygens used very modern approach of relativity according to which he describes the same event, collision of two bodies, as seen by the observer of a boat where the experiment is conducted and also by another observer standing still on the bank.

Figure 1.44 shows a collision experiment conducted by one standing on a boat moving with a uniform speed v_0 towards right. Let the conductor of the boat experiment be called A. A brings the balls (identical) by moving his hands. Let the right hand be moved at speed u_1 and the left hand be moved at speed u_2, so that $u_1 > u_2$. Thus, the balls 1 and 2 collide moving at speeds u_1 and u_2, respectively, as observed

[21]J. Barbor—'Absolute or Relative Motion', Cambridge University Press, 1989.

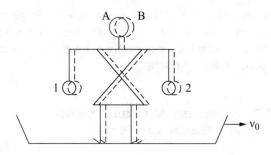

Fig. 1.44   Collision experiment as thought of by Huygens

Fig. 1.45   Huygens' thought experiment on the centre of gravity of two bodies

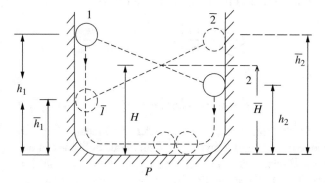

Huygens used this result with his centre of gravity law for having a complete solution for the collision problem. Huygens earlier showed that if through a mechanical phenomenon involving gravity the centre of gravity of a system of bodies could be raised it could lead to the development of a perpetual motion machine; i.e., work could be done without any input energy. Since from the antiquity it was not considered possible, Huygens suggested that at best the heights of the centre of gravity before and after the mechanical process could be same. Huygens did not have

by A. As observed by B, standing still on the bank, balls 1 and 2 move with speed $u_1 + v_0$ and $u_2 + v_0$. Speed towards right in Fig. 1.44 is considered to be positive. If A observes the collision and finds the post-collision speeds of 1 and 2 to be v_1 and v_2 (+ve or −ve depending on their directions), observer B will find the post-collision speeds to be $v_1 + v_0$ and $v_2 + v_0$, obviously. Next come the ingenuity of Huygens and the demonstration of the basic Principle of Relativity. Observer B, standing on the bank, conducts a separate experiment with spheres 1' and 2', identical to 1 and 2, by moving them with speeds $u_1 + v_0$ and $u_2 + v_0$. It is a different physical experiment, but if the two experiments are conducted simultaneously at adjacent locations the balls will appear to move together before experiment. Huygens asks 'what should be the post-collision speeds of spheres 1' and 2'?' Since the motions before the collision were identical in the two experiments, observer B cannot distinguish between their movements after the collision also and those will be $v_1 + v_0$ and $v_2 + v_0$.[22] Thus, if a collision can be described as

$$u_1, u_2 \rightarrow v_1, v_2$$

then any experiment with initial speeds $u_1 + v_0$ and $u_2 + v_0$ should yield the resulting speeds as $v_1 + v_0$ and $v_2 + v_2$.

Newton's concept of mass but what he did was far superior to Descartes' concept of quantity of matter. He considered weight of a body to present the quantity of matter it contains. If one considers the acceleration due to gravity, g, to be a constant on the surface of the earth, then weight is proportional to mass.

So, instead of using the sign w for mass, m can be used to match with the modern convention. It was also shown by Galileo that the speed, u, gained by a body after a free fall through a height 'h' is given by

$$u = \sqrt{2gh} \qquad (1.12.1)$$

He also showed that when the body again goes up it can reach up to height h. Now, let two bodies 1 and 2 of masses m_1 and m_2 are allowed to fall from original heights of the two bodies h_1 and h_2. They come down and collide at the location P at speeds u_1 and u_2, and reversing their directions starts climbing with speeds v_1 and v_2, reaching the final heights $\bar{h}_1$ and $\bar{h}_2$, respectively, as shown in Fig. 1.45.[23] So, at the start of the mechanical process, the height of the centre of gravity is

---

[22]The observer B could then go on to the boat and conduct the experiment with pre-collision speeds $u_1 + v_0$ and $u_2 + v_0$, and the result will be same (i.e. post-collision speeds $v_1 + v_0$, $v_2 + v_0$).

[23]One need not bother to find out if they reach their respective maximum heights simultaneously or not when they are dropped simultaneously. Once a body reaches the maximum height it can be locked at that position and wait for the other to reaches its topmost location.

$$H = \frac{m_1 h_1 + m_2 h_2}{m_1 + m_2} \qquad (1.12.2\text{a})$$

The final height of the centre of gravity at the end of the mechanical process is given by

$$\bar{H} = \frac{m_1 h_1 + m_2 h_2}{m_1 + m_2} \qquad (1.12.2\text{b})$$

Huygens stated that when the collision was elastic

$$\bar{H} = H \qquad (1.12.2\text{c})$$

Using (1.12.1) in (1.12.2a) and (1.12.2b)

$$H = \frac{m_1 u_1^2 + m_2 u_2^2}{2g(m_1 + m_2)} = \bar{H} = \frac{m_1 v_1^2 + m_2 v_2^2}{2g(m_1 + m_2)}$$

Hence

$$m_1 u_1^2 + m_2 u_2^2 = m_1 v_1^2 + m_2 v_2^2 \qquad (1.12.3)$$

So this quantity (known as twice the kinetic energy in modern language) is conserved in all elastic collisions. It has been shown earlier in this section that if the pre-collision speeds of these bodies be $u_1 + v_0$ and $u_2 + v_0$ the post-collision speeds have to be $v_1 + v_0$ and $v_2 + v_0$. According to (1.12.3), one gets

$$m_1(u_1 + v_0)^2 + m_2(u_2 + v_0)^2 = m_1(v_1 + v_0)^2 + m_2(v_2 + v_0)^2$$

or,

$$m_1 u_1^2 + m_1 v_0^2 + 2m_1 u_1 v_0 + m_2 u_2^2 + m_2 v_0^2 + 2m_2 u_2 v_0$$
$$= m_1 v_1^2 + m_1 v_0^2 + 2m_1 v_1 v_0 + m_2 v_2^2 + m_2 v_0^2 + 2m_2 v_2 v_0$$

Using (1.12.3) in the above relation, one gets

$$m_1 u_1 + m_2 u_2 = m_1 v_1 + m_2 v_2 \qquad (1.12.4)$$

This is the correct form of Descartes' conservation of quantity of motion relation. Equations (1.12.3) and (1.12.4) can be used to solve elastic collision problems of any two bodies. Huygens proposed in 1669 three fundamental laws of nature related to science of motion.

(i)   Momentum conservation
(ii) Kinetic energy conservation in no loss interactions,
(iii) The centre of gravity law mentioned above.

Huygens should be considered to be the pioneer in effectively using the Principle of Relativity and proposing the correct momentum conservation law. His energy conservation principle became very important at a later stage.

Before the section on Huygens' work on elastic collision problem is closed, it should be mentioned that he emphatically stated many times that the relative speed with which two impacting bodies approach before collision is the same with which the two bodies separate after the collision. Thus, the actual physical happenings at the point of contact depend only on the relative velocity.

1.12.2   Kinematics of Circular Motion and 'Centrifugal Force'

Study of motion related to rotation was a very important activity during the seventeenth century and eighteenth century. It was extremely relevant to the concept of a rotating earth. A major objection to the earth's rotation was that if the earth is spinning about its axis once every 24 h all objects on its surface, those were not rigidly attached to the earth, should be flung away. It was observed that when a wet wheel spins at high speed water particles fly away from the rim along tangential directions. Galileo suggested that the gravity is stronger than the tendency that tries to throw away the objects. The question of this tendency of objects on rotating bodies was also a serious topic taken up by Copernicus. In Descartes' works 'The World' and 'The Principles of Philosophy', this subject takes a significant part. He attempted to explain the terrestrial gravity with an idea of a kind of centrifuge effect of the cosmic vortex.

Huygens' main idea, to begin with, was to follow Descartes along the line showed by him. This led him to investigate the motion of rotating objects. The term 'centrifugal force' was first used by Huygens in his paper 'De Vi Centrifuga', and he proposed a number of theorems related to, what is called today, centripetal acceleration. He was the first to correctly determine the magnitude of centripetal acceleration.

In his first theorem, Huygens proposed on the subject, he proves that the centripetal acceleration[24] was proportional to the radius 'r' when the speed of rotation remains constant. In his original form, the statement is as follows[25]:

> If two equal bodies pass around unequal circles in equal times the ratio of the centrifugal force on the larger circle to that on the smaller is equal to the ratio of the circumferences, or the diameters.

Figure 1.46 shows the situation as depicted in Huygens' work. If a particle moves along the larger circle and starts at the point A after a time t, it comes to the point P. But if the particle gets detached from the rotating body at A according to Descartes and Huygens, it should move uniformly along

---

[24]In fact used 'centrifugal force' as a measure of the acceleration. He already came to the great revolutionary idea that force and acceleration are proportional. The next section will take up the topic in details.
[25]Barbor, J.—'Absolute or Relative Motion', Cambridge University Press, page 491.

Fig. 1.46 Dependence of centripetal acceleration on radius

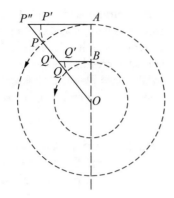

the tangent and reach the point P' (so that $AP' = \widehat{AP}$) in time t. If AP' is extended and the radius OP is also extended, they intersect at the point P''. Now, for very small values of t (i.e. when the phenomenon is restricted to the immediate vicinity of A), $AP' \approx AP''$ and PP' curve will be indistinguishable from the line PP''. Huygens conceived that the point reaches the location P, instead of the location P'' ($\equiv P'$) because of acceleration, a_A, directed towards the centre. He reasoned out that

$$P''P = 1/2\, a_A t^2 \qquad (1.12.5a)$$

Next, he considered a particle to move on the same rotating body and started at point B. Following the same logic elaborated above, Huygens found

$$Q''Q = 1/2\, a_B t^2 \qquad (1.12.5b)$$

where a_B represents the centripetal acceleration of particle B. Next, he realized that

$$AP'' = AP' = AP = \omega \cdot AO$$

and

$$BQ'' = BQ' = BQ = \omega \cdot BO$$

Hence,

$$\frac{AP''}{BQ''} = \frac{AO}{BO} \qquad (1.12.6)$$

Next, he found that

$$PP'' = OP'' - OP = OP'' - OA$$

and

$$OP'' = (OA^2 + AP''^2)^{\frac{1}{2}} \approx OA\left(1 + \frac{AP''^2}{2 * OA^2}\right)$$

as $AP''/OA \ll 1$, t being very small as mentioned above. From the above, one finds

$$PP'' \approx \frac{1}{2}\frac{AP''^2}{OA} \qquad (1.12.7a)$$

Similarly

$$QQ'' \approx \frac{1}{2}\frac{BQ''^2}{OB} \qquad (1.12.7b)$$

From the above two relations

$$\frac{PP''}{QQ''} \approx \left(\frac{AP''}{BQ''}\right)^2 \left(\frac{OB}{OA}\right) \qquad (1.12.8)$$

Combining (1.12.5a), (1.12.5b), (1.12.6) and (1.12.8), one gets

$$\frac{a_A}{a_B} = \frac{PP''}{QQ''} = \frac{AO}{BO} \qquad (1.12.9)$$

Hence, centripetal acceleration is

$$a \propto r \qquad (1.12.10)$$

for a given rotational speed.

Huygens' second theorem on centripetal acceleration states that it is proportional to the square of the rotational speed for a given value of the radius. In his language, 'If two equal bodies move around the same circle or wheels with unequal speeds, though uniformly, the ratio of the centrifugal force of the faster to the slower is equal to the ratio of the squares of the velocities'.

One should note that Huygens used the term 'two equal bodies' in both theorems. Thus, the force will be proportional to acceleration (or, in another way, 'force' represents 'acceleration'). Figure 1.47 is used for proving the second theorem.

The bodies describe the same circle with radius OA, but with different rotational speeds. One body moves a distance

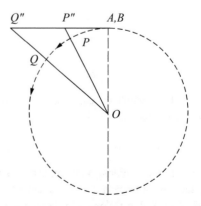

Fig. 1.47 Dependence of centripetal acceleration on speed

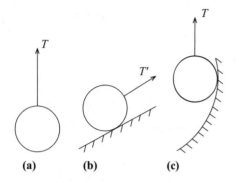

Fig. 1.48 Huygens' experiments of tug and acceleration

$\widehat{AP}$ in time t. The other one starts at the same point $A\ (\equiv B)$ and reaches Q after time t.

Therefore, $\widehat{AP} = \omega_A t$ and $\widehat{AQ} = \omega_B t$. Following the logic already explained, Huygens arrived at the relation

$$\frac{a_A}{a_B} = \frac{PP''}{QQ''} \qquad (1.12.11)$$

Now

$$PP'' = OP'' - OP = OP'' - OA \approx \frac{1}{2}\frac{AP''^2}{OA}$$

and

$$QQ'' = OQ'' - OQ = OQ'' - OA \approx \frac{1}{2}\frac{AQ''^2}{OA}$$

From the above two relations, one gets

$$\frac{PP''}{QQ''} = \frac{AP''^2}{AQ''^2} \approx \frac{\widehat{AP}^2}{\widehat{AQ}^2} = \frac{\omega_A^2}{\omega_B^2}$$

Hence, combining the above result with (1.12.11), Huygens arrived at his final result

$$\frac{a_A}{a_B} = \frac{\omega_A^2}{\omega_B^2}$$

or,

$$a \propto \omega^2 \qquad (1.12.12)$$

Combining the results of the two theorems, Huygens correctly arrived at the final result

$$a \propto \omega^2 r \qquad (1.12.13)$$

It should be remembered that modern language and symbolism have been used in the above derivations. But the logic is the same as that followed by Huygens.

Using the above results, Huygens was able to prove quantitatively that the centrifugal force on an object on the earth's surface is very small compared to that exerted by the earth's gravity, and therefore, bodies do not fly away even though the earth spins around its axis once every 24 h. The correct expression for the centripetal acceleration as derived by Huygens played a very important role in developing the science of motion as will be seen later in the volume.

1.12.3 Modern Concept of Force and the 'Force–Acceleration' Relation: Second Law of Motion in Primitive Form

Huygens made the real breakthrough in the science of motion and paved the way for the final development. Huygens accomplished this feat through a series of intense reasoning when trying to solve the 'centrifugal force' problem and an attempt to understand the phenomenon of gravity. This led him to study Galileo's free fall experiments. He found that the most important characteristic feature of the earth's gravity is a motion with constant acceleration. Slowly, it emerged in his mind that dynamically the most significant quantity could be acceleration. He studied the 'tug' on a string attached to a ball under different conditions as indicated in Fig. 1.48.

He noticed that the 'tug' was preventing the body to accelerate down. Comparing the magnitudes of the 'tugs' and the corresponding accelerations the ball could have when the 'tugs' were removed, he found that the magnitudes of the 'tugs' were correlating to the subsequent motion. He further observed that the 'tugs' did not have any relation to the displacement of the object. This he demonstrated using the cases shown in Fig. 1.48a, c. At the moment of release, the accelerations are identical though the subsequent gross motions are different. He made a revolutionary idea—'the tug represents what the bodies acceleration would be at the instant when the tug is released, not its subsequent overall motions'. Huygens concluded with the following statement:

> It is therefore clear that when we wish to determine the force we must consider, not what happens during a length of time after the body has been released, but rather what happens in an arbitrarily small amount of time at the beginning of the motion.

Thus, the nature of the relationship of the 'force' and 'motion' as an instantaneous proportionality between 'force' and 'acceleration' was discovered. This was definitely the most important (and most difficult too) discovery in the history of science of motion. Huygens also had the realization that a 'tug' (or force) is experienced only when constraints are imposed on to the motion of a body. When the tug is released, the body will be subjected to an immediate acceleration in the direction of the tug, and the acceleration will be proportional to the tug. In modern terminology, it can be written as $T \propto a$.

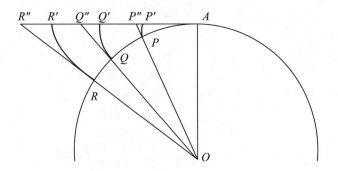

Fig. 1.49 Acceleration of a point in circular motion

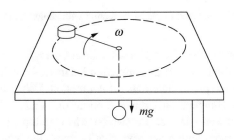

Fig. 1.50 Huygens' experiment to show equivalence of gravity and inertia force

This was a major breakthrough, and using this result along with his expression for the centripetal acceleration, the planetary motion could be solved by Newton and others. Of course, the 'force–acceleration' relationship could not be framed in the form of an equation. Only when Newton introduced the concept of 'mass', the proportionality sign could be replaced by an equality sign yielding the modern form of the second law of motion.

1.12.4 Early Concept of the Principle of Equivalence

Another great discovery was made by Huygens during the process of studying rotational motions. He considered a thought experiment as described below.

Figure 1.49 shows a particle at A describing circular motion with its centre at O. The particle takes the position P, Q and R at equal intervals of time so that $\widehat{AP} = \widehat{PQ} = \widehat{QR}$. On the other hand, it would have reached positions P', Q' and R', if released at A, instead of reaching positions P, Q and R. When the time intervals are very short and the whole experiment is confined to a region very near to A, the curves PP', QQ' and RR' become indistinguishable from the sections of the radial lines PP'', QQ'' and RR''. If the circumferential speed of the particle be v and the time intervals be 'τ', then $AP'' = v\tau$, $AQ'' = 2v\tau$ and $AR'' = 3v\tau$. Huygens considered the displacement $P''P$, $Q''Q$ and $R''R$ (all directed

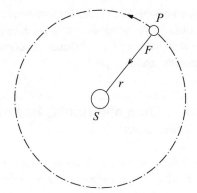

Fig. 1.51 Kepler's third law and law of gravitation

towards the centre O) to be the effect of centripetal action. He found that

$$P''P = P''O - OA = (AP''^2 + OA^2)^{\frac{1}{2}} - OA \approx \frac{1}{2}\frac{AP''^2}{OA}$$
$$\approx \frac{v^2}{2OA}\tau^2$$

Similarly,

$$Q''Q \approx \frac{v^2}{2OA}4\tau^2 \quad \text{and} \quad R''R \approx \frac{v^2}{2OA}9\tau^2$$

Thus, Huygens obtained the result that as in case of a free fall under gravity the displacement due to centripetal acceleration also is increasing as square of the time. Or in other words, the particle is falling towards O with a motion that resembles free fall under gravity. This led Huygens to the revolutionary realization that the effects of centrifugal force and the gravity are indistinguishable in a localized sense. One is amazed to note that this profound realization waited for two and half centuries and then was taken up by Einstein for developing his General Theory of Relativity.

To demonstrate the nature of equivalence between gravity and inertia force (centrifugal force is nothing but an inertia force in modern language), Huygens conducted experiments in which a rotating body could balance the tug due to another body's gravity as shown in Fig. 1.50. Huygens did not stop at the geometric analysis represented by Fig. 1.49, but he also showed that the path of the particle with respect to the point A when it is detached from the rotating wheel (and executes a uniform straight line motion) is an involute that starts at A in the radially outward direction. This way he could prove that the direction of centrifugal force will be in the radially outward direction.

Before ending the section on Huygens' contributions to the development of the science of motion, it is important to mention how a mindset can prevent one from making the final synthesis of a science in spite of developing all the basic elements. Newton's success in solving the problem of planetary motion and unifying the rules dictating the motions in the heavens with those for terrestrial movements became possible for his acceptance of the forces that can act without physical contact—a fundamental departure from Galileo, Descartes and Huygens.

1.13 Halley, Wren and Hooke: Rudiments of Gravitation

Once a discussion on Newton's contribution to the development of mechanics is initiated, one becomes so overwhelmed with his brilliance that the contributions of other contemporary scientists fade out into the background. Nevertheless, it must be remembered that there were a few contemporary scientists (most noteable among them being Halley, Wren and Hooke) who did make definite contributions towards the development of gravitational theory. As it happened with the science of motion, in case of gravitation also masterly final stroke by Newton almost obliterated others' useful additions. But they are there and should be acknowledged. It is quite possible that their work even helped Newton to arrive at his final results. It is Hooke who first proposed that the planets are subjected to an attractive force directed towards the sun.

Wren and Halley were interested in investigating the nature of the force that keeps the planets in their orbits. They used Kepler's third law along with Huygens' expression for centripetal acceleration to show that gravitational attraction diminishes as square of the distance.

Figure 1.51 shows a planet P in its orbit around the sun S. The eccentricities of most orbits being very small, the orbit can be approximated as a circle with radius r. Kepler's third law states that the square of the orbital period T is proportional to the cube of the orbital size, r. Thus

$$T^2 \propto r^3 \qquad (1.13.1)$$

Now, if the v be the orbital speed

$$T = \frac{2\pi r}{v} \qquad (1.13.2)$$

Using (1.13.2) in (1.13.1)

$$\frac{r^2}{v^2} \propto r^3$$

or,

$$v^2 \propto r^{-1} \qquad (1.13.3)$$

By using the concept of composition of motion developed by Galileo, scientists could figure out that the orbital motion was a combination of uniform inertial motion in the transverse direction and a free fall towards the sun due to its attractive force. So it was known that the gravitational force,

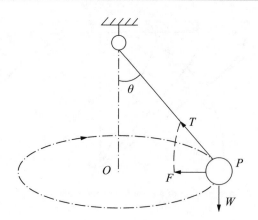

Fig. 1.52 Simulation of planetary motion with conical pendulum

F, acting on the planet is always directed towards the sun. They figured out that this attraction balances the centrifugal force on the planet due to its orbital motion. Using the work done by Huygens on circular motion, one finds the acceleration towards the centre (which when multiplied by the planets' mass with a negative sign becomes the 'centrifugal force')

$$a \propto \omega^2 r$$

$$a \propto \frac{v^2}{r} \qquad (1.13.4)$$

Huygens also demonstrated that

$$F \propto a$$

Using (1.13.3) and (1.13.4) in the above relation, it was possible to show that

$$F \propto \frac{1}{r^2} \qquad (1.13.5)$$

This way Halley and others came to the conclusion that gravitational attraction is inversely proportional to the square of the distance. Around the same time, Hooke conducted many experiments with conical pendulums to simulate planetary motions.

Figure 1.52 shows a conical pendulum. The component of the tension T in the direction PO can represent an attractive force always directed towards O (the vertical component of T balances the weight of the bob representing a planet). Hooke demonstrated various aspects of planetary motions (both circular and elliptical).[26] Using a double pendulum, Hooke was even successful in simulating the motion of the earth–moon system round the sun. Though the results were of purely qualitative nature, Hooke posed the

---

[26]The simulation is actually not correct as the attractive force F in these experiments is not inversely proportional to PO^2.

problem in a very focused manner. Hooke did make some valuable contributions which were, of course, of qualitative nature. However, the first focused presentation and clear statement of the planetary motion dynamics were undoubtedly made by Hooke.

1.14 Newton and the Final Synthesis

Traditionally, it is mentioned that Isaac Newton was born on the Christmas Day in the year 1642, the year Galileo died. According to the new calendar, his date of birth was 4 January 1643. There are a vast number of books and articles on his life and work. It is entirely justified keeping in view of his enormous contributions to the development in science in general and the science of motion in particular. Newton made many breakthroughs in different branches of physics, astronomy, mechanics and also was one of the creators of a new branch of mathematics—calculus.

Newton used many developments in his research on the motion of bodies those were already accomplished by Huygens about a decade or so before, but many theories were rediscovered by Newton himself. It will be attempted here to present the final synthesis of the results by him leading to the formation of the most fundamental branch of science—dynamics.[27] The main topics are the development of the concept of mass and the second law of motion, discovery of the third law, general solution of collision problem, law of universal gravitation and the solution of the problems of planetary motion. Using his second law of motion along with the law of universal gravitation, Newton completely solved the two-millennia old problem of planetary motions.

1.14.1 Concepts of Mass, Momentum, Force and the Second Law of Motion

Once the earth was dethroned from its fixed position at the centre of the universe, a serious difficulty was faced by the philosophers and scientists. The relational nature became apparent, and the point of contention was how such relative motion can be considered to follow some set of universal rules. The matter is still not fully settled till today. Newton based his whole science of motion on the basis of a concept of absolute space and universal time. Newton developed his whole mechanics using the absolute space as the fame of reference for all motions.

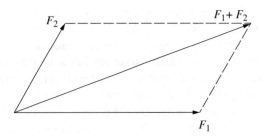

Fig. 1.53 Law of addition of motive force

Newton's concept of inertia was more or less the same as developed by Descartes and Huygens, but he was more systematic in analysing the situation. He had primarily two types of 'forces' that are linked to the inertial property of moving bodies. The first type of force that he called 'vis insita' or 'innate force' of matter keeps a body moving uniformly when imparted with the motion at the start. This 'force' or 'impetus' for motion is proportional to the body. Newton conceived the idea of 'inertial mass' as a quantitative measure of the content of matter in a body from this definition. Newton, unlike his predecessors, made a clear distinction between 'mass' and 'weight' and mentioned that these two quantities always maintain a fixed ratio. This was a great achievement that helped to transform the rules of dynamics to become applicable beyond the terrestrial region free from its gravity. Weight is a quantity that depends on the earth's gravity, but mass is an invariant quantity for a particular body.[28] Newton polished the statement of the law of inertia of motion (proposed by Descartes and Huygens) as his first law in the following form:

Every body continues in its state of rest, or of uniform motion in a right line, unless it is compelled to change that state by forces impressed upon it.

Kepler's original concept of inertia as an inherent laziness of matter is accepted by Newton, and he extends it to the situation where continuing in a particular state of motion is also included in the definition of 'laziness'.

Unlike Descartes, Newton recognized from the very beginning the importance of the direction of motion beside its magnitude, the speed. Thus, the quantity of motion, the product of the quantity of matter with speed, was also a directional quantity. This was the modern concept of momentum, and Newton recognized that even to change the direction without causing any change in speed (i.e. the quantity of motion) an external 'motive force' has to act on the body. His other concept of 'force' was this 'motive force', and instead of philosophizing too much about its true status, very intelligently he decided that whatever the real

---

[27]This word 'dynamics' was first coined by Leibniz in relation to his work on energy-based mechanics. The word was finally given its current meaning by D'Alembertt in his famous book 'Traite de Dynamique' published in 1743.

[28]So a ball may appear very light on moon's surface, but it will generate the same impact on the toe (as on the earth) when kicked.

nature of this 'motive force' may be its quantitative measure must be through the change in the 'innate force' (momentum in today's language) it produces. He also conjectured that the direction of change of the momentum has to be the same as that of the impressed 'motive force'. He identified this force with change in directional speed (velocity) multiplied by the bulk (mass) of the body. Thus, a matured dynamics started crystallizing to handle motion in three-dimensional space.

Another advance Newton made was to free the concept of force from its Cartesian status of 'instantaneous physical impact'. Instead, he conceived a continuously acting force as a series of impacts spaced with infinitesimally small time intervals. The parallelogram law of addition of motive force (Fig. 1.53) was clearly stated by him.

If a body is acted upon by two forces, F_1 and F_2 simultaneously, the change of motion of the body (i.e. the instantaneous acceleration) will be along the diagonal and its magnitude will be proportional to the force with a magnitude given by the length of the diagonal. Newton's concept of a continuously active force as the result of innumerable infinitesimally small impacts slowly led to the need to handle such situations mathematically. This could be the impetus for Newton to develop calculus. He formally defined 'motive force' (or, impressed force) as follows:

> An impressed force is an action exerted upon a body, in order to change its state, either of rest, or of uniform motion in a right line.

The proportionality of 'tug' (or force) was already demonstrated by Huygens. It was also known to the scientists in UK as Wren and Halley used that to deduce the inverse square law of the gravitational attraction from the third law of Kepler. Hence, the second law of motion, that gave a quantitative description between the causes (force) and effect (change in motion), finally came from Newton in the following form:

> The change of motion is proportional to the motive force impressed, and is made in the direction of the right line in which that force is impressed.

One should note that Newton's original formulation of the second law was also in the form of proportionality. Using mass as a quantity of matter and its product with velocity to yield the quantity 'momentum', the second law can be in the form of an equality, by suitably choosing the unit of mass, in the modern form:

$$\overline{F} = \frac{d}{dt}(m\overline{v}) \qquad (1.14.1)$$

$$\overline{F} = m\overline{a}$$

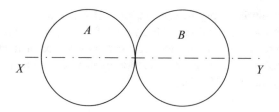

Fig. 1.54 Newton's study of collision problem

1.14.2 Collision Problem and the Discovery of the Third Law

It is evident that the first law of motion was discovered by Newton's predecessors almost to its final form. The basic tenant of the second law was also unravelled to a great extent by Huygens. However, the credit for discovering the third law of motion totally goes to Newton.

Newton solved the collision problem in a far more elegant manner after almost a decade of Huygens' solution. It appears that Newton was not familiar with Huygens' work. It is noteworthy that Newton's solution to collision problem was applicable to both elastic and inelastic impacts. While solving the collision problem, Newton received a number of deep understanding those led him to the development of a mature and elegant dynamics. While studying the collision problem, the precise idea of 'force' crystallized in Newton's mind and he realized that 'force' never appears in isolation. A force is always in pairs comprising of the impressed force and its equal and opposite reaction. Using Newton's language in studying the collision problem, the following statement indicates how the idea of third law emerged.

Figure 1.54 shows two bodies A and B in collision along the line XY. Newton's statement could be written as follows:

(a) If A presses B towards Y, then B presses A towards X (i.e. in the opposite direction). It is obvious.
(b) A body must move that way in which it is pressed.
(c) If two bodies A and B meet the one and the other, the resistance in both is the same for so much as A presses upon B so much B presses on A. And, therefore, they both must suffer in equal changes in their motion.

This third law that the force on B due to A is equal and opposite to the force that acts on A due to B provides the necessary condition for solving collision problem, both elastic and inelastic. Newton attempted to justify the law using the results of a significant number of experiments on collisions. The scheme of the experiments was based on that demonstrated by Wren in the Royal Society. Pendula of different weights were suspended from points A and B (Fig. 1.55).

Fig. 1.55 Wren's collision experiment demonstration in the Royal Society

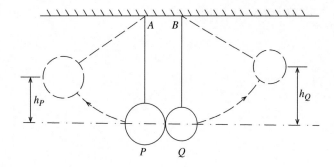

The speeds of the two spheres P and Q with which they collided could be controlled (and also determined) by adjusting the heights from where they were allowed to ascended, using Galileo's formula ($v_Q = \sqrt{2gh_Q}$). Again, the velocities after the collision could be estimated from the heights they ascended to. Newton used the length of the pendula as 10 ft and adjusted h_P and h_Q to 8, 12 and 16 ft. He always found (within an error of 3 inches, in the reascended heights) that the forces[29] acting on P and Q during collision were equal and opposite. Newton, of course, used the language '… the action and the reaction were always equal'.

Once the equality of action and reaction is realized, the solution becomes simple. It is true the scientists and philosophers were aware that every action involves some kind of reactions. But Newton's credit goes in discovering the magnitudes to be equal. This Newton did when he was around 23–24 years old. Much later, at the time of formalizing his work on the science of motion, he proposed this discovery as the third law of motion by generalizing it so that the rule could be applicable to all types of forces.

1.14.3 Law of Universal Gravitation and Planetary Motion

Newton's most spectacular success was in finally discovering the law of universal gravitation and solving the two-millennia old planetary motion problem. He not only solved the problem of the orbital motion of the planets but also solved a number of other finer and more subtle problems (like precision of the equinox). Of course, one should remember that his approach was more of geometrical nature rather than the elegant treatments found in the currently used textbooks.

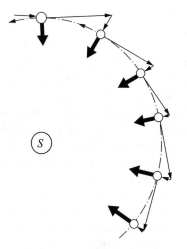

Fig. 1.56 Orbital motion as a composite motion

As mentioned earlier, the concept of gravitation as an attractive force between two bodies was first started by Kepler. He also proposed that the sun's gravitation keeps the planets orbiting around the sun. However, it was primarily Hooke (and also Halley and Wren) who successfully showed the gravitational force to be an attractive force towards the sun. It was also recognized that the orbital motion is nothing but a composite motion composed by inertial motion superimposed on the constant falling motion towards the sun as indicated in Fig. 1.56.

The credit for the concept of composite motion goes obviously to Galileo who derived the trajectory of a projectile using the concept. The fact that the attractive force diminishes as square of the distance was also figured out by using Kepler's third law and Huygen's formula for centripetal acceleration. Hooke first posed the problem of gravitation to Newton when Hook was the curator of the 'English Academy of Sciences (Royal Society)'. In a reply to a letter from Hooke regarding the trajectory of a falling body (due to the earth's attraction), Newton proposed the trajectory as shown in Fig. 1.57a. Hooke corrected Newton and suggested that in the absence of any resistance due to air the trajectory should be as shown in Fig. 1.57b.

---

[29]However, it is clear that the studied quantity was impulse (in modern language) and not force. But considering the fact that the duration of the impact was very small, the equality of average forces could be represented by the equality of the impulses.

Fig. 1.57 Trajectory of a falling body

(a)   **(b)**

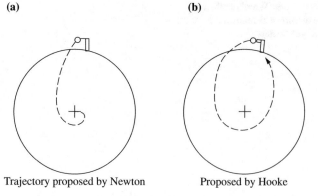

Trajectory proposed by Newton   Proposed by Hooke

Hooke also tried to derive the shape of an orbit due to gravitational attraction (inverse square law); he did not possess the necessary skill, but Hooke found approximate solutions of the 'equation of motion' using Huygen's formula. He drew the orbits and found those to be somewhat like ellipses. He called them as 'ellipsoids'. Not being able to mathematically prove, he suggested to Newton to take up the problem.

Newton took up the problem of gravitation and proved that Kepler's second law (equal area in equal time) is valid for any central force, not necessarily for the inverse square force due to gravitational attraction. Like most other contemporary scientists, he also used geometry for analysing the problems of motion. Newton replaced the continuous attraction of the sun into a series of closely placed impulses causing the change in inertial motion.

If the trajectory is locally approximated by a circle with O, as, the centre (Fig. 1.58) a planet occupies positions B', C', D', ... (starting from A) in times t, $2t$, $3t$, ... In the absence of any attractive force, the planet would have moved in a straight line (first law of motion) uniformly and would have occupied positions B, C, D, ... at times t, $2t$, $3t$, ... (starting the clock when the planet was at A). Since the whole analysis is being done in the immediate neighbourhood of A, $A - B' - C' - D'$ can be approximated as a circle and the lengths of arcs $\widehat{AB'}$,

$\widehat{B'C'}$, $\widehat{C'D'}$ are equal. Similarly, $AB = BC = CD$ ... and points $OB'B$, $OC'C$, $OD'D$, ... are collinear. Now, Newton showed that

$$BB' = \left[OA^2 + AB^2\right]^{1/2} - OA$$
$$\approx \frac{1}{2} \times \frac{AB^2}{OA} = \frac{1}{2} \times \frac{v^2}{OA} \times t^2 = \frac{1}{2}ft^2$$

$$CC' = \left[OA^2 + AC^2\right]^{1/2} - OA$$
$$\approx \frac{1}{2} \times \frac{AC^2}{OA} = \frac{1}{2} \times \frac{v^2}{OA^2} \times (2t)^2 = \frac{1}{2}f(2t)^2$$

Since $AB = vt$, $AC = 2vt$, $AD = 3vt$ with v as the speed at A, the fall towards O given by BB', CC', DD', ... increases as square of the time—a result of free fall.[30] Then, Newton proceeds to prove the second law as described below.

The motion of the orbiting body about a fixed centre of attraction O is divided into small time intervals. The resultant motion (giving the position of the body at the end of each interval) is due to an inertial uniform rectilinear motion compounded by a displacement due to the attraction towards O. Figure 1.59 shows the scheme. Let P_1 be the starting position, and the orbiting body has a velocity in the direction 1. In a small interval of time, it will reach the position P_2' as shown. The sun exerts an attractive force in a direction AO where A is at the middle of the segment P_1P_2'. If the resultant effect of the attractive force during the period τ (when the body goes from P_1 to P_2') be represented by a displacement in a direction parallel to AO (average direction during the whole time interval), the final location the body will reach is a point P_2 so that P_2P_2' is parallel to AO.

After moving from A to P_2, the body would have moved along the same direction and would have reached position P_3' after a period of time τ in the absence of any attraction towards O. However, if the resultant effect of the attractive

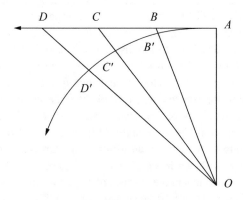

Fig. 1.58 Motion due to an attractive force

---

[30]This analysis is similar to that obtained by Huygens earlier while deriving the expression for centripetal acceleration.

Fig. 1.59 Newton's proof of Kepler's second law

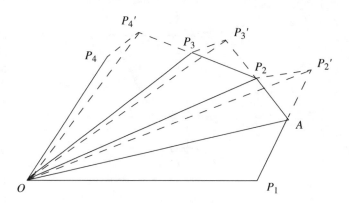

force be a displacement along the mean direction of attraction given by P_2O, it will reach the position P_3 as shown so that P'_3P_3 is parallel to P_2O. Now, the following conditions are satisfied.

$$P_1A = AP_2; P'_2P_2 \text{ is parallel to } AO$$
$$AP_2 = P_2P'_3; P'_3P_3 \text{ is parallel to } P_2O$$
$$P_2P_3 = P_3P'_4; P'_4P_4 \text{ is parallel to } P_3O$$

The time taken to travel the segments P_1A, AP_2, P_2P_3, P_3P_4 is same and equal to $\tau/2$. Again, the following relations among the areas of the triangles are also satisfied.

$$\text{As } P_1A = AP'_2, \Delta P_1OA = \Delta AOP'_2;$$
$$\Delta AOP'_2 = \Delta AOP_2 \text{ as } P'_2P_2 \text{ is parallel to } AO$$
$$\text{As } AP_2 = AP_2P'_3, \Delta AOP_2 = \Delta P_2OP'_3;$$
$$\Delta P_2OP'_3 = \Delta P_2OP_3 \text{ as } P'_3P_3 \text{ is parallel to } P_2O$$
$$\text{As } P_2P_3 = P_3P'_4, \Delta P_2OP_3 = \Delta P_3OP'_4;$$
$$\Delta P_3OP'_4 = \Delta P_3OP_4 \text{ as } P'_4P_4 \text{ is parallel to } P_3O$$

The following relations yield the following area conditions

$$\Delta P_1OA = \Delta AOP_2 = \Delta P_2OP_3 = P_3OP_4 = \cdots$$

Since the time taken to traverse the distances P_1A, AP_2, P_2P_3, P_3P_4 is equal, the second law of Kepler is proved.

Of course, Newton conceptualized that when the time intervals become infinitesimally small the trajectory will be a smooth continuous curve. He also emphasized that the second law is valid for any central force as no condition on the nature of force is assumed in the proof above.

1.14.4 Universality of Gravitational Force

Newton demonstrated that the inverse square law of gravitational attraction is not of local character but valid universally. That is why the name 'universal gravitation' was coined by him. At present, it is generally understood that the

laws of nature are valid universally and at all times. However, that was not the concept generally prevalent among philosophers. According to Aristotle's philosophy, the rules for the heavenly bodies were different from those followed on the earth. Thus, the idea of a law valid all over was a very novel concept in the early periods of the modern science.

Newton demonstrated this with the example of a falling body on the earth's surface and the moon. He was successful in demonstrating that the moon is continually falling towards the earth by the same gravitational attraction that causes a body to fall on the earth's surface. This wonderful demonstration was done by Newton around 1665 when he was just a 21-year-old young person! His analysis is presented below.

Newton first found the distance through which moon falls towards the earth in 1 s. From the work done by the Hellenistic astronomers, Newton was aware that the distance to moon from the earth is approximately 60 times that of the radius of the earth. Since the distance of an attracted object has to be measured from the centre of the earth, the strength[31] of the gravitational force on moon will be 1/3600th of that on a falling body near the earth's surface. Again, from the work of Eratosthenes, the earth's radius was known to be 6400 km. Therefore, the distance to the moon was 384,000 km. Since the moon takes 27.3 days to make one complete round of its orbit, the speed at which it moves along the orbit is given by

$$v = \frac{2\pi \times 384,000}{27.3 \times 24 \times 3600} \text{ km/s}$$
$$= 1.023 \text{ km/s} \approx 1000 \text{ m/s}$$

From Fig. 1.60, it is seen that if there had been no attraction by the earth the moon would have gone to position A (starting from position M) in 1 s. But it goes to the position M' due to the gravitational attraction of the earth. Hence, the fall of the moon towards the earth in 1 s is given by the

---

[31]Strength of gravitation can be indicated either by the force on a body of unit mass or by the acceleration due to gravity at the location.

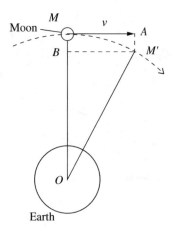

Fig. 1.60 Falling moon

height AM'. Now, $AM' = MB$ when B is the point (on the radial line OM) where a perpendicular line to OM from point M' meets OM. Again by geometry

$$M'B^2 = MB(2OM - MB)$$
$$\approx 2 \times MB \times OM \quad \text{(neglecting } MB^2 \text{ as } MB \ll OM)$$

Hence,

$$M'B = AM \approx \sqrt{2 \times MB \times OM}$$
$$= \sqrt{2 \times MB \times 3.84 \times 10^5} \text{ km}$$

But AM is nothing but the distance travelled by the moon in 1 s, i.e. its velocity. Converting all quantities in m, Newton obtained[32]

$$2 \times MB \times 3.84 \times 10^8 \approx AM^2 = 1000^2$$

or,

$$MB = AM' \approx \frac{1}{2 \times 3.84 \times 10^2} \text{ m} = \frac{1}{768} \text{ m}$$

On the surface of the earth, an apple falls $\frac{1}{2} \times g$, i.e. $\frac{1}{2} \times$ 9.81 m in 1 s. Thus, the ratio of the distance of fall by moon to that of the apple in 1 s is given by

$$\frac{\frac{1}{768} \text{ m}}{\frac{9.81}{2} \text{ m}} = \frac{1}{3767}$$

That is close to 1/3600 as predicted. Hence, Newton concluded that it is the same force that acts on both the apple and the moon.

---

[32]Of Course, Newton did not use metre as the length dimension. However, it must be mentioned here that this idea that celestial motions are governed by the same forces which control the motion of bodies on the earth was not original to Newton. Before him, the same idea was found in the propositions of Kepler, Borelli and Hooke.

Of course, the matter was not simple for Newton. The distance to moon from the earth was less ambiguous. But Newton had to first prove that for the apple the distance had to be taken from the centre of the earth. It took him almost 20 years to prove that, and he published his work only in 1685!

1.14.5   Orbit for Inverse Square Law

It has been shown in Sect. 1.13 that gravitational pull was known to decrease as the square of the distance from Kepler's third law and Huygen's formula for centripetal acceleration. However, the task to prove that the resulting orbit due to such a force is an ellipse remained as a challenge.

The discovery of the relationship of the inverse square law of gravitational attraction with orbit shape has an interesting history. The idea of explaining the orbital motion of the planets by combining the inertial motion and the attraction by the sun was first hinted by the Italian philosopher Giovanni Alphonso Borelli (1608–1679) in the year 1666. But his exposition was somewhat nebulous, and a better expression of this revolutionary idea was presented by Robert Hooke (1635–1703), secretary of the Royal Society, in the year 1674 in his book 'An Attempt to Prove the Motion of the Earth'. Even before this publication, Hooke gave a lecture at Gresham College in London in the year 1670, a part of the lecture dealt with the problem of planetary motions. He proposed very clearly what in modern scientific language one can mention as 'orbital motions are combinations of tangential inertial rectilinear motions and continuous free fall towards the centre of attraction'. Many historians of science feel that Hooke should be credited with the first unambiguous and correct statement of the dynamic nature of orbital motion. However, a quantitative treatment of the problem was beyond Hooke's capability. Realizing this, Hooke corresponded with Newton and requested him to take up and solve the problem in the year 1679.

This problem of planetary motion also interested Edmond Halley and Christopher Wren. They had a meeting with Hooke in January 1684 in which this problem was discussed. The problem of deriving the laws of planetary motion from the law of gravitation could not be solved by them. In Halley's own words, 'Sir Christopher, to encourage the Inquiry said, that he would give Mr. Hooke, or me, two months time, to bring him a convincing demonstration thereof; and besides the honour, he of us, that did it, should have from him a present of a book of 40 shillings'. Of course, neither Hooke nor Halley succeeded in solving the problem. In August 1684, about 7 months after this meeting, Halley visited Cambridge and happened to meet Newton. When he asked Newton what would be the shape of a

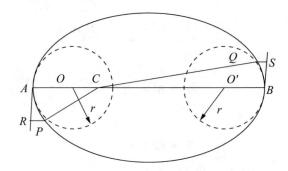

Fig. 1.61 Proof of proposition 2

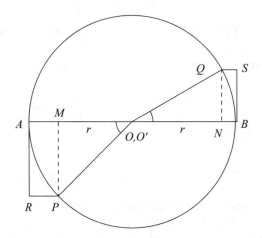

Fig. 1.62 Circle representing the sections of the ellipse at the two ends

planet's orbit under the action of a gravitational force that decreases according to an inverse square law, Newton's immediate reply was that the orbit would be an ellipse with one of the foci being the centre of attraction. Unfortunately, Newton could not show his derivation and suggested that the paper was misplaced and could not be found. But after another three months, in November 1684 Newton sent a nine page manuscript containing a proof of Kepler's first law from an inverse square law of gravitation. In what follows, Newton's proof is presented in an abridged form.

Newton solved the problem in two stages using a purely geometric approach. In his first proposition, Newton proved Kepler's second law as described in the previous section. His second proposition states:

'If a body be attracted towards either focus of an Ellipsis and the quantity of the attraction be such as suffices to make the body revolve in the circumference of the Ellipsis; the attraction at the two ends of the Ellipsis shall be reciprocally as the squares of the body in those ends from that focus'. Following Newton's lines of argument, the proof is as follows:

Figure 1.61 shows an ellipse with AB as the major axis with C as one of the foci which is the centre of attraction.

The objective is to prove that the ratio of the forces of attraction at A and B is equal to CB^2/CA^2.

If the planet moves to point P from A in a small interval of time, the planet in the same period will move from B to Q so that the area ACP is equal to area CBQ. It should be remembered that Newton proved in Proposition 1 'equal area in equal time' law for all central forces. For very small intervals of time, ACP and CBQ can be considered to be right-angled triangles. Hence, the area law suggests

$$\frac{1}{2} \times AP \times AC = \frac{1}{2} \times BQ \times BC$$

Now, in Newton's own language 'because the Ellipsis is alike crooked at both ends', the areas AP and BQ can be assumed to be on the same circle (Fig. 1.62) of radius r as explained in Fig. 1.61.

Now, PR and QS are perpendiculars dropped on the tangents drawn at A and B, respectively. PM and QN are normal drawn on line AB. As

$$PM^2 = AM(2r - AM)$$
$$\approx 2r \times AM$$

whereas AM^2 can be neglected for very small time interval. Similarly,

$$QN^2 \approx 2r \times BN$$

Hence,

$$\frac{PM^2}{QN^2} \approx \frac{AM}{BN}$$

For very small time intervals, $PM \to AP$ and $QN \to BQ$. Thus,

$$\frac{PR}{QS} = \frac{AM}{BN} \approx \frac{PM^2}{QN^2} \approx \frac{AP^2}{BQ^2}$$

But from Kepler's second law, already it has been shown that $AP \times AC = BQ \times BC$, or,

$$\frac{AP}{BQ} = \frac{BC}{AC}$$

i.e.

$$\frac{PR}{QS} = \frac{AP^2}{BQ^2} \approx \frac{BC^2}{AC^2}$$

Now, PR and QS represent the amount of fall of the planet (from its uniform inertial linear motions) due to the attractive force directed towards the centre of attraction. But the length of fall for a given time is proportional to the acceleration which is again proportional to the force as demonstrated by Huygens.

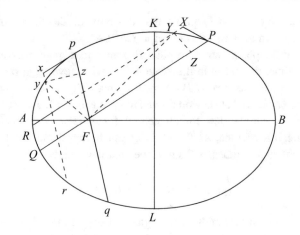

Fig. 1.63 Newton's geometrical proof of Kepler's first law

(PR/QS)

= (Force of attraction at A/Force of attraction at B)

$$= \frac{1}{AC^2} \times \frac{BC^2}{1}$$

This proves that the force of attraction at the ends in inversely proportional to the distance from the centre of attraction.

In his Proposition 3, Newton proved that for an elliptic orbit the force of attraction at any point on the orbit to be inversely proportional to the square of the distance from the centre of attraction. Proposition 3 states 'if a body is attracted towards either focus of any Ellipsis and by that attraction is made to revolve in the Perimeter of the Ellipsis: the attraction shall be reciprocally as the square of the distance of the body from that focus of the Ellipsis'.

The proof by Newton is purely geometrical and will show his mathematical competence of extraordinary nature. The proof is presented in a compact form without too much explanation of the standard geometrical results (Fig. 1.63).

Let P and p be two positions of a planet on its elliptic orbit under the influence of an attractive force directed towards the focus F, the centre of attraction. In a small interval of time, the planet would have moved to X (due to inertial motion). But due to the attraction towards F, it falls through the distance XY and the planet reaches point Y on the orbit. Line XR is parallel to line PFQ through F. Z is the point where the normal on PQ from point Y meets. In a similar way, the planet moves to point y (instead of x) in the same time interval. Line xr is parallel to pFq. Purely from geometric theories, Newton derived the following relation:

$$XY = \frac{AB \times PQ \times YZ^2}{XR \times KL^2}$$

For very small time intervals, PY and py can be treated as straight lines. According to the second law of Kepler, the area of ΔFPY and ΔFpy is equal. Hence,

$$\frac{1}{2} \times FP \times YZ = \frac{1}{2} \times Fp \times yz$$

or,

$$FP \times YZ = Fp \times yz$$

or,

$$FP \times YZ = Fp \times yz$$

Now, from the first relation

$$\frac{XY}{xy} = \frac{PQ}{pq} \times \frac{xr}{XR} \times \frac{YZ^2}{yz^2}$$

because $PQ \rightarrow XR$ and $pq \rightarrow xr$ when the time interval becomes infinitesimally small. Since XY and xy represent the fall towards F and proportional to the force of attraction

$$(\text{Force at } P)/(\text{Force at } p) = \frac{YZ^2}{yz^2} = \frac{Fp^2}{FP^2}$$

Thus,

$$\text{Force of attraction} \propto \frac{1}{\text{Distance from the focus}^2}$$

Of course, using modern mathematical tools, the proof is far more elegant and familiar to all. But the above proof of Kepler's first law is of historical importance. It also demonstrates the importance of geometry in the development of science in the pre-calculus era.

1.15 Newtonian Dynamics in Matured State

Newton is justly credited for the final synthesis of all the prevailing and discovered (by him) concepts that resulted in his famous (but least read) book 'Principia'. By now, it is clear that the first law of motion was not a discovery by Newton and the credit for the second law of motion should go to both Huygens and Newton. However, it was Newton's insight that there is an entity that is required to alter inertial motion. The quantitative measure of this entity is what was called by Newton as 'force'. Till his time, the idea of force was somewhat imprecise and nebulous. It was purely Newton's contribution that force always appears in pair. Each force has an equal and opposite reaction which is known as the third law of motion. This discovery was made by Newton while solving the collision problems. Newton's solution to the collision problems became far more general and elegant than those obtained by Huygens. Newton also was aware that there are two forms of third law—the weak form and the strong form. In the weak form, it is stated that there is an equal and opposite force to every impressed force.

In the strong form, the action and reaction are not only equal and opposite, but they act along the same line of action. The third law of motion—'Lex Tertia'—as mentioned by Newton provided the final link to complete a comprehensive science of motion. Newton also recognized the extremely important fact that the total quantity of motion (i.e. 'momentum' in modern language) is preserved in any collision as a consequence of the third law.

Newton also realized that the force depends on not only the change in inertial motion (i.e. acceleration) but also the 'bulk' of the body. In fact, this type of force, that changes the inertial motion of a body, was called by Newton as 'motive force'. In contrast, Newton also used a term 'innate force' (i.e. inherent force) to describe the capacity of a body to continue in its inertial motion. Of course, to any careful reader, it becomes immediately clear that such 'innate force' has to depend on the frame of reference and is not an inherent property of the body. But Newton continued to believe throughout in the existence of an 'absolute frame of reference'.

1.15.1 Concept of Mass

Newton's next (and perhaps the most significant one) conceptual contribution was the concept of 'mass' as a measure of the bulk of a body. He clearly distinguished between 'weight' and 'mass' of a body. As per Newton's law of universal gravitation, if the earth has a gravitational mass μ_e then an object with gravitational mass μ on the surface of the earth will be pulled by a force F towards the centre of the earth where

$$F = G\frac{\mu_e\mu}{R^2} = g\mu$$

R is the radius of the earth, and G is the constant of universal gravitation. This force is the weight of the body. If the same force is applied to the body, it is accelerated at the rate a and

$$F = ma$$

where m is another property of the body, called the 'inertial mass'. But first, it is necessary to make sure that the inertial mass of a body, defined above, is a 'property' of that body. This is essential as both the definitions of force and mass are linked through only one equation. If experiments are conducted on different bodies with inertial masses m_1, m_2, m_3, ... by applying the same force F (ensured by a spring balance), then the resulting accelerations of these bodies will be a_1, a_2, a_3, ... From the second law

$$F = m_1a_1 = m_2a_2 = m_3a_3 = \cdots$$

or,

$$\frac{m_2}{m_1} = \frac{a_2}{a_1}; \quad \frac{m_3}{m_1} = \frac{a_3}{a_1}; \cdots$$

If any one of these bodies, say m_1, be considered as the unit, all other masses can be quantitatively measured in terms of the standard mass.

Of course, experiments have been conducted along the line described above using different magnitudes of the applied force. Though the acceleration magnitudes are different, the ratios in the above equation have been found to remain unchanged. Therefore, inertial mass of an object is independent of the force and acceleration and can be considered as a property of the body concerned. Since the gravitational pull and accelerated motion are two different phenomena, there is no guarantee, per se, that μ and m be related. Using a suitable unit, these two can be made equal. When experiments are conducted, it is found that

$$m \propto \mu$$

and with suitable choice of units, $m = \mu$. This equivalence of the inertial and gravitational masses of a body is called the 'principle of equivalence' and represents one of the most mysterious and deep-rooted phenomena in the physical world.

1.15.2 Principia and Subsequent Development

No book in the whole history of science can match the fame of Newton's Principia. During the interaction with Halley once, Newton wrote a paper presenting his results related to mechanics in details. He titled the paper as 'Philosophic Naturalis Principia Mathematica' and submitted it to the Royal Society in 1686. Unfortunately, the Society had no funds for the publication, and Newton was unwilling to spend any money on this endeavour. So, finally, Halley took the responsibility of publishing the book and even read the proofs himself.

In Principia, all the basic elements of science of motion were encapsulated. However, the book was not easy to read, and popularization of the subject 'mechanics' had to wait till the works of Bernoulli and Euler. In the whole book, there was no mention of entities like energy and Newton wanted to resolve all problems using $F = ma$.

It was Leibniz who used the term 'Vis Viva' as another type of force. This is nothing but the kinetic energy (without the term ½) in modern terminology. It is interesting to note that the term 'dynamics' was first coined by Leibniz in relation to his treatment of the science of motion. Much later, the word transformed to mean 'dynamics' the way it is done in modern science. Freeing 'dynamics' from the narrow Leibnizian concept was done by D'Alembert in 1743.

As has been revealed through research on the historical developments in mechanics after the publication of Principia, not more than six to seven persons understood the book during Newton's lifetime. However, Newton's doctrine did not face any problem in becoming accepted in the academic world of England, but there was not any further creative development in mechanics as follow-up action. In the continental Europe, the popularization of Principia was slow because of the quarrel between Newton and Leibniz regarding the priority of discovering calculus. The first book that systematically presented Newton's mechanics after Principia was Phoronomia by Jacob Bernoulli, in the year 1716. This book for the first time presented the laws of motion in differential form. The first response to Principia that had original and creative elements contained therein came from Daniel Bernoulli who learnt mathematics and physics from his father Johann Bernoulli. Daniel was also guided by the philosophy of Descartes, Huygens and Leibniz. However, as his father considered Newton to be in the enemy camp because of the fight between Leibniz and Newton, Daniel's work on Principia to take it to higher level of perfection enraged his father. Nevertheless, Daniel's work 'Hydrodynamica' in 1738 and his paper on tides in 1741 led to the triumph of Newton's doctrine in the continental Europe. More or less contemporaneously Leonhard Euler published his book 'Mechanica' in 1736. This book opened up the world of mechanics accessible to general readers for the first time which helped to establish mechanics as an identified domain of knowledge in the university curricula. Only after a total synthesis, the works done by Descartes, Huygens, Newton, Bernoulli, Leibniz and Euler gave the subject its modern look.

Following the line of thought started by Leibniz, Euler gave importance to energy in developing his mechanics which later became known as analytical mechanics after Lagrange gave the final shape to the subject.

2.1 Nature of Motion and Space

Some of the deepest concepts in science include the nature of motion and space. From the very beginning of modern science, the matter has been examined by many great scientists and thinkers. One can safely remark that the pursuit of deeper understanding of the nature of motion and space continues even today and many points remain unresolved. The difficulty arises because these two aspects of the science of motion are at the very root of man's understanding of the universe. Motion is defined as the change of place as time progress, and this led the philosophers to ponder over the question 'what is space?' and 'what is time?' Though it is next to impossible to arrive at definite answers to the above questions, it is not very difficult to recognize 'space' as the 'arena of motion'. Any further philosophical discussion may not yield any fruitful results except framing of flowery sentences demanding a good vocabulary in English. Therefore, now onwards further discussions will be devoted to space and time relevant to better understanding of Newtonian dynamics.

2.1.1 Newton's Concept of Absolute Space and Time

Newton took a more straightforward route to the understanding of the nature of the space and time to describe his dynamics. As dynamics deals with motion that is defined as the change of position with time, it is obvious that the fundamental characteristics of space have to be linked with geometry. Newton believed in the existence of an entity called 'absolute space' that exists by itself on its own right independently. Similarly, he considered 'time' to be also an entity that is universal and flows uniformly independent of space. It may be worthwhile to quote Newton from his Scholium where he defined 'absolute time', 'absolute space' and 'absolute motion':

Hitherto I have laid down the definitions of such words as are less known, and explained the sense in which I would have them to be understood in the following discourse I do not define time, space, place, and motion, as being well known to all. Only I must observe that the common people conceive those quantities under no other motions but from the relation they bear to sensible objects. And thence arise certain prejudices, for the removing of which it will be convenient to distinguish them into absolute and relative, true and apparent, mathematical and common.

I. Absolute, true, and mathematical time, of itself, and from its own nature, flows equably, without relation to anything external, and by another name is called duration: relative, apparent and common time, is some sensible and external (whether accurate or unequable) measure of duration by the means of motion, which is commonly used instead of true time; such as an hour, a day, a month, a year.

II. Absolute space, in its own nature, without relation to anything external remains always similar and immovable. Relative space is some movable dimension or measure of the absolute spaces; which our senses determine by its position to bodies; and which is commonly taken for immovable space; such is the dimension of a subterraneous, an aerial or celestial space, determined by its position in respect of the earth. Absolute and relative space are the same in figure and magnitude; but they do not remain always numerically the same. For if the earth, for instance, moves, a space of our air, which relatively and in respect of the earth remains always the same, will at one time be one part of the absolute space into which the air passes; at another time it will be another part of the same, and so, absolutely understood, it will be continually changed.

III. Place is a part of space which a body takes up, and is according to the space, either absolute or relative.

IV. Absolute motion is the translation of a body from one absolute place to another; and relative motion, the translation from one relative place into another.

Newton ascribed the absolute space with three-dimensional Euclidean geometry. It will be shown later how this geometrical structure of absolute space gives rise to the conservation principles. It is desirable that the primary characteristics of the absolute space and time of Newton are compiled here.

Important properties of Newtonian space and time: (i) Absolute space and time are passive; i.e., these do not get influenced by external agency. These do not depend on the presence of matter and energy in the universe. (ii) The

© Springer Nature Singapore Pte Ltd. 2018
A. Ghosh, *Conceptual Evolution of Newtonian and Relativistic Mechanics*,
Undergraduate Lecture Notes in Physics, https://doi.org/10.1007/978-981-10-6253-7_2

behaviour of space remains same at all times. Similarly, time is also independent of the location where it is measured. (iii) Time interval between two events is same to all observers, and the order of events is independent of the observer and appears same to all. (iv) The absolute space follows Euclidean geometry.

Though all the above properties appear to be obvious to us as our senses have developed in a very small region of space surrounded by non-massive objects over a very short span of time, all the above points have been refuted by the philosophers and scientists of a later period. In more concrete terms, the following properties of absolute space lead to the observed rules of Newtonian dynamics. These are as follows: (i) the homogeneity of space, (ii) the isotropy of space, (iii) the continuity of space, (iv) the completeness of space, (v) the homogeneity of time, (vi) the continuity of time, (vii) the independence of space and time. These points will be again revisited in a later section to demonstrate the origin of the conservation principles of Newtonian dynamics.

The Newtonian dynamics hinges on one kinematic parameter, i.e. acceleration. Newton believed that this acceleration of an object, which is equal to the impressed force divided by the mass of the object, is with respect to the absolute space. So an observer who is moving with a uniform velocity with respect to the absolute space finds the acceleration to remain unchanged and the second law of motion works. Thus, the second law of motion remains valid in all frames of reference which are moving with uniform velocity with respect to the absolute space. It should be noted that the first and third laws also remain valid in these frames of reference which Newton called as 'inertial frames of reference'.

To demonstrate the fact that acceleration of a particle with respect to the absolute space is relevant in dynamics and its relative motion with respect to other objects does not play any role, Newton devised his famous bucket experiment described in the following section.

2.1.2 Newton's Bucket Experiment

This experiment, which was devised by Newton to prove his point that motion with respect to the absolute space is relevant, is quite simple and anyone can perform it. One has to hang a bucket, with some water in it, with the help of a rope (Fig. 2.1a). To start with, both the water and the bucket are stationary. Then, the bucket is spun as shown in Fig. 2.1b, but initially, the water is still stationary as it will take some time to pick up the rotational motion from the inner wall of the spinning bucket. Thus, it is observed that the bucket is rotating but the water is stationary with its surface flat as before. Next, the water in the bucket gradually picks up the rotational motion from the bucket, and at a stage, both the bucket and the water rotate and the water surface becomes concave as

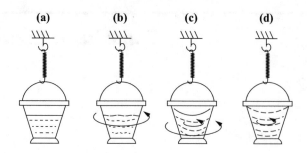

Fig. 2.1 Newton's bucket experiment

expected from our common knowledge (Fig. 2.1c). At the final stage of the experiment, the bucket is stopped by hand, but the water still continues to rotate due to inertia, and one finds that the surface of the rotating water is concave as shown in Fig. 2.1d. Newton's logic in support of the existence of absolute space was along the following lines.

Initially, the bucket and water are both stationary and the water body does not possess any acceleration with respect to the absolute space. Hence, the water surface is flat. In the next stage, the bucket body is rotating, but the water body is still stationary. So, the water body has motion relative to the bucket but has no acceleration with respect to the absolute space. Therefore, the water surface is flat. In the third stage, both the bucket and the water rotate and there is no motion of the water with respect to the bucket but has motion with respect to the absolute space. Hence, according to Newton's logic water surface is curved. Finally, again the water is moving with respect to the absolute space and the bucket. As expected, the water surface is curved because of its motion with respect to the absolute space.

The above experiment demonstrates (according to Newton) that the relative motion between the bucket and the water in it does not generate any force (due to relative acceleration of the water molecules) on the water. But when the water body is accelerating with respect to the absolute space, its molecules are subject to forces. Therefore, Newton concluded motion relative to absolute space has relevance to dynamics; relative motion between bodies develops no dynamical effect.

Newton suggests another experiment to establish his idea of absolute space using two globes in the following manner:

> For instance, if two globes, kept at a given distance one from the other by means of a cord that connects them, were revolved about their common centre of gravity, we might, from the tension of the cord, discover the endeavour of the globes to recede from the axis of this motion, and from thence we might compute the quantity of their circular motions. And then if any equal forces should be impressed at once on the alternate faces of the globes to augment or diminish their circular motions, from the increase or decrease of the tension of the cord, we might infer the increment or decrement of their motion; … And thus we might find both the quantity and the determination of this circular motion, even in an immense vacuum, where there was

nothing external or sensible with which the globes could be compared.

However, it is not difficult to recognize the difficulty in conducting the above-mentioned globe experiment in an otherwise empty universe.

At this, it is quite pertinent to ask why only acceleration with respect to absolute space leads to dynamical effect, i.e. development of force. Uniform non-accelerated motion with respect to the absolute space should also lead to the development of force. This matter will be taken up for further discussion in the later sections. However, at this stage, it is enough to point out that all experiments conducted so far have not indicated the existence of any inertial resistive force to counter uniform motion with respect to the absolute space. Either it does not exist or the current level of technology is not capable of detecting such forces those being extremely small. The accuracy that can be achieved in the K-meson decay experiments may not be high enough to detect any ultra-small force, if there is any.

2.1.3 Newton's Bucket Experiment Follow up

Notwithstanding the logic and suggested experiments by Newton to establish the concept of absolute space and absolute nature of motion, his philosophy did not go unchallenged. Two contemporary scientist philosophers questioned the validity of the concept of absolute space and motion. G. W. Leibniz (1646–1716) was introduced to science of motion by none other than Huygens. Leibniz did not accept Newton's idea of absolute space and absolute time. His contention was that it is impossible to feel space without the presence of any matter, its existence can be identified as a 'system of relationships', and it is meaningless to endow 'space' with absolute existence. It is worthwhile to quote from his letter to S. Clarke, one of Newton's disciples, on this subject.

> As for my opinion, I have said more than once, that I hold space to be something merely relative, as time is; that I hold it to be an order of coexistences, as time is an order of successions. For space denotes, in terms of possibility, an order of things which exist at the same time, considered as existing together; without enquiring into their manner of existing. And when many things are seen together, one perceives that order of things among themselves.

But Leibniz was unable to provide a satisfactory explanation to Newton's bucket experiment considering his concept of reality of relative motion only. Interestingly, the seed of a truly relational mechanics was sowed by Newton's disciple, Clarke, in his last letter to Leibniz that reached him on 29 October 1716. No reply was ever written as Leibniz died on 14 November 1716. It is appropriate to quote the relevant part of the letter.

> It is affirmed,[1] that motion necessarily implies a relative change of situation in one body, with regard to other bodies; and yet no way is shown to avoid this absurd consequence, that then the mobility of one body depends on the existence of other bodies; and that any single body existing alone, would be incapable of motion; or that the parts of a circulating body (suppose the sun) would lose the vis centrifuga arising from their circular motion, if all the extrinsic matter around them were annihiliated.

Thus, without realizing, Clarke hit the bull's eye by conjecturing that if the matter in the rest of the universe is annihilated effects of motion, like curving of the water surface in a rotating bucket or the tension in the cord connected to the globes, would vanish. Hence, the relative motion of the water is not important as the matter content in the bucket is small to produce any visible effect. Though Clarke raised this point to establish the futility of a 'relative motion' theory, the hint got lost. Otherwise, both Leibniz and Berkeley, at a later time, could answer the question raised by Newton's bucket experiment.

Soon after Leibniz, Newton's concept of absolute space and absolute motion was criticized by Bishop G. Berkeley (1685–1753) following the line laid down by Leibniz. The criticism is found in his two books—'The Principle of Human Knowledge' published in 1710 and 'Of Motion'— published in 1721. His opinion on motion can be quoted from his first book as follows:

> But, notwithstanding what has been said, I must confess it does not appear to me that there can be any motion other than relative; so that to conceive motion there must be at least conceived two bodies, where of the distance or position in regard to each other is varied. Hence, if there was one only body in being could not possibly be moved. This seems evident, in that the idea I have of motion do the necessarily include relation.

Similarly, in his other book also the criticism is quoted as follows:

> No motion can be recognized or measured, unless through sensible things. Since this absolute space in no way affects the senses, it must necessarily be quite useless for the distinguishing of motions. Besides, determination or direction is essential to motion; but that consists in relation. Therefore, it is impossible that absolute motion should be conceived.

But, like Leibniz, Berkeley also failed to grasp the unintentional hint from Clarke that in the absence of matter in the rest of the universe, no effect of any motion can be sensed.

> Both Leibniz and Berkeley were somewhat confused but Berkeley took one step further in the right direction. He tried to replace Newton's absolute space by the conglomeration of the fixed stars that do not have motions relative to one another and can be conceived to form a frame of reference that can be used to describe motion. He mentions in his book "On Motions –", "-- - - for the removal of ambiguity and for the furtherance of the

---

[1]By Leibniz.

mechanics of these philosophers who take the wider view of the system of things, it would be enough to bring in, instead of absolute space, relative space as confined to the heavens of the fixed stars, considered as at rest - - - - - -.'"

Anyhow, the tremendous success of Newtonian mechanics in explaining all motions—celestial and terrestrial—did not allow much scope for further debate on the issue and Newton's ideas continued without any challenge for a century and a half.

2.1.4   Ernst Mach and Mach's Principle

The relative–absolute debate was resurrected by the nineteenth-century scientist philosopher Ernst Mach (1838–1916). He wanted to clear the nebulosity surrounding the nature of space and provide a solid concept devoid of any ambiguity. In the preface of his famous book 'The Science of Mechanics', he clearly stated in the preface that 'The present volume is not a treatise upon the application of the principles of mechanics. Its aim is to clear up ideas, expose the real significance of the matter and get rid of metaphysical obscurities'. He was the first to take up the challenge posed by Newton's thought experiment with two globes tied up by a chord and executing rotation and the other famous experiment with the bucket full of water. To explain the tension in the chord and the curvature of the water surface without bringing in the concept of absolute space was at a task with insurmountable difficulty for the Leibnizian relativists. Mach solved the difficulty with one simple stroke—by rejecting the experiments. According to him, one can say what will happen to the globe experiment when the matter from the rest of the universe vanishes. According to him it would be impossible to see any rotation of the pair of globes. So far as the bucket experiment is concerned, his view is quoted as follows:

Newton's experiment with the rotating vessel of water simply informs us, that the relative rotation of the water with respect to the sides of the vessel produces no noticeable centrifugal forces, but that such forces are produced by its relative rotation with respect to the mass of the earth and other celestial bodies. No one is competent to say how the experiment would turn out if the sides of the vessel are increased in thickness and mass till they were ultimately several leagues thick. The one experiment only lies before us, and our business is, to bring it into accord with the other facts known to us, and not with the arbitrary fictions of our imagination.

Thus, according to Mach, dynamical effect is produced by relative motion between bodies and there is no role for the existence of a real absolute space. If the bucket would have been sufficiently massive, a rotating bucket could produce curved surface in the water even if it were not rotating with respect to the fixed stars (representing the absolute space). Hence, the problem is with the magnitude. Had our technology been advanced enough, such effects could be observed.

Mach did not prescribe a specific physical theory, but his proposition was also in the form of a principle without any quantitative model. Utilizing this principle, M. Schlick was the first to interpret in 1915 that 'the case of inertia must be assumed to be an interaction of masses'. Thus, the first hypothesis of the origin of inertia as the interaction of a body moving with respect to the matter present in the rest of the universe was born and became widely known as the 'Mach's principle'.

This concept greatly impressed Einstein, and in his paper of 1918, he clearly indicates 'Mach's principle' as a founding pillar of the General Theory of Relativity. To quote him

c) Mach's Principle: The G-field is completely determined by the masses of the bodies. Since mass and energy are identical in accordance with the results of the Special Theory of Relativity and the energy is described formally by means of the symmetric energy tensor ($T_{\mu\nu}$), this means that the G-field is conditioned and determined by the energy tensor of the matter.

When Mach was 75 years old and became immobile because of a paralytic attack, Einstein's last letter from Zurich on 25 June 1913 included the following:

Recently you have probably received my new publication on Relativity and Gravitation which I have at last finished after unending labour and painful doubt. Next year at the solar eclipse it will turn out whether the light rays are bent by the sun, in other words whether the basic and fundamental assumption of the equivalence of the acceleration of the reference frame and of the gravitational field really holds. If so, then your inspired investigations into the foundations of mechanics – despite Planck's unjust criticism – will receive a splendid confirmation. For it is a necessary consequence that inertia has its origin in a kind of mutual interactions of bodies, fully in the sense of your critique of Newton's bucket experiment.

Subsequently, many scientists and philosophers have provided support to the concept. The 'Mach's principle' is also considered important for providing explanation to the origin of inertial effect. According to Thomas Phipps, 'when the subway jerks, it's the fixed stars that throw you down'. With time, the term 'fixed stars' has been replaced by the term 'matter present in the rest of the universe'. Many scientists consider that inertia is nothing but the manifestation of a body accelerating with respect to the rest of the universe.

2.1.5   Quantification of Mach's Principle— Concept of Inertial Induction

Although the Mach's principle is very attractive as a concept, the primary difficulty was the absence of a quantitative model of the theory. Without such a quantitative model, not much further progress was possible. To remedy the situation, D.W. Sciama, the Cambridge University Professor of Physics, proposed the concept of 'inertial induction' as a quantitative description of Mach's principle. According to Sciama, a major advantage of Mach's principle over the 'absolute space' concept of Newton is that the primary role played by 'acceleration' can be explained. According to the conventional doctrine of Mach's principle, the interactive force between two material objects depends on the relative 'acceleration' between the two bodies and the relative velocity does not lead to the development of any interactive force. To establish this point (that the velocity of a particle with respect to the matter present in the rest of the universe is not detectable), experiments have been conducted on the decay of K meson into π meson. The rate of K meson decay into two π mesons was expected to depend on its velocity with respect to the matter in the rest of the universe if the interactive force were dependent on velocity. But no such dependence was found from the experiments, and the existence of an interactive force depending on the relative velocity between two objects was disproved.

However, this could be a matter of accuracy. If the interactive force due to relative velocity be too small to be detected by the K meson decay experiment, then the question remains open. Anyhow Sciama considered the interactive force between two particles to depend only on their relative acceleration and named the phenomenon as 'inertial induction'. According to Sciama, the total force between two particles of mass[2] m_1 and m_2 has two parts—one static and the other dynamic as follows:

$$F = F_s + F_i \tag{2.1.1}$$

The static part of the force is the usual Newtonian gravitational attraction

$$G \times \frac{m_1 m_2}{r^2}$$

where r is the distance between the two particles and G is the constant of universal gravitation (Fig. 2.2a). The dynamic part of the total interactive force is due to the inertial induction effect due to the relative acceleration and is given by

$$G \times \frac{m_1 m_2}{cr^2} \times a$$

where c is the speed of light in a vacuum and a is the acceleration of m_2 with respect to m_1 as shown in Fig. 2.2b. Thus,

$$F = G \frac{m_1 m_2}{r^2} + G \frac{m_1 m_2}{cr^2} a \tag{2.1.2}$$

If a is reversed, the direction of F_i will also reverse. Of course, this is a very simplistic situation and the magnitude

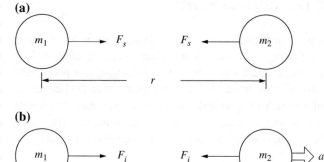

Fig. 2.2 Concept of inertial induction

of a can be either positive or negative. Sciama arrived at this concept after considering the force between two electric charges. Figure 2.3a shows two charges $+e_1$ and $-e_2$ at a distance r. Coulomb's law of electrostatic force suggests that the attractive force is given by

$$F = \varepsilon \frac{e_1 e_2}{r^2} \tag{2.1.3}$$

where ε is the dielectric constant of the free space. The similarity of the law with the law of universal gravitation is quite obvious. Now, if $-e_2$ is moved away with an acceleration a with respect to $+e_1$, then the total force between the two charges is given by

$$F' = \varepsilon \frac{e_1 e_2}{r^2} + \varepsilon \frac{e_1 e_2}{cr^2} a \tag{2.1.4}$$

Sciama attributed a similar effect of a in the case of gravitational interaction also in an intuitive manner. Of course, the model is quite simple and the question whether relative velocity generates any force or not remains unresolved. But this was the first attempt to quantify Mach's principle.

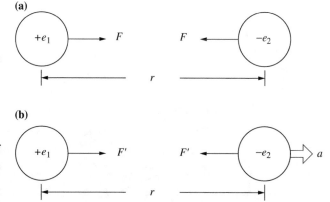

Fig. 2.3 Force between electric charges

---

[2]Actually, this is the gravitational mass.

2.1.6 Origin of Inertia

Once a quantitative model is available, it is possible to use it for finding the source of the inertial property of an object. Using Sciama's model of inertial induction, it is possible to have a rough estimate of the total force due to the interaction of a particle of (gravitational) mass m accelerating at the rate a with respect to the rest of the matter in the universe. If M_j be the mass of the jth object in the universe (Fig. 2.4)

$$F_i = \sum_j^{\text{ObservableUniverse}} \left(\frac{GM_j}{c^2 r}\right) ma \qquad (2.1.5)$$

In the very large scale, it may be assumed that the matter is uniformly distributed all over and the matter in the universe does not possess any particular acceleration. Thus, for a stationary, homogenous and isotropic universe, the total force on the particle due to inertial induction from the rest of the universe can be expressed as follows:

$$F_i = \frac{Gm\rho}{c^2}\left[\iiint \frac{dv}{r}\right] a \qquad (2.1.6)$$

where dv is the volume of a thin spherical elemental cell as shown in Fig. 2.5 and ρ is the uniform density of matter in the universe. The radius of the cell is r with the test particle at the centre.

Now, the elemental volume dv can be expressed as follows:

$$dv = 4\pi r^2 dr$$

Using this in (2.1.6), the total force on m due to inertial induction becomes

$$F_i = \frac{4\pi Gm\rho}{c^2}\left(\int_0^{R_0 = c/H} \frac{1}{r} r^2 dr\right) a \qquad (2.1.7)$$

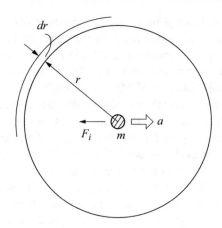

Fig. 2.5 Interaction of a particle with an element of the universe

where R_0 is the radius of the observable universe given by c/H with H as the Hubble's constant. From (2.1.7), one obtains

$$F_i = \frac{2\pi G\rho R_0^2}{c^2} ma$$
$$= \frac{2\pi G\rho}{H} ma \qquad (2.1.8)$$

Using the currently estimated density ρ as 10^{-26} kg m^{-3} and H as 50 km s^{-1} Mpc^{-1} ($\approx 1.5 \times 10^{-18}$ s^{-1}), (2.1.8) yields

$$F_i \approx 1.8 ma \qquad (2.1.9)$$

Ideally, F_i (i.e. the force to be exerted for accelerating a mass m at the rate a) should be equal to exactly ma. However, considering the approximate nature of the inertial induction model and the uncertainties involved in estimating the values of ρ and H, the close agreement is remarkable. This provides enough hint that Mach's principle may be the real cause of inertia as suggested by some scientists. It is felt that a more accurate modelling could lead to more accurate results.

It should be mentioned here that considering the theory of Weber's electrodynamic force law

$$F = \frac{e_1 e_2}{4\pi\varepsilon r^2}\left[1 - \frac{1}{c^2}\left(\frac{dr}{dt}\right)^2 + \frac{2r}{c^2}\frac{d^2 r}{dt^2}\right] \qquad (2.1.10)$$

Tisserand proposed a modified law for gravitational interaction between two masses in the following form as early as 1872:

$$F = \frac{Gm_1 m_2}{r^2}\left[1 - \frac{1}{h^2}\left(\frac{dr}{dt}\right)^2 + \frac{2r}{h^2}\frac{d^2 r}{dt^2}\right] \qquad (2.1.11)$$

where h was considered to be the speed of propagation of gravitation. It is interesting to note that besides the Newtonian static term, the force law contains a velocity-dependent term and an acceleration-dependent term very similar to that proposed by Sciama. However, Tisserand proposed the

Fig. 2.4 A particle accelerating in the universe

model to explain the anomaly observed in the advance of the perihelion of Mercury. Since this was adequately explained by Einstein's General Theory of Relativity, Tisserand's model of dynamic gravitational interaction as a possible quantitative model of Mach's principle remained unnoticed till very recently.

Though it is expected that a more accurate modelling following Sciama's logic will give the desired result, there is a serious philosophical issue involved. The exactness of the equation $F = ma$ cannot be dependent on the fine tuning of the values of G, ρ, H and c. So, there must be some other matter of very fundamental nature to take care of such exactness. This problem will be taken up again in a later chapter.

2.2   Relative–Absolute Duality of Nature of Motion

The debate between the supporters of Newton's concept of motion (i.e. acceleration) with respect to an absolute space and the continental scientists' philosophy of relative (with respect to other objects in the universe) motion continued for a long time. However, as the progress of science did not get affected by the philosophical nature of the disagreement the debate is not given any serious consideration by the practising scientists; the subject has been gradually pushed to a dark corner of the world of philosophy of science.

One may think that the matter is of purely academic nature without any relevance to down-to-earth science but it is not exactly so. In the Newtonian concept of absolute space as the background for motion to take place 'inertia' is an intrinsic property of mass and the second law of motion is a 'law'. But it will be seen later in this volume that 'inertia' can be shown to be manifestation of an object's motion relative to the matter present in the rest of the universe. This solves also a number of paradoxes and mysteries in Newtonian scheme of things. These will be taken up in later sections of this book. In this section, it will be attempted to demonstrate that a sense of absolute motion can emerge from the results obtained through the observation of an object's relative motion with respect to other objects present in the universe. To keep the treatment uncomplicated, only a two-dimensional situation will be analysed though the results for a three-dimensional situation will remain the same.

2.2.1   The Nature of the Universe

It will be necessary first to identify the nature of the universe. From all observations, the universe appears to be infinite, homogenous, and isotropic. The objects in the universe have random motions, but there is no systematic universal motion. Locally, there are evolutions of systems and destructions are also of local nature. Such a universe is considered to satisfy the 'perfect cosmological principle'. In the nineteenth century, it was thought that the universe is filled up by stars. But now it is known that the main constituent members of the universe are the galaxies along with some other types of objects.

There are a number of cosmological models. According to the currently popular 'Big-Bang' theory, the universe is finite in size and possesses a universal expansion and the universe is also evolving as a whole. In the 'steady-state' theory the universe is infinite but possesses a universal expansion. The steady state is maintained by continuous creation of matter to keep the matter density unchanged. However, according to the less-known 'stationary universe' model the universe satisfies the 'perfect cosmological principle'. The expanding universe model is favoured because that provides a simple and straightforward explanation for the observed cosmological redshift without invoking new physics. But there is no direct evidence of universal expansion so far. So, to illustrate the absolute nature of motion using the relative motion of a body with respect to other bodies present in the universe, a model of infinite, homogenous, isotropic non-evolving stationary universe model will be chosen.

Figure 2.6 shows a plot of a million galaxies in the observable universe. It is clear that in the very large scale, the lumpiness of the matter content in the universe gradually reduces and the homogenous and isotropic nature of the universe becomes apparent. Apart from this, all contemporary observations also point towards the absence of any curvature of the universal space–time continuum and the geometry is strictly Euclidean. There is no systematic motion of the bodies in the universe though all objects possess some finite motion. In such a scenario, it is possible to conceive a mean rest frame of this quasistatic universe. When very large sections are considered, the centre of mass of each of these sections is almost without any perceptible motion. An imaginary network connecting these centres of mass constitutes a frame of reference embedded in the universe that may be treated as an absolute frame of the quasistatic infinite universe. Even the cosmic microwave radiation that fills the whole universe may be treated as a background for detecting absolute motions, i.e. motions with respect to the mean rest frame of the universe.

2.2.2   Absolute Motion in Terms of Relative Motion

In this section, an attempt will be made to show a methodology by which the displacement of an object in the universe can be provided with some scientific meaning.

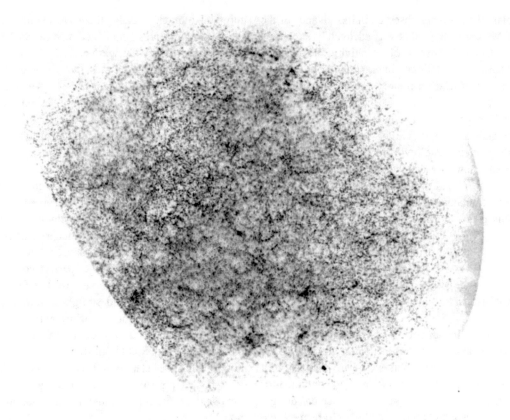

Fig. 2.6 A million galaxies

Figure 2.7a shows a section of the universe containing the targeted object P and other bodies. If another snapshot of the same region of the universe is taken at another instant of time, it may look as shown in Fig. 2.7b. Now as all objects in this universe are in motion, it is not possible to find out the amount of displacement of the object P. Even it may be, at a first glance, meaningless to talk about the displacement of the objects. This is the fundamental problem of motion; many philosophers and scientists feel that it is impossible to have any sense in talking about the motion of an object unless it is defined in relation to other objects. But as all objects are in motion, there cannot be any sense in contemplating motion of a body. But in an infinite, non-evolving homogenous and isotropic universe, the

relative displacements of a body in relation to other bodies can lead to a sense of absolute motion.

The problem will be dealt with in a reverse order. First, let it be assumed that there is an absolute frame, and finally, the validity of this assumption is established. At the first instant, when a snapshot is taken, let the position of the body P with respect to all the other bodies is given by

$$\boldsymbol{r}_1; (i = 1, 2, 3, \ldots, n, \ldots, \infty)$$

When the next snapshot is taken after an interval of time Δt, let these positions be $\boldsymbol{r}'_1$; ($i = 1, 2, 3, \ldots, n, \ldots, \infty$) when the body P has moved to position P' (with respect to the assumed mean rest frame of the universe). The other bodies at positions 1, 2, 3, … have moved to positions 1′, 2′, 3′, … (as indicated in the figure) in this assumed mean rest frame of the universe. Thus, the displacements of these bodies are given by $\Delta\boldsymbol{\rho}_1$; ($i = 1$, 2, 3, …, n, …, ∞), whereas body P moves through the displacement Δs (all in the imagined mean rest frame of the universe). The quantities which are directly observed are the position vectors $\boldsymbol{r}_1$ and $\boldsymbol{r}'_1$; ($i = 1, 2, 3, \ldots, n, \ldots, \infty$). It can be shown that Δs can be determined from these quantities. From Fig. 2.8, the following equations are obtained:

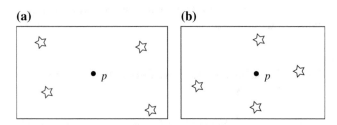

(a)   **(b)**

Fig. 2.7 Snapshots of a region at the two different instants

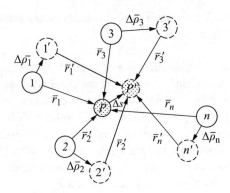

Fig. 2.8 Relative displacement of bodies

$$r_1 + \Delta s = \Delta \rho_1 + r'_1$$
$$r_2 + \Delta s = \Delta \rho_2 + r'_2$$
$$r_3 + \Delta s = \Delta \rho_3 + r'_3$$
$$\dots\dots\dots\dots\dots$$
$$r_n + \Delta s = \Delta \rho_n + r'_n$$

$$\dots\dots\dots\dots\dots$$

Summing up both sides of the above vector equations for N terms (when N is very large tending to infinity), the following equation is obtained:

$$\sum_{i=1}^{N} r_i + N \times \Delta s = \sum_{i=1}^{N} \Delta \rho_i + \sum_{i=1}^{N} r'_i \qquad (2.2.1)$$

To proceed further, let the displacements $\Delta \rho_i$ be considered to be equal in the magnitude, $\Delta \rho_{av}$ but randomly oriented. In reality, the magnitudes of the displacements of the objects satisfy a normal distribution with an average value $\Delta \rho_{av}$. But the final outcome of this argument remains the same. From the random walk theorem,

$$\sum_{i=1}^{N} \Delta \rho_i = \Delta \rho$$

where $|\Delta \rho| = \sqrt{N} \times \Delta \rho_{av}$. If $\hat{k}$ be a unit vector along the direction of the resultant vector $\Delta \rho$,

$$\sum_{i=1}^{N} \Delta \rho_i = \hat{k} \sqrt{N \Delta \rho_{av}} \qquad (2.2.2)$$

Substituting this resultant in (2.2.1) and dividing both sides by N

$$\frac{1}{N} \sum_{i=1}^{N} r_i + \Delta s = \frac{\hat{k}}{\sqrt{N}} \Delta \rho_{av} + \frac{1}{N} \sum_{i=1}^{N} r'_i$$

when N becomes very large, $\frac{\Delta \rho_{av}}{\sqrt{N}}$ tends to zero, and the above equation becomes after rearrangement

$$\Delta s = \frac{1}{N} \left(\sum_{i=1}^{N} r'_i - \sum_{i=1}^{N} r_i \right) \qquad (2.2.3)$$

Therefore, it is seen that if the initial and final positions of an object with respect to a very large number of other objects in the universe are known, the displacement of the object with respect to the mean rest frame of the universe can be determined from (2.2.3). Thus, relative position can yield the absolute motion. This is a fine example of relative–absolute duality in the science of motion.

As already mentioned earlier in this section, the concept of absolute motion is not absurd as it sounds. In fact, the cosmic microwave background radiation provides such a notion to absolute nature of motion. The motion of the earth through this 'soupe' of background radiation can be determined by observing the dipole anisotropy in the Doppler effect. The magnitude of the earth's absolute velocity (with respect to this radiation) has been found out to be 260 km s^{-1}.

2.3 Inertial and Gravitational Mass

The concept of mass was first proposed by Newton to quantify the quantity of matter. The definition of mass given by Newton was 'quantity of matter', and he quantified it by the product of density and volume. In earlier times, the 'bulk' or the 'quantity of matter' used to be recognized by the weight of an object. Newton recognized mass as an intrinsic property of matter and different from weight which is nothing but the force of gravitational pull by the earth. So weight is not a property of an object as the gravitational pull does not remain constant. In Newton's scheme of things, 'mass' comes into picture in two ways. In the first case, the mass represents the inertia to change of motion. Quantitatively, this is manifested in the second law

$$F = ma$$

On the other hand, the law of universal gravitation proposed by Newton also incorporates a quantity called 'mass'. The force of gravitational attraction between two bodies with masses m_1 and m_2 is given by

$$|F| = G \frac{m_1 m_2}{r^3} |r|$$

where r is the position vector of one body with respect to the other. It should be noted that in Newtonian mechanics, the phenomenon of accelerating a body with an impressed force has no relation to the gravitational attraction between two bodies. So the properties designated by the term 'mass' in the two cases need not be the same. In modern times, two different names are used to denote this difference: in the case of a body's resistance (or inertia), change in the state of motion is called 'inertial mass' (m_i), and in the case of gravitational attraction, the property is defined as the 'gravitational mass' (m_g). Over and above, to make further finer distinction, the gravitational mass of a body on which the gravitational pull acts is often termed as 'passive gravitational mass' and the mass of the body exerts the pull is called the 'active gravitational mass'. However, according to Newton's third law, the situation is symmetric and there is no important difference between the 'active' and 'passive' gravitational mass of a body.

Newton did not make any special effort, to differentiate between the 'inertial' and 'gravitational' mass though he was aware of the difference. Perhaps, the primary reason for this was the traditional way of considering the weight of an object as its bulk (representing the mass). But, nevertheless, he proposed to demonstrate the equality by more accurate experiments than the standard 'free fall' experiments of Galileo. This point will be taken up later in this section.

2.3.1 Inertial Mass

To concretize the concept of inertial mass, a methodology needs to be developed. A thought experiment can be conceived in which different bodies are impressed upon the same force.

The scheme is indicated in Fig. 2.9. To ensure the constancy of the impressed force F, the compression of the spring is used. A definite force compresses a spring by a definite predetermined amount. If the experiments are conducted in an inertial frame of reference, the second law of motion stipulates that

$$F = m_1 a_1 = m_2 a_2 \qquad (2.3.1)$$

where m_1 and m_2 are the inertial masses of two bodies being accelerated at the rates a_1 and a_2, respectively. Conducting the experiments in an inertial frame also ensures that the results will be agreed upon by all inertial observers. Equation (2.3.1) yields

$$\frac{m_1}{m_2} = \frac{a_2}{a_1} \qquad (2.3.2)$$

This is an unambiguous way of defining the ratio of inertial masses of two bodies. So, if one of the bodies, say body 2, is assumed to have mass equal to 1 (as the standard), then the mass of any other body is given by

$m = m/1 = $ (acceleration of the standard mass)/(acceleration of the body)

where the above relation is also used to define the unit of force. In the SI unit, the mass of the standard mass is 1 kg, and when a force acting on the body produces unit acceleration, 1 m s^{-2}, the force is defined to be 1 N. In Newtonian mechanics, the inertial (and also gravitational) mass remains conserved. The mass also shows additive property. If two objects have masses m_1 and m_2, they accelerate at the rate a when subjected to forces F_1 and F_2,

$$F_1 = m_1 a \quad \text{and} \quad F_2 = m_2 a$$

When the two bodies are combined and subjected to a force ($F_1 + F_2$), the following result is obtained:

$$F_1 + F_2 = (m_1 + m_2)a$$

So the combined bodies as a single body have mass $m_1 + m_2$.

2.3.2 Gravitational Mass

The fact that all bodies fall with equal acceleration when no resistance acts was well established by Galileo first and other subsequent scientists later. If an object has a gravitational mass m_g and inertial mass m_i, the force of attraction due to gravity on the earth's surface is given by Newton's law of universal gravity[3] as follows:

$$F = \frac{G m_{E_g} m_g}{R_E^2} \qquad (2.3.3)$$

when m_{E_g} represents the gravitational mass of the earth and R_E is the earth's radius. This force acting on the object produces an acceleration according to the second law as follows:

$$a = \frac{F}{m_i} = \left(\frac{G m_{E_g}}{R_E^2}\right)\frac{m_g}{m_i} = g \qquad (2.3.4)$$

where g is the acceleration due to gravity on earth's surface. Since 'g' is found to be same for two objects 1 and 2,

$$\frac{m_{g_1}}{m_{i_1}} = \frac{m_{g_2}}{m_{i_2}} = \alpha \text{ (const)} \qquad (2.3.5)$$

which implies that the ratio of the gravitational mass and the inertial mass is same for all objects. Using the known value

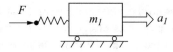

Fig. 2.9 Experiment to determine inertial mass

---

[3]In fact, Newton did not formulate his law of universal gravitation in the familiar form.

$m_{E_g} = 5.976 \times 10^{24}$ kg, $R_E = 6.378 \times 10^6$ m   and
$g = 9.80665$ m s^{-2}

$$\frac{m_g}{m_i} = \frac{1}{G} \times 9.80665 \times \frac{6.378^2 \times 10^{12}}{5.976 \times 10^{24}} = \frac{6.6754 \times 10^{-11}}{G} = \alpha$$

Taking $G = 6.67 \times 10^{-11}$ m^3 kg^{-1} s^{-2}, α becomes equal to unity rendering the gravitational and inertial masses equal. This equality of gravitational and inertial masses intrigued all scientists as there has been no theory relating the phenomena of inertia and gravitation till Mach propose his very qualitative philosophy towards the end of the nineteenth century.

2.3.3 Equivalence of Inertial and Gravitational Mass

Traditionally, all users of Newtonian mechanics generally do not ponder over the matter of using 'mass' without making any distinction whether it is inertial or gravitational. This is so because of the exact equivalence of the inertial and gravitational mass of an object (i.e. the ratio of these two properties is a constant for all objects). Though the observed results of the free fall experiments lead to this understanding, the accuracy of these results was not high and many other experiments have been devised to test this equivalence. Even at present some billion-dollar experiments are being planned to test the accuracy of this equivalence—the matter is of so profound importance to scientists and philosophers.

Huygens was the first to derive in 1673 the expression for the period of oscillation of a simple pendulum as follows:

$$T = 2\pi \sqrt{\frac{l}{g}}$$

Using this formula, the value of g was determined with a higher degree of accuracy than that obtained from the free fall experiments. By conducting experiments with the length l, as 11 ft and using the oscillating mass made of different materials, Newton was able to prove

$$(m_i - m_g)/m_i \approx \pm 10^{-3}$$

meaning that the inertial and gravitational masses of a body are equal within one part in thousand. Using a spherical pendulum, Newton was able to improve the accuracy further.

Figure 2.10 shows the arrangement. The equilibrium configuration of the pendulum, when rotated at an angular speed ω, is given by the balancing act of the deflecting moment $m_i\omega^2 l \sin\theta$. $l \cos\theta$ and the restoring moment $m_g gl \sin\theta$. The obtained relation is

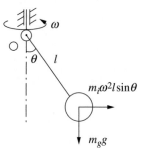

Fig. 2.10 Newton's conical pendulum experiment

$$m_i\omega^2 l^2 \sin\theta \cos\theta = m_g gl \sin\theta$$

or,

$$(m_i/m_g) = g/(\omega^2 l \cos\theta)$$

As the pendulum mass rotates at speed ω, the period of rotation is given by

$$T = \frac{2\pi}{\omega} = \frac{2\pi}{\left[g/\left\{(m_i/m_g)l\cos\theta\right\}\right]^{1/2}}$$

Newton equated this period with the time period of a pendulum with length $l \cos\theta$ given by

$$T = \frac{2\pi}{\left[g/\left\{(m_i/m_g)l\cos\theta\right\}\right]^{1/2}}$$

and used the time period expression for a simple pendulum to determine the speed ω of a conical pendulum. Using the experiments, Newton established the equivalence of m_i and m_g with a much higher level of accuracy.

In 1880, Baron Eötvös from Budapest conducted an experiment to demonstrate the equivalence with an accuracy level of 10^{-8}. In 1964, Dicke and, in 1971, Braginski and Panov verified the equivalence to a limit of 10^{-12}. In more recent experiments using space technology, the accuracy of one part is 10^{18} has been achieved! The basic principle followed in Eötvös experiment is explained through the simplified diagram in Fig. 2.11.

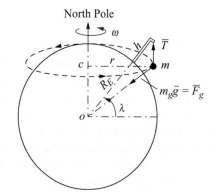

Fig. 2.11 Principle of Eötvös experiment

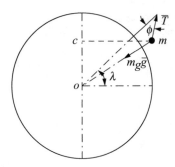

Fig. 2.12 Forces on mass in Eötvös experiment

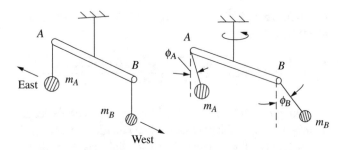

Fig. 2.13 Development of an unbalance moment when the equivalence is not valid

Because of the earth's rotation, the mass has acceleration towards the centre C of the circle. The magnitude of the centripetal acceleration directed towards C is $\omega^2 r \approx {}^2R\cos\lambda$ as $h \ll R$, where h is the height of the post from which the pendulum suspended. The force required for this acceleration is equal to $m\omega^2R\cos\lambda$. This force is the resultant of the tension T and the gravitational pull F_g on the mass. The force towards C acting on the bob (Fig. 2.12) is given by

$$m_g g \cos\lambda - T\cos(\lambda+\varphi) = m_i\omega^2 R\cos\lambda$$

or,

$$(m_g g - m\omega^2 R)\cos\lambda = T\cos(\lambda+\varphi) \qquad (2.3.6)$$

Since the mass is not subjected to any acceleration in the direction of the local vertical

$$m_g g \sin\lambda = T\sin(\lambda+\varphi) \qquad (2.3.7)$$

Eliminating T from the above two equations

$$\left(1 - \frac{m_i\omega^2 R}{m_g g}\right)\cot\lambda = \cot(\lambda+\varphi)$$

Simplifying after approximation

$$\varphi \approx \left(\frac{m_i}{m_g}\right)\frac{\omega^2 R}{g}\sin\lambda\cos\lambda \qquad (2.3.8)$$

Since φ depends on the ratio (m_i/m_g), it will not be same for different materials if the equivalence of m_i and m_g is not true. To check this, two bobs of same weight but made of different materials are hung from a suspended lever AB as shown in Fig. 2.13. If the angles φ_A and φ_B are different, then the suspended lever will be subjected to a resultant moment causing it to rotate as indicated. However, if $m_i \equiv m_g$, φ will be same for different material bobs and no rotation of the suspended lever will be noticed. In the experiments conducted by Eötvös, no rotation of the suspended lever was found.

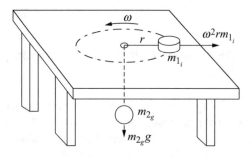

Fig. 2.14 Balancing of gravitational pull by inertial centrifugal force

Before Newton, Huygens recognized the equivalent nature of the inertial centrifugal force and gravitational pull. He suggested an experiment through which this could be verified.

Figure 2.14 shows the scheme of this experiment. When the mass m_1 is rotated at the appropriate angular velocity ω keeping the length of the rotating part of the rope r, the centrifugal force $m_{1_i}\omega^2 r$ due to the inertial mass can balance the hanging weight. By suitable adjustment, it can be shown that $m_i = m_g$ when $m_1 = m_2$. Only two and half centuries later this equivalence of gravity and inertia became the foundation for Einstein's General Theory of Relativity. It is also interesting to note this is at the root of all planetary motions.

2.4 Space–Time and Symmetry in Newtonian Mechanics

It may occur to careful readers if there are definite reasons for the laws of motion to be as they are. As the laws of the game of chess are closely linked to the structure of a chessboard, the laws of motion are also linked to certain properties of the space and time. In fact while presenting the laws of motion such special properties are implicitly taken for granted. Of course the laws of mechanics cannot be directly derived only from the structure of space and time,

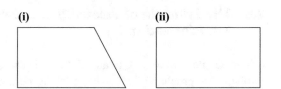

Fig. 2.15 Symmetry of geometric shapes

but understanding their close inter dependence provides a deeper understanding of the concepts.

Three important properties have been *a priory* assumed while formulating the laws of motion. These are as follows:

i. Homogeneity of space,
ii. Isotropy of space and
iii. Homogeneity of time

The homogeneity of space implies that the results of a dynamics experiment do not depend on the location where the experiment is conducted. Similarly, the isotropic nature of space ensures that the results of a dynamics experiment are not dependent on the orientation. Similarly, the outcome of a dynamical lest is independent of the time when it is conducted suggesting the homogeneity of time.

Another important aspect that plays a very profound role in mechanics is 'symmetry'. To understand the meaning, it may be helpful to present the concept of symmetry through geometrical shapes.

Figure 2.15 shows five geometric shapes with varying degrees of symmetry. The shape shown in (i) does not possess any symmetry as the appearance changes when it is rotated in the range 0 < angle of rotation < 360°. The rectangle shown in (ii) preserves its appearance when rotated through 180°. The square and the equilateral triangle shown in (iii) and (iv) possess increasingly higher degree of symmetry. The square and the triangle maintain the shape through a rotation of 90° and 60°, respectively. Finally, the circle shown in (v) is perfectly symmetric as any amount of rotation does not alter the shape. Thus, symmetry is signified by the 'absence of change', and in case of geometrical shapes, it is recognized by the lack of change through rotation or reflection.

Figure 2.16 shows the reflections in the cases of the shapes (i) and (ii). It is shown that the shape is preserved in the case of a rectangle but not for the shape shown in (i). Any amount of rotation of the reflected shape does not produce the original shape.

In case of geometrical figures, the symmetry is depicted by the absence of change when it is rotated or reflected. In the case of dynamics, the symmetry is identified by the lack of change in the results of an experiment when one goes from one inertial frame to another or when such an experiment is moved through space or conducted at a different time

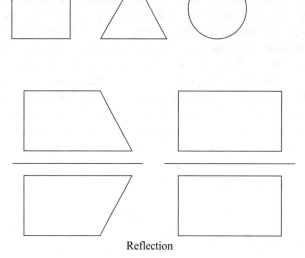

Reflection

Fig. 2.16 Reflection of shapes

or it is rotated through space. How the symmetry properties are linked to various laws of motion and conservation principle will be discussed in the later section.

2.5   Early Concept of Energy

At this point, it is desirable to focus the attention to the development in science of motion that was taking place in the continent at a time when Newton was developing his mechanics. After Rene Descartes and Christian Huygens, it was Gottfried Wilhelm Leibniz (1646–1716) who was busy in developing this branch of science. One should not forget that it was Leibniz who first introduced the word 'dynamics' to describe this branch of natural philosophy. Newton used to refer to this discipline as 'rational mechanics' and considered it to be a mathematical discipline. However, Leibniz considered his 'dynamics' to be a physical discipline. Of course, Leibniz's treatment of the subject was predominantly metaphysical and, quite often, lost its meaning in the modern context.

The primary difference in the approach followed by Leibniz was because of his refusal to accept the concept of force as per the Newtonian doctrine. At the same time, being a follower of Descartes and Huygens, Leibniz did not agree to the 'action-at-a-distance' approach for gravitational interaction. Apart from this the basic idea, behind his dynamics were conceived some years before the publication of Principia. So, the explanations for the planetary motion given by Leibniz depended heavily on metaphysics. To him, 'motion' has no reality and only relative motion is meaningful. He also totally rejected Newton's concept of 'absolute space'.

Leibniz proposed two types of forces—'passive force' and 'active force'. The 'passive force' is somewhat akin to the inertial effect in bodies, and the 'active force' is directly related to the motion of bodies. Thus, passive force represents the mass of a body, whereas the active force relates to the mass and speed of a body. In Leibniz's opinion, the motion of a body has no reality; it is possible to conceive motion as change of position relative to other bodies. Active forces, according to the scheme proposed by Leibniz, again could be classified into two types—'living' forces and 'dead' forces. Those which are associated with only the tendency of motion were considered as 'dead' forces, and those which were associated with the motion were called 'living'. Without entering into the metaphysical discussion on Leibniz's science of motion, in modern language, one can understand 'dead' forces as the product of mass and an infinitesimal change is speed, i.e. $m \cdot dv$. This is nothing but an infinitesimal change in momentum and is same as the concept of impetus or impulse used by Newton in his analyses. (It should be remembered that Newton did not use the second law in a form it is used in modern times. He solved his mechanics problems using the concept of impulse that is equal to the change in momentum in very short durations.) The 'living' force in Leibniz's mechanics is proportional to the product of mass and the square of the speed, i.e. mv^2. Leibniz used the term 'vis viva' for this quantity. A considerable degree of nebulosity exists in Leibniz's writings, and maybe the true nature of these quantities was not very clear to Leibniz himself. When translated into the modern language, 'dead' force bears a close relationship with the concept of force or, more strictly speaking, with the idea of 'virtual work'. But the living force is merged with the modern concept of kinetic energy with some subtle differences. According to Leibniz, the living force had both direction and magnitude, whereas kinetic energy is a purely scalar quantity.

Using examples from collision problems, Leibniz also claimed that his 'motive action' (meaning living force, most probably) remains conserved. Using this principle, he solved collision problems with elastic bodies. Thus, the concept of an entity that is conserved took shape in continental mechanics that laid the foundation of the modern concept of mechanical energy. More elaborate discussion on energy and related issues will be taken up in the later section.

2.6 The Principle of Relativity and Galilean Transformation

It was Galileo who first removed the biggest hurdle in resolving the problem of a moving earth by introducing the concept of relativity and the indistinguishability of inertial frames from one another. At the present time, it is next to impossible to understand the tremendous difficulty in incorporating this idea into the thought process of not only the common men but even the philosophers.

Since mass is an invariant quantity in Newtonian mechanics and the acceleration of a body is observed to be the same by the observers in all inertial frames, the second law ensures that the observers will also agree on the force to be the same. Thus, two inertial observers agree about the cause of a given motion to be the same though the description of motion will be different. To elaborate this point, let the following example be considered.

An observer A sitting in a moving train throws a ball in the vertical direction and catches it back. Another observer B stands on the platform and observes the motion of the ball. Their descriptions are indicated in Fig. 2.17a, b. But both can predict their respective observations by using the second law of motion employing the respective initial conditions. The implication of the above experiment is that the laws of motion are the same in all inertial frames. This is an example of the Principle of Relativity, the general statement of which is that the laws of physics do not distinguish one inertial frame from another. Though this idea was used by Einstein for developing his Special Theory of Relativity, it was first proposed by Galileo in his famous book 'Dialogue Concurring Two Chief World Systems—Ptolemaic and Copernican'.

Though the text is widely read, it may be useful to present here the dialogue of two imaginary characters (one was, of course, Galileo himself), Sagredo and Salviati.

Salviati: Shut yourself up with some friend in the main cabin below decks on some large ship, and have with you there some flies, butterflies, and other small flying animals. Have a large bowl of water with some fish in it; hang up a bottle that empties drop by drop into a wide vessel beneath it. With the ship standing still, observe carefully how the little animals fly with equal speed to all sides of the cabin. The fish swim indifferently in all directions; the drops fall into the vessel beneath; and, in throwing something to your friend, you need to throw it no more

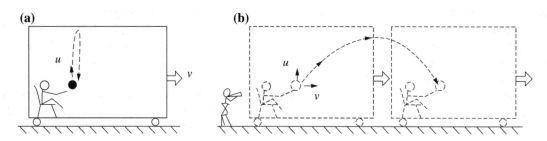

Fig. 2.17 **a, b** A ball's motion observed by two observers, **c** Two inertial observers looking at the motion of a particle

(c)

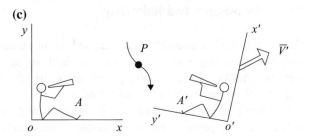

Fig. 2.17 (continued)

strongly in one direction than another, the distances being equal; jumping with your feet together, you pass equal spaces in every direction. When you have observed all these things carefully (though there is no doubt that when standing still everything must happen in this way), have the ship proceed with any speed you like, so long as the motion is uniform and not fluctuating this way and that. You will discover not the least change in all the effects named, nor could you tell from any of them whether the ship was moving or standing still....

The above statement of Salviati suggests that it is not possible to detect uniform motion from observing internal motion phenomena. Thus, Galileo was the first proponent of 'The Principle of Relativity' that states that 'all inertial frames are equivalent so far as the laws of motion are concerned'. This principle which is within the domain of Newtonian mechanics is also termed as Galilean Relativity to distinguish it from Einstein's Special Theory of Relativity. The transformation of kinematic quantities from one inertial frame observer to another follows the rules which are known as Galilean transformation. If A and A' be two observers in inertial frames S and S', respectively, and, if the O_x and $O'_{x'}$ are aligned with the direction of the velocity V with which S' moves with respect to frame S, then the velocity components of a particle follow the transformation rules as given below:

$$t' = t$$
$$u'_x = u_x - V$$
$$u'_y = u_y$$
$$u'_z = u_z$$

The transformation of coordinates follows the following rules:

$$t' = t$$
$$x' = x - Vt$$
$$y' = y$$
$$z' = z$$

It is also clear that acceleration components transform as follows:

$$a_{x'} = \frac{\mathrm{d}u'_x}{\mathrm{d}t'} = \frac{\mathrm{d}(u_x - V)}{\mathrm{d}t} = \frac{\mathrm{d}u_x}{\mathrm{d}t} = a_x$$

Similarly,

$$a_{y'} = a_y \quad \text{and} \quad a_{z'} = a_z$$

Hence,

$$\boldsymbol{a'} = \boldsymbol{a}$$

It will be seen in what follows is that many interesting features of Newtonian mechanics can be derived through the application of the symmetry concepts and the Principle of Relativity.

2.6.1 The Principle of Relativity

Before proceeding further, it is desirable to examine the Principle of Relativity in a more mathematical form to concretize the concepts.

Figure 2.17c shows two different inertial observers A and A' taking note of the motion of a particle P as indicated. The inertial frame $o'-x'-y'$ has a uniform velocity $\boldsymbol{V'}$ with respect to the other inertial frame $o-x-y$ as shown in the figure. Although, A and A' will have different descriptions of the motion of P as $y(x)$ and $y'(x')$, both of them will agree on the acceleration of P as $\boldsymbol{a} = \boldsymbol{a'}$. The Principle of Relativity in Newtonian Mechanics can be expressed in a more elaborate manner as described below.

Let S be an isolated system in which the particles interact with one another but not with the rest of the universe. Observer A marks the motion of all particles (with specific initial positions and velocities at $t = 0$) as time progresses. The positions and velocities of all particles are recorded as

$$\boldsymbol{r}_i = \boldsymbol{r}_i(t); \quad i = 1, 2, \ldots, N$$
$$\boldsymbol{v}_i = \boldsymbol{v}_i(t); \quad i = 1, 2, \ldots, N$$

Now, observer A' conducts an experiment with the same particles with the same initial positions and same initial velocities in A' frame of reference, and then, the description of motion with the progress of time will be the same as that marked by A. Thus, with

$$\boldsymbol{r}_i(o) = \boldsymbol{r}'_i(o); \quad i = 1, 2, \ldots, N$$

and

$$\boldsymbol{v}_i(o) = \boldsymbol{v}'_i(o); \quad i = 1, 2, \ldots, N$$

the subsequent motion's description will be the same or mathematically

$$\boldsymbol{r}_i(t) = \boldsymbol{r}'_i(t'); \quad i = 1, 2, \ldots, N$$

and

$$\boldsymbol{v}_i(t) = \boldsymbol{v}'_i(t'); \quad i = 1, 2, \ldots, N$$

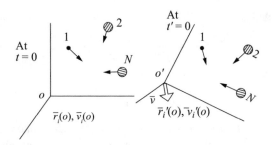

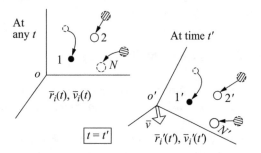

Fig. 2.18 Explicit illustration of the Principle of Relativity

with a further stipulation that

$$t = t'$$

Figure 2.18 illustrates the above diagrammatically.

One important point needs to be carefully noted. The Principle of Relativity also ensures three important properties of the space and time. The homogeneity of space is ensured as the inertial frames are located at different locations. The orientation of the inertial frames does not affect the laws of motion indicating the isotropy of space. Since the principle is valid not depending on when the experiments are conducted, it ensures the homogeneity of time.

The profoundness of the principle is evident in the fact that according to the Principle of Relativity, all inertial frames are equivalent for *all* physical phenomena not only the dynamical ones.

2.6.2 Symmetry and Relativity

The Principle of Relativity, when applied to mechanics problems involving some symmetry, can lead to many important results without solving the problems. Some examples presented below illustrate the point.

The first example is based on a collision problem of two isolated identical particles. Before the collision, the situation is 'perfectly' symmetric when viewed by an inertial observer *A* as shown in Fig. 2.19a. The possibilities of the state of motion after the collision could be as indicated in Fig. 2.19b. To determine the possibility without solving the collision problem, let the same phenomenon be observed by *A'* as shown in Fig. 2.19a. At the beginning, both *A* and *A'* found the situation to be symmetric. According to the Principle of Relativity, the two observers must find the description of motion after the collision to be indistinguishable. A little careful study shows that the situation indicated in (i) of Fig. 2.19b is the only case that satisfies the Principle of Relativity. Cases (ii) and (iii) look different to the two observers, and the descriptions differ. If a third observer *A''* is considered in the frame $x'' - y'' - z''$, the possibilities shown in (iv) and (v) look different and the descriptions differ after the collision. Therefore, the only possible situation is as shown in (i). In general, if a situation is symmetric, after a dynamical interaction the situation must maintain the symmetry according to the Principle of Relativity.

If the case of a head-on symmetric collision of two identical particles appears to be too restrictive, a little more generalized case can be considered where two identical particles collide in an unsymmetrical fashion as indicated in Fig. 2.20.

Figure 2.20 shows the phenomenon. The figure shows the initial positions and velocities of the particles, and let the particles be initially free to move (as they are far off from each other and the system is an isolated one from the rest of the universe). Therefore, in the inertial frame $x-y-z$, both will move with no acceleration. Next, let the collision be

Fig. 2.19 Presentation of symmetry

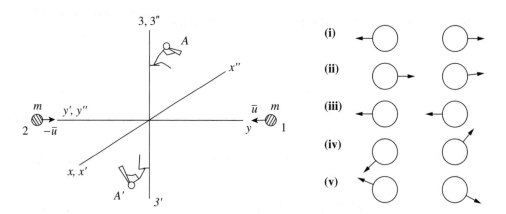

Fig. 2.20 Unsymmetrical collision of two identical particles

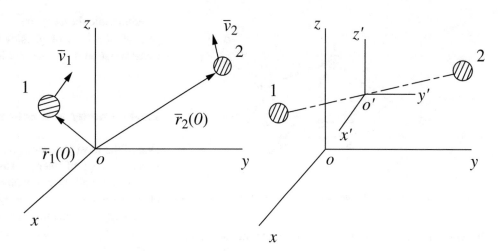

observed from a frame $x'-y'-z'$ whose origin is at the midpoint of the line joining the two particles, O'. Initially, the velocity of the point O' is

$$\frac{\mathrm{d}}{\mathrm{d}t}\frac{1}{2}[\boldsymbol{r}_1(o)+\boldsymbol{r}_2(o)]=\frac{1}{2}[\boldsymbol{v}_1(o)+\boldsymbol{v}_2(o)]=\boldsymbol{v}_{o'}(o)$$

i.e. a constant. Therefore, if the axes of the primed frame of reference be always kept parallel to those in the inertial frame $x-y-z$, the primed frame moves with respect to the inertial frame with a constant velocity. Thus, the frame $x'-y'-z'$ is also an inertial frame if the velocity is kept constant. Now, when observed from this second inertial frame, the collision is a symmetrical one. The initial velocities of the two particles in this frame are found out as follows:

$$\boldsymbol{v}_1'(o)=\boldsymbol{v}_1(o)-\boldsymbol{v}_{o'}(o)=\boldsymbol{v}_1(o)-\frac{1}{2}[\boldsymbol{v}_1(o)+\boldsymbol{v}_2(o)]$$

$$=\frac{1}{2}[\boldsymbol{v}_1(o)+\boldsymbol{v}_2(o)]=\boldsymbol{u}(o)$$

and

$$\boldsymbol{v}_2'(o)=\boldsymbol{v}_2(o)-\boldsymbol{v}_{o'}(o)=\boldsymbol{v}_2(o)-\frac{1}{2}[\boldsymbol{v}_1(o)+\boldsymbol{v}_2(o)]$$

$$=-\frac{1}{2}[\boldsymbol{v}_1(o)+\boldsymbol{v}_2(o)]=-\boldsymbol{u}(o)$$

As this frame is also an inertial frame, the Principle of Relativity has to be satisfied; i.e., the initial symmetry has to be maintained. So the positions of particles 1 and 2 in the new frame must also remain symmetrical, implying that during the subsequent motion also $\boldsymbol{r}_1(t) = -\boldsymbol{r}_2(t)$. It ensures that even when the particles start interacting the origin O' remains at the middle of the two particles. This also ensures that the velocities and accelerations of the particles are always equal and opposite when the collision is viewed from the frame $x'-y'-z'$. Thus,

$$\boldsymbol{v}_1'(t)=-\boldsymbol{v}_2'(t)\quad\text{and}\quad\boldsymbol{a}_1'(t)=\boldsymbol{a}_2'(t)$$

Therefore, $\boldsymbol{a}_1'(t)+\boldsymbol{a}_2'(t)=0$ in the $x'-y'-z'$ frame, and since accelerations are same in all inertial frames, all inertial observers conclude that

$$\boldsymbol{a}_1'(t)+\boldsymbol{a}_2'(t)=0$$

or,

$$\frac{\mathrm{d}}{\mathrm{d}t}[m\boldsymbol{v}_1(t)+m\boldsymbol{v}_2(t)]=0$$

or,

$$m\boldsymbol{v}_1(t)+m\boldsymbol{v}_2(t)=\text{constant}$$

meaning that the total momentum of the system is conserved. Therefore, a considerable amount of information can be extracted by applying the symmetry concept and the Principle of Relativity.

2.6.3   Form Invariance of Physical Laws

Physical laws represent the generalized patterns of physical phenomena. The Principle of Relativity states that no distinctions can be made among the inertial frames by conducting internal experiments only. It means that the observers in different inertial frames must arrive at identical conclusions about the laws of physics from the results of the internally conducted experiments. Or in other words, the form of the laws of physics must be invariant. This very important property dictated by the Principle of Relativity is called the 'form invariance of physical laws'.

This property of physical laws puts very severe restrictions on the allowed forms. A very suitable example can be taken to illustrate this as follows. Let two particles 1 and 2 interact gravitationally, and let the phenomenon be observed

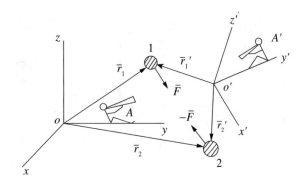

Fig. 2.21 Gravitational interaction as viewed from two inertial frames

by two inertial observers A and A' sitting in the inertial frames $x-y-z$ and $x'-y'-z'$ as indicated in Fig. 2.21.

Observer A notices the positions of the two particles 1 and 2 as r_1 and r_2. Similarly the observer A' records the positions of the same particles with respect to the primed frame of reference $x'-y'-z'$ as r_1' and r_2'. Now let a law of gravitation is suggested in the following form. The gravitational force is given by

$$|F| = \frac{Gm_1m_2}{|r_1 + r_2|^2}$$

Hence, the accelerations of the particles should be

$$a_1 = \frac{F}{m_1} \quad \text{and} \quad a_2 = -\frac{F}{m_2}$$

The force of acceleration according to the observer A', according to this suggested force law, will be

$$|F'| = \frac{Gm_1m_2}{|r_1' + r_2'|^2}$$

as mass is an invariant quantity. Thus, the acceleration of the particles according to A' will be

$$a_1' = \frac{F'}{m_1} \quad \text{and} \quad a_2' = -\frac{F'}{m_2}$$

But both A and A' being inertial observers they must agree that $a_1 = a_1'$ and $a_2 = a_2'$. This requires the two quantities

$$\frac{Gm_2}{|r_1 + r_2|^2} \quad \text{and} \quad \frac{Gm_2}{|r_1' + r_2'|^2}$$

must be equal which is not true as $|r_1 + r_2| \neq |r_1' + r_2'|$. Hence, the suggested form is not admissible. On the other hand, a force law

$$|F| = \frac{Gm_1m_2}{|r_1 - r_2|^2}$$

is admissible because $|r_1 - r_2| = |r_1' - r_2'|$. Therefore, in general, the laws of physics must remain invariant under transformation from one inertial frame to another.

2.6.4 Energy and Energy Function

One of the most profound concepts in the science of motion is the concept of energy. However, an attempt to define energy is futile (like many other very important quantities in physics such as charge, force). Qualitatively, one can state that energy is something that is required to do work or get things done. In the whole of Principia, Newton did not refer to this concept even once. It was the contemporary continental scientists like Huygens and Leibniz who considered a quantity like energy (Leibniz's vis viva) to play an important role in dynamics.

The significance of 'energy' is in its conservation property. The 'energy', i.e. the capacity to do work in a dynamical system, is conserved. Of course, energy can be stored (or possessed by something) in various forms, viz. chemical, heat, sound, electromagnetic, nuclear and mechanical. The sum total of energy in an isolated system is always conserved though there can be transformation from one form into another. So far as dynamics is concerned, the only form of energy that is relevant is mechanical energy. In dynamics, if mechanical energy is transformed into another non-mechanical form like heat or chemical, then it is considered to be lost. Thus, in mechanics, energy conservation of a system implies that the total mechanical energy remains constant.

Mechanical energy can be in two forms. When the capacity to do work of a body is by virtue of its motion (like a bullet), the energy is called 'kinetic energy' (it is very similar to Leibniz's 'vis viva'). Mechanical energy can exist in another form, i.e. 'potential energy'. A body with potential energy can do work not by virtue of its motion but because of its internally stored energy. Best examples of a body possessing potential energy are that which is raised to a height against gravity, or a spring that is deformed.

Figure 2.22 shows the various examples of a mechanical system (or body) to possess mechanical energy in different manners. In the first case, the capacity to do work comes from the speed of the body, whereas in the second and third examples, the capacity to do work is in potential form. Quantitative measures of energy can be done by equating it to the amount of work done. So it is important to understand what is meant by work. Work is done either by a force or against a force, and the amount of work is the product of the force and the displacement of the point of application of the force A as illustrated in Fig. 2.23.

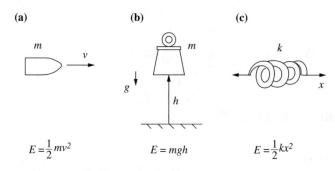

$$E = \frac{1}{2}mv^2 \qquad\qquad E = mgh \qquad\qquad E = \frac{1}{2}kx^2$$

Fig. 2.22 Examples of mechanical energy

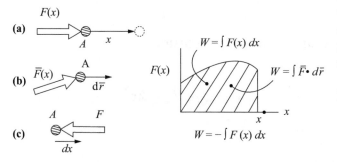

Fig. 2.23 Work done by a force

If the force and displacement of the point of application of the force A are vectors, then the scalar product summed over the whole process is taken as the work done by the force. The scalar quantity, work, goes into the system (to which the point A belongs) to be stored as energy. When the displacement of the point A is against the direction of application of the force (Fig. 2.23c), the work is done by the system against the force. Hence, the work done is negative and is achieved through the expenditure of energy stored in the system. In case of the vector nature of the force and the displacement, the sign of the scalar product automatically decides the situation. Using this concept, it is seen that the kinetic energy is given by $\frac{1}{2}mv^2$ and the gravitational or strain potential energy is given by mgh and $\frac{1}{2}kx^2$ where 'h' is the height raised and x is the deformation of a spring from its natural length, k being the stiffness of the spring given by the force required to produce unit deformation (elongation or compression). The law of conservation of mechanical energy of a system states that the sum total of the kinetic and potential energy remains constant. Figure 2.24 shows a number of different types of mechanical systems and their respective total energy expression.

2.6.5   Energy Function

The concept of energy function is very important. A particular mechanical system can be disturbed by putting it in motion. A system undergoes infinite states of motion depending on the initial disturbance. But is there something that describes all possible states of motion of the system and not a particular state? If such a 'thing' exists, then that describes the 'system' consisting of all the 'possible' states of motion. In fact, there is such a thing that describes a 'system' instead of a particular motion executed by it. This 'something' can be called the 'energy function' of a system. Each mechanical system possesses a unique 'energy function'. The positions and velocities of the individual particles constituting a system determine the value of its 'energy function'. An example can illustrate the importance of energy function concept.

Let the first case be that of a simple spring mass system. Its energy function is given as follows:

$$E = \frac{1}{2}m\dot{x}^2 + \frac{1}{2}kx^2$$

Since it is a conservative isolated system, E must be conserved:

$$\text{or,} \quad \frac{dE}{dt} = 0$$
$$\text{or,} \quad m\dot{x}\ddot{x} + kx\dot{x} = 0$$
$$\text{or,} \quad \dot{x}(m\ddot{x} + kx) = 0$$

Hence, two possible situations are permitted. Either the body is at rest, i.e. $\dot{x} = 0$, or the following relation is satisfied:

$$m\ddot{x} + kx = 0$$

The example is simple but shows how the dynamical behaviour expressed in the form of a relationship among various quantities (in modern language this is called the equation of motion). It can be stated that the dynamical characteristics of a mechanical system are governed by its 'energy function'. It should be carefully noted that the 'energy function' of a system is not explicitly dependent on time. For an isolated conservative system, the energy function remains constant implying

$$E[\boldsymbol{r}_i(t); i = 1, 2, \ldots, N, \boldsymbol{v}_i(t); i = 1, 2, \ldots, N] = \text{constant}$$

or

$$\frac{dE}{dt} = 0$$

Fig. 2.24 Examples of mechanical system and their total energy

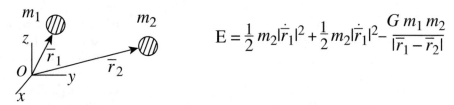

$$E = \frac{1}{2} \sum_{i=1}^{n} m_i |\overline{v}_i|^2$$

(a) A system of noninteracting particle

$$E = \frac{1}{2} m_2 |\dot{\overline{r}}_1|^2 + \frac{1}{2} m_2 |\dot{\overline{r}}_1|^2 - \frac{G\, m_1\, m_2}{|\overline{r}_1 - \overline{r}_2|}$$

(b) A pair of gravitationally interacting particles

$$E = \frac{1}{2} m \dot{x}^2 + \frac{1}{2} k x^2$$

Equilibrium position

(c) A spring-mass system

$$E = \frac{1}{2} m l^2 \dot{\theta}^2 + mgl(1 - \cos\theta)$$

(d) An oscillating pendulum

which signifies that the results of an experiment are independent of the time when it is conducted. Thus, homogeneity of time is a direct consequence of energy conservation.

It is interesting to study what is the situation when a mechanical phenomenon is observed in a non-inertial frame. It is quite obvious that the velocities of the bodies seen from a non-inertial frame do not remain constant. Thus, the energy of a mechanical system does not remain conserved. Since energy conservation is not valid in a non-inertial frame of reference, homogeneity of time is not satisfied in such frames. As just mentioned, the kinetic energy of a particle (or, a system of particles) is not same in different inertial frames (though it remains conserved in any particular frame) as the velocities are different in different inertial frames of reference. However, the potential energy of a system depends on the interaction among particles. Considering a case of a two-particle system, it is clear that the potential energy depends on the distance between two particles. Hence, as the distance remains unchanged to different

observers, the potential energy is an invariant quantity independent of the observer's frame of reference.

2.7 Laws of Motion and the Properties of Space and Time

In this section, the laws of motion will be studied in association with the properties of space and time, and from the Principle of Relativity. Let the case of an isolated single particle moving freely in an inertial frame be taken up. The energy function of this particle will be of the form $E|v|^2$. It cannot depend on time as time is homogenous; similarly, the energy function cannot depend on either the position r or the direction of v since the space is both homogenous and isotropic. Next, let the y and y' axes of two inertial observers A and A' be aligned with the direction of $-v$ as indicated in Fig. 2.25 without affecting the generality of the result.

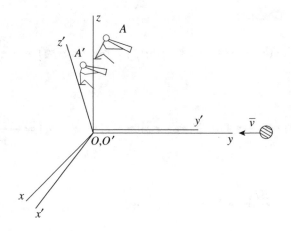

Fig. 2.25 Motion of a free particle with respect to two inertial frame with different orientation

To begin with the motion possesses symmetry in both the frames as x, z, x' and z' components of v are zero. Following the Principle of Relativity, both observers must not be able to detect any break of symmetry during the subsequent motion. So all the above four components of v must continue to remain zero. Hence, the particle moves along a straight line. Furthermore as $E|v|^2$ remains constant, the speed $|v|$ must remain constant. These are the two predictions of the first law.

2.7.1 The Second Law of Motion

Extracting the second law of motion from the properties of the space and time along with the application of the Principle of Relativity can be achieved as demonstrated below.

Figure 2.26 shows two particles 1 and 2 that are interacting with each other but are isolated from the rest of the universe. Let the particles be in motion with velocities v_1 and v_2 as shown, their instantaneous locations being r_1 and r_2. Let U be the potential energy of the system because of the

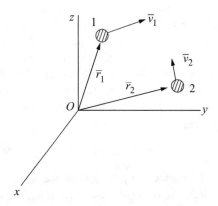

Fig. 2.26 System of two interacting particles

interaction of the particles, and it is a function of their instantaneous distance $|r_1 - r_2|$ only. If the kinetic energies of the individual particles be T_1 and T_2, depending on the square of their speeds v_1 and v_2, the total mechanical energy of the whole system can be expressed as follows:

$$E = T_1 v_1^2 + T_2 v_2^2 + U[(r_1 - r_2) \cdot (r_1 - r_2)]$$

Since the system is an isolated one and neither of the above energy terms being explicitly dependent on time (as time is homogenous as mentioned earlier), the total energy is conserved. Hence, writing $v_1^2 = v_1 \cdot v_1$ and $v_2^2 = v_2 \cdot v_2$

$$\frac{dE}{dt} = \frac{d}{dt}[T_1(v_1 \cdot v_1) + T_2(v_2 \cdot v_2) \qquad (2.7.1)$$
$$+ U[(r_1 - r_2) \cdot (r_1 - r_2)]] = 0$$

Using the identities

$$\frac{d(v_1 \cdot v_1)}{dt} = 2v_1 a_1, \quad \frac{d(v_2 \cdot v_2)}{dt} = 2v_2 a_2$$

and

$$\frac{d[(r_1 - r_2) \cdot (r_1 - r_2)]}{dt} = 2(r_1 - r_2) \cdot (v_1 - v_2)$$

(2.7.1) can be written in the following form:

$$\frac{dT_1}{dv_1^2}(2v_1 \cdot a_1) + \frac{dT_2}{dv_2^2}(2v_2 \cdot a_2)$$
$$+ \frac{dU}{d[(r_1 - r_2) \cdot (r_1 - r_2)]} 2(r_1 - r_2) \cdot (v_1 - v_2) = 0$$

$$(2.7.2)$$

Let the following quantities are defined as given below:

$$m_1(v_1)^2 = 2\frac{dT_1}{dv_1^2}$$
$$m_2(v_2)^2 = 2\frac{dT_2}{dv_2^2}$$

and,

$$f(|r_1 - r_2|) = 2\frac{dU}{d[(r_1 - r_2) \cdot (r_1 - r_2)]}$$

Incorporating these definitions in (2.7.2), one obtains

$$\frac{dE}{dt} = m_1(v_1^2)v_1 \cdot a_1 + m_2(v_2^2)v_2 \cdot a_2$$
$$+ f(|r_1 - r_2|)(r_1 - r_2) \cdot (v_1 - v_2) = 0$$

Separating the kinetic and potential energy terms

$$m_1(v_1^2)v_1 \cdot a_1 + m_2(v_2^2)v_2 \cdot a_2$$
$$= -f(|r_1 - r_2|)(r_1 - r_2) \cdot (v_1 - v_2) \qquad (2.7.3)$$

Now, it should be remembered that all equations are form invariant and valid in all inertial frames. The right-hand side of (2.7.3) has the same value in all inertial frames as the quantities $(|\boldsymbol{r}_1 - \boldsymbol{r}_2|)$ and $(|\boldsymbol{v}_1 - \boldsymbol{v}_2|)$ have the same values in all inertial frames. Hence, the quantity

$$f(|\boldsymbol{r}_1 - \boldsymbol{r}_2|)(\boldsymbol{r}_1 - \boldsymbol{r}_2) \cdot (\boldsymbol{v}_1 - \boldsymbol{v}_2) = \frac{dU}{dt}$$

is an invariant quantity[4] and, therefore, the term

$$m_1(v_1^2)\boldsymbol{v}_1 \cdot \boldsymbol{a}_1 + m_2(v_2^2)\boldsymbol{v}_2 \cdot \boldsymbol{a}_2 = \frac{dT}{dt}$$

is also invariant. Therefore,

$$\frac{dT}{dt} = m_1(v_1^2)\boldsymbol{v}_1 \cdot \boldsymbol{a}_1 + m_2(v_2^2)\boldsymbol{v}_2 \cdot \boldsymbol{a}_2$$
$$= m_1(v_1'^2)\boldsymbol{v}_1' \cdot \boldsymbol{a}_1' + m_2(v_2'^2)\boldsymbol{v}_2' \cdot \boldsymbol{a}_2' = \frac{dT'}{dt}$$

This invariance restricts the function $m_1(v_1^2)$ and $m_2(v_2^2)$. It can be shown that one choice of m_1 and m_2 satisfies the condition of invariance as demonstrated below. Let the following forms be assumed

$$m_1(v_1^2) = m_1(\text{constant}), \quad m_2(v_2^2) = m_2(\text{constant})$$

and,

$$m_1\boldsymbol{a}_1 = -m_2\boldsymbol{a}_2 \tag{2.7.4}$$

with above choices

$$\frac{dT}{dt} = m_1\boldsymbol{v}_1 \cdot \boldsymbol{a}_1 + m_2\boldsymbol{v}_2 \cdot \boldsymbol{a}_2 = m_1\boldsymbol{v}_1 \cdot \boldsymbol{a}_1 - m_2\boldsymbol{v}_2 \cdot \boldsymbol{a}_2$$
$$= m_1\boldsymbol{a}_1 \cdot (\boldsymbol{v}_1 - \boldsymbol{v}_2) = m_1\boldsymbol{a}_1' \cdot (\boldsymbol{v}_1' - \boldsymbol{v}_2') = \frac{dT'}{dt}$$
$$\tag{2.7.5}$$

by using the Galilean transformation rules. It can be proved that the choices given in (2.7.4) are the only way to maintain invariance of $\frac{dT}{dt}$. It is left as an exercise for the reader. Next, let the following definitions be introduced:

$$\boldsymbol{F}_1 = -f(|\boldsymbol{r}_1 - \boldsymbol{r}_2|)(\boldsymbol{r}_1 - \boldsymbol{r}_2) \quad \text{and}$$
$$\boldsymbol{F}_2 = -f(|\boldsymbol{r}_1 - \boldsymbol{r}_2|)(\boldsymbol{r}_2 - \boldsymbol{r}_1) \tag{2.7.6}$$

---

[4]This can be demonstrated in a more direct way using the laws of Galilean transformation. Let x–y–z be two inertial frames with the following transformation equations $\boldsymbol{r}' = \boldsymbol{r} - \boldsymbol{u}t$, $\boldsymbol{v}' = \boldsymbol{v} - \boldsymbol{u}$, $\boldsymbol{a}' = \boldsymbol{a}, t' = t$; then, the following equalities are obtained: $\boldsymbol{r}_1' - \boldsymbol{r}_2' = \boldsymbol{r}_1 - \boldsymbol{r}_2$ and $\boldsymbol{v}_1' - \boldsymbol{v}_2' = \boldsymbol{v}_1 - \boldsymbol{v}_2$; and so, $\frac{dU}{dt} = f(|\boldsymbol{r}_1 - \boldsymbol{r}_2|)(\boldsymbol{r}_1 - \boldsymbol{r}_2) \cdot (\boldsymbol{v}_1 - \boldsymbol{v}_2) = f(|\boldsymbol{r}_1' - \boldsymbol{r}_2'|)(\boldsymbol{r}_1' - \boldsymbol{r}_2') \cdot (\boldsymbol{v}_1' - \boldsymbol{v}_2') = \frac{dU'}{dt}$.

Therefore,

$$\boldsymbol{F}_1 \cdot (\boldsymbol{v}_1 - \boldsymbol{v}_2) = -f(|\boldsymbol{r}_1 - \boldsymbol{r}_2|)(\boldsymbol{r}_1 - \boldsymbol{r}_2) \cdot (\boldsymbol{v}_1 - \boldsymbol{v}_2) = -\frac{dU}{dt}$$

and

$$\boldsymbol{F}_2 \cdot (\boldsymbol{v}_2 - \boldsymbol{v}_1) = -f(|\boldsymbol{r}_1 - \boldsymbol{r}_2|)(\boldsymbol{r}_2 - \boldsymbol{r}_1) \cdot (\boldsymbol{v}_2 - \boldsymbol{v}_1) = -\frac{dU}{dt}$$

so,

$$\frac{dU}{dt} = -\boldsymbol{F}_1 \cdot (\boldsymbol{v}_1 - \boldsymbol{v}_2) = -\boldsymbol{F}_2 \cdot (\boldsymbol{v}_2 - \boldsymbol{v}_1) \tag{2.7.7a}$$

Again from (2.7.5)

$$\frac{dT}{dt} = m_1\boldsymbol{a}_1 \cdot (\boldsymbol{v}_1 - \boldsymbol{v}_2) = m_2\boldsymbol{a}_2 \cdot (\boldsymbol{v}_2 - \boldsymbol{v}_1) \tag{2.7.7b}$$

Combining (2.7.7a) and (2.7.7b)

$$\frac{dT}{dt} + \frac{dU}{dt} = (m_1\boldsymbol{a}_1 - \boldsymbol{F}_1) \cdot (\boldsymbol{v}_1 - \boldsymbol{v}_2)$$
$$= (m_2\boldsymbol{a}_2 - \boldsymbol{F}_2) \cdot (\boldsymbol{v}_2 - \boldsymbol{v}_1) = \frac{dE}{dt} = 0 \tag{2.7.8}$$

Unless $\boldsymbol{v}_1 = \boldsymbol{v}_2$ or the vector $(\boldsymbol{v}_1 - \boldsymbol{v}_2)$ is perpendicular to the vector $(m_1\boldsymbol{a}_1 - \boldsymbol{F}_1)$, (2.7.8) can be satisfied if and only if

$$\boldsymbol{F}_1 = m_1\boldsymbol{a}_1$$

and,

$$\boldsymbol{F}_2 = m_2\boldsymbol{a}_2 \tag{2.7.9}$$

which are nothing but the second law of motion and $\boldsymbol{F}_1$ and $\boldsymbol{F}_2$ are the forces. Thus, the second law of motion can be derived from the properties of the space, time and the Principle of Relativity.

2.7.2 The Third Law of Motion

This law is a contribution of Newton alone unlike the first and second laws which had major contributions from others before Newton. The profound importance of this law is often missed. The basic content of the law has two forms—weak form and strong form. The weak form of the third law states that the forces between two interacting particles are always equal and opposite. Thus if the forces on particles 1 and 2 be $\boldsymbol{F}_1$ and $\boldsymbol{F}_2$, then

$$\boldsymbol{F}_1 = -\boldsymbol{F}_2$$

Figure 2.27 shows two particles and the forces acting on those. As per the third law, the forces are equal and opposite. However, the strong form of the third law states that these

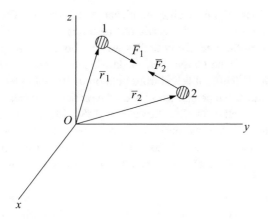

Fig. 2.27 Third law of motion

forces are not only equal in magnitude and opposite in direction, but act along the same line of action.

Thus, using the quantities shown in Fig. 2.27

$$F_1 = -F_2$$

and

$$F_1 \times (r_1 - r_2) = F_2 \times (r_1 - r_2) = 0$$

and

$$F_1 \times r_1 = -F_2 \times r_2 \qquad (2.7.10)$$

In fact from the previous section, it has been found that

$$\frac{dU}{dt} = f(|r_1 - r_2|)(r_1 - r_2) \cdot (v_1 - v_2) = -F_1 \cdot (v_1 - v_2)$$
$$= -F_2 \cdot (v_2 - v_1)$$

or,

$$-F_1 \cdot (v_1 - v_2) = F_2 \cdot (v_1 - v_2)$$

or,

$$F_1 = -F_2$$

Again,

$$\frac{dU}{dt} = -F_1 \cdot (v_1 - v_2) = -F_1 \cdot v_1 - F_2 \cdot v_2$$
$$= -F_1 \cdot \frac{dr_1}{dt} - F_2 \cdot \frac{dr_2}{dt}$$

Hence,

$$\partial U = -F_1 \cdot \partial r_1 - F_2 \cdot \partial r_2 \qquad (2.7.11)$$

Using $\partial U = 0$, since $dU/dt = 0$

$$F_1 \cdot \partial r_1 - F_2 \cdot \partial r_2 = 0 \qquad (2.7.12)$$

Now, considering the homogeneity of space a pure translation of the pair of particles should not result in any change. Under pure translation, both the particles undergo equal displacement

$$\partial r_1 = \partial r_2 = \partial r$$

Hence,

$$F_1 = -F_2$$

as a result of the homogeneity of space. When a pure rotation is imparted, the displacement of the two particles are equal and opposite, as shown in Fig. 2.28.

So,

$$\partial r_1 = -\partial r_2 = \partial r$$

Using the above condition in (2.7.12), one obtains

$$F_1 \cdot \partial r_1 - F_2 \cdot \partial r_2 = 0$$

or,

$$2F_1 \cdot \partial r_1 = 0$$

But for infinitesimal displacement, δr is perpendicular to $(r_1 - r_2)$. Hence, F_1 and F_2 are parallel to the line joining the particles. This is a consequence of the isotropy of space. The above two results complete the third law of motion.

The third law of motion also leads to two important conservation principles as explained below. For the pair of particles under consideration, the two important quantities are linear momentum and the angular momentum. These are defined as follows:

P = Linear momentum = $m_1 v_1 + m_2 v_2$

l_o = Angular momentum about O = $r_1 \times m_1 v_1 + r_2 \times m_2 v_2$

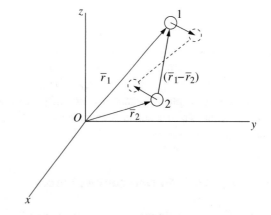

Fig. 2.28 Rotation of a pair of particles

When the third law is employed, the following results are obtained:

$$\frac{d\boldsymbol{P}}{dt} = \frac{d}{dt}(m_1\boldsymbol{v}_1 + m_2\boldsymbol{v}_2) = m_1\boldsymbol{a}_1 + m_2\boldsymbol{a}_2 = \boldsymbol{F}_1 + \boldsymbol{F}_2 = 0$$

So $\boldsymbol{P}$ is conserved. Again

$$\frac{d\boldsymbol{l}_o}{dt} = \frac{d}{dt}(\boldsymbol{r}_1 \times m_1\boldsymbol{v}_1 + \boldsymbol{r}_2 \times m_2\boldsymbol{v}_2)$$
$$= \boldsymbol{v}_1 \times m_1\boldsymbol{v}_1 + \boldsymbol{r}_1 \times m_1\boldsymbol{a}_1 + \boldsymbol{v}_2 \times m_2\boldsymbol{v}_2 + \boldsymbol{r}_2 \times m_2\boldsymbol{a}_2$$

as O is a point fixed in the frame of reference meaning $\dot{\boldsymbol{r}}_1 = \boldsymbol{v}_1$ and $\dot{\boldsymbol{r}}_2 = \boldsymbol{v}_2$. So,

$$\frac{d\boldsymbol{l}_o}{dt} = \boldsymbol{r}_1 \times \boldsymbol{F}_1 + \boldsymbol{r}_2 \times \boldsymbol{F}_2 = \boldsymbol{r}_1 \times \boldsymbol{F}_1 - \boldsymbol{r}_2 \times \boldsymbol{F}_2$$

by the first part of the third law of motion.

Thus,

$$\frac{d\boldsymbol{l}_o}{dt} = (\boldsymbol{r}_1 - \boldsymbol{r}_2) \times \boldsymbol{F}_1 = 0$$

because $\boldsymbol{F}_1$ acts along the vector $(\boldsymbol{r}_1 - \boldsymbol{r}_2)$ by the second part of the third law.

Finally, the properties of the space and time, i.e. the homogeneity of space, the isotropy of space and the homogeneity of time, result in:

- The conservation of linear momentum,
- The conservation of angular momentum and
- The conservation of energy.

So, these conservation principles and the laws of motion are the manifestations of the properties of the space and time and the Principle of Relativity.

2.8 Action-at-a-Distance and Spatiotemporal Locality

Accepting the existence of non-contact force in the science of motion was a serious problem to a large number of scientists in the seventeenth century. Frankly speaking, it is not very easy to grasp the reality behind such 'action-at-a-distance' phenomenon. It is not very easy even at the present time. This section is devoted to a discussion on this very important issue.

2.8.1 Early Work on Non-contact Forces

The earlier diffused ideas and nebular concepts regarding the motion of planets led to the introduction of metaphysical

treatment like providing the planets with 'soules'. The first work on physical force without requiring physical contact between the interacting bodies was the book 'De Magnete, Magnetieisque Corporibus, et de Magno Magnete Tellure' by W. Gilbert published from London in 1600. Kepler, who was the prime person working on the motion of planet, could free himself from J.C. Scaliger's (1484–1558) concept of 'souls' and 'spirits' once he came across Gilbert's book. In the earlier days, he used to call the gravitational force also as a 'magnetic force'. The matter has been already discussed in Sect. 1.7.3.

It was not very easy to accept the gravitational interaction as in the continent the influence was of Descartes' physics that was based upon the impact between particles. Kepler was an exception; he not only conceived of a force like gravitation by the sun on the planets controlling their motions, but he correctly attributed the attraction by the moon to explain the phenomenon of tide. But Galileo was also a disbeliever in such 'occult' non-contact forces. Galileo proposed a theory of the tides which was obviously wrong. According to Galileo's theory, the accumulation of ocean water in certain regions of earth was related to the speed of the earth relative to the distant stars based on a combination of the orbital motion and the daily rotation. Even during the later period, scientists like Descartes and Huygens were totally opposed to the idea of non-contact 'gravitational force'. A meeting to discuss the nature of 'gravity' was organized in 1669 at Paris Académie. Two scientists, Robervel and Frenicle proposed the gravity to be a force of attraction by the earth on objects near the earth's surface—a mutual force between like bodies. But Huygens opposed the proposition vehemently as is clear from a translation of his statement given below:

> To discover a cause of weight that is intelligible, it is necessary to investigate how weight can come about while assuming the existence only of bodies made of one common matter in which one admits no quality or inclination to approach each other but solely different sizes, figures and motions.

In fact, according to many historians of science, this total rejection of the concept of a non-contact gravitational attraction denied Huygens in developing a complete synthesis of the science of motion. Although he developed most of the basic elements much before Newton.

To begin with, Newton also attempted many models to explain gravitational attraction through mechanical means. The story of apple falling giving the idea of explaining the motion of the moon to young Newton in 1665 may contain a certain element of truth, but Newton's attempt at that time was to try for a mechanical explanation. Such models based on Cartesian principles of collision of fundamental particles filling the universe could somehow generate a force towards the earth. But such models could not result in an equal and opposite force on the earth that attracted it to the other interacting bodies, viz. the apple and the moon. The correct idea of universal gravitation took shape in Newton's mind much later, perhaps, in 1680. Thus, Newton transformed dynamics from its restricted domain of 'contact force' only to forces extending over long distances. Thus, the concept of the 'action-at-a-distance' achieved acceptability among the scientists.

2.8.2 Spatiotemporal Locality and Action-at-a-Distance

In general, it is easy to conceive that a 'cause' can give rise to a 'result' only at that location in space and precisely at that instant of time. Of course, there are instances where, apparently, the result takes some time to show up, but a little thinking reveals that in all such cases there are innumerable 'mini-intermediate results' which connect the original 'cause' with the final 'result'.

In a loose sense, one can logically accept that a particle can influence the motion of another one when it is in physical contact. It can be stated that two bodies can be considered to be in physical contact when no region of space, however small, can be placed between them. Similarly when two events are separated by a gap of interval, however small, can be considered to be temporally local. Very precise and scientific definitions of 'spatial', 'temporal' and 'spatiotemporal locality' have been given by Marc Lange in 'An Introduction to the Philosophy of Physics'. These are quoted below:

> Spatial locality: For any event E and for any finite distance $\delta > 0$, no matter how small, there is a complete set of causes of E such that for each event C in this set, there is a location at which it occurs that is separated by a distance no greater than δ from a location at which E occurs.
>
> Temporal locality: For any event E and for any finite temporal interval $\tau > 0$, no matter how short, there is a complete set of E's causes such that for each event C (a cause) in this set, there is a moment at which it occurs that is separated by an interval no greater than τ from a moment at which E occurs.
>
> Spatiotemporal locality: For any event E, any finite temporal interval $\tau > 0$, and any finite distance $\delta > 0$, there is a complete set of causes of E such that for each event C in this set, there is a location at which it occurs that is separated by a distance no greater than δ form a location at which E occurs, and there is a moment at which C occurs at the former location that is separated by an interval no greater than τ from a moment at which E occurs at the latter location.

Even without the precise and robust definitions quoted above, it is not difficult to conceive the violation of spatiotemporal locality in the physical world. 'Action-at-a-distance' is something that violates this condition; precisely because of this, many philosophers and scientists in the past could not accept the concept of gravitational interaction between two bodies at a distance. It is also the reason why Mach's principle has remained unacceptable to many. Gravitational action violates 'spatial locality' condition as the interacting bodies are separated by finite space. On the other hand, Mach's principle violates 'spatiotemporal locality' condition as illustrated through Fig. 2.29.

Two bodies A and B, as shown in Fig. 2.29 are located at a distance r. At $t = t_1$, body A is given an acceleration a with respect to body B. According to Mach's principle, A experiences a force F resisting the acceleration because of body B at the same instant. The information about A's acceleration takes a finite time to reach B, and its effect should take a finite time to reach back A in the form of the resisting force F_i. This is a kind of a 'reverse' violation of spatiotemporal locality. The bodies at separate locations cannot develop 'instantaneous' relationship for any 'cause–result' phenomenon. It is an insurmountable hurdle in accepting and understanding such 'action-at-a-distance' phenomenon.

2.8.3 The Concept of Field

Apart from the phenomenon of gravitation, very serious problems involving violation of spatiotemporal locality condition are exhibited by the phenomena of electricity and magnetism. Therefore, from the mediaeval period, philosophers and scientists continuously attempted to provide acceptable physical solutions to the observed 'action-at-a-distance' phenomena.

The first clear and concrete evidence of a body influencing another at a distance was the phenomenon of tide. In the ninth century AD, an Arabian philosopher, Abu Mashar published a book in 850 AD in which he clearly emphasized the relationship between the moon's position in the sky and the state of the tide. Of course, his primary aim was to establish 'astrology' on scientific basis. But it was Kepler who first realized the need for an action without physical contact to explain the motions of planets. Gilbert's

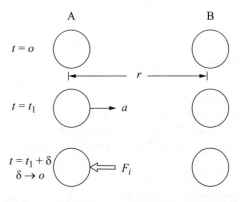

Fig. 2.29 Violation of spatiotemporal locality by Mach's principle

book also brought the existence of magnetic force as another example of action-at-a-distance. Though Galileo and Huygens were against any such 'action-at-a-distance' phenomena and associated gravitational attraction, Newton's development of the celestial mechanics stopped all further hostilities. However, both Newton and Huygens had to explain the phenomena involving transfer of influence through open space. In case of Newton, it was the phenomenon of gravitation and Huygens had to explain his theory of light in which the light propagated as the wave-like oscillations in a medium called 'aether'.[5] Thus, the concept of a continuous medium filling all space became essential to physically explain the observed phenomena involving 'action-at-a-distance'. The presence of a medium helped the scientists to accept such phenomena, and the violation of 'spatiotemporal locality' condition was eliminated. Roger Boscovitch (1711–1787), who worked upon the mechanics of both Newton and Leibniz, was perhaps the first to propose a theory in which the intermediate space between two interacting bodies (without any direct physical contact) is filled up by points associated with local properties that could interact with other points and develop either attraction or repulsion. This kind of a hypothetical model may be considered as the forerunner of the later mathematically developed theory of field. Though science historians like Mary Hesse attribute 'field' concept to the hydrodynamics developed by Leonhard Euler in the true sense, the 'velocity field' belonging to a flowing fluid is very different from the concept of gravitational or electromagnetic fields. The velocity field represents the property of the material particles occupying the space. The concept of an aether filling all space was also used by the philosopher Kant (Immanuel Kant: 1723–1804).

The foundation stone for the modern field concept was first laid down by Michael Faraday (1791–1867). He could understand the phenomenon of electrical induction only by assuming the propagation of the effect was by forces filling the intervening space between the interacting bodies. Initially, he was hesitant to assert his assumption as a real physical entity; however, gradually he became more confident about the existence of the field and the associated lines of forces. Finally, it was James Clerk Maxwell who published his seminal paper in 1864 and put the field concept on a rigorous mathematical foundation. In the earlier times, the field at a point used to be considered to be a condition of the all pervading 'aether' at that location. The subject received more impetus from the urge to understand electrostatic and electromagnetic forces. The force between two stationary electric charges $+q_1$ and $+q_2$ is given by Coulomb's law as follows:

$$F = \frac{\varepsilon q_1 q_2}{r^2}$$

when r is the distance between the charges and ε is a universal constant. Augustin de Coulomb (1736–1806) framed this law in 1785. Though the law of electrostatic force is very similar to that of universal gravitation proposed by Newton almost a century before Coulomb, the gravitational force is extremely weak in comparison with electrostatic force. Thus, the gravitational attractive force between two electrons is 10^{-42} times that of the electrostatic force between them. If $+q_1$ be a stationary charge and $+q_2$ be the test charge, then the electrostatic interaction is represented by Fig. 2.30.

The electrostatic field is represented by the radially outflowing 'lines of forces' from $+q_2$, and the intensity of the field at a location drops as inversely proportional to the square of the distance (qualitatively indicated by the size of the arrow heads). Force on the test particle is given by the product of the test charge and the intensity of the field at the location of the test charge. Thus, the intensity of the electrostatic field at location r is given by

$$E_e = \frac{\varepsilon q_1}{r^3} r \qquad (2.8.1)$$

In a similar way, the gravitational field of a stationary mass m_1 can be expressed as follows:

$$E_g = -\frac{G m_1}{r^3} r \qquad (2.8.2)$$

The diagrammatic representation of (2.8.2) is shown in Fig. 2.31.

Thus, the 'field lines' or the 'lines of forces' characterize the field and may be considered to be the diagrammatic representation of a field. If the test charge be considered to be always positive, then the field due to two equal charges with opposite signs becomes as shown in Fig. 2.32.

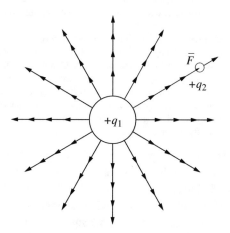

Fig. 2.30 Electrostatic field of a stationary charge

[5]Also spelled as 'ether'.

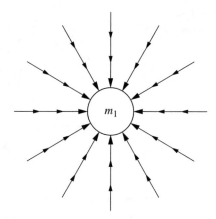

Fig. 2.31 Gravitational field of a stationary mass

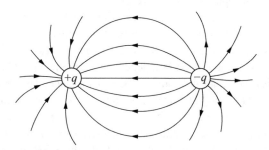

Fig. 2.32 Electrostatic field due to two equal and opposite charges

It should be noted that unlike electric charge, gravitational mass does not have any polarity and the interaction between two mass particles is always attractive. The gravitational field due to two equal masses can be represented by the 'field lines' or 'lines of forces' as shown in Fig. 2.33.

The idea of 'lines of forces' first came to Faraday's mind from his observation of the patterns generated by iron filings around a magnet (familiar with most students of science in high schools). Just like the electrostatic case, magnets also produce magnetic fields. As magnets also possess two poles —north and south—the lines of forces around a bar magnet (or a solenoid with current passing through the coil) are similar to those developed due to two equal and opposite charges as shown in Fig. 2.32. This scheme satisfactorily explains 'action-at-a-distance' phenomena. However, it is still unclear if the 'lines of forces' are real entities or not.

Experiments with gravitational field in a laboratory are difficult. But extensive research has been conducted on the electric lines of forces. If one takes a bar magnet (or a solenoid with current passing through the coil), its surrounding space is expected to be filled by the lines of forces of the associated magnetic field. If a conductor is moved through these hypothesized lines of forces keeping the magnet stationary, a current is generated due to electromagnetic induction. The cutting of the lines of forces by the conductor is considered to be the cause of the current developed. Now, if the magnet is moved keeping the conductor stationary, a current is introduced in the conductor. It looks obvious if one assumes that the lines of forces are rigidly attached to the source, i.e. the magnet. However, the results from experiments with 'Faraday's disc' based on a report by Faraday in 1832 lead to a serious problem as explained below.

Figure 2.34 shows the scheme diagrammatically. A metallic disc is attached to the end of a bar magnet with the help of a pin so that the disc can be rotated keeping the magnet still. The lines of forces coming out of the bar magnet are also indicated to pass through the metallic disc. A galvanometer can detect any current developed in the wire whose one end is connected to the central pin and the other end touches the disc's periphery. If the magnet (and the conductor with the galvanometer) is kept still and the disc is rotated, a current develops in the conducting wire. This is easy to understand as the rotating conducting disc cuts the lines of forces which are stationary along with their source, the bar magnet. But when the disc and the magnet are rotated together (so that there is no relative motion between the disc and the magnet), a current is induced. Again if the magnet is rotated keeping the rest of the system stationary, no current is induced. If the lines of the forces are assumed to be rigidly connected to their source, both the above results defy explanation. This poses a serious question on the reality of the lines of forces. However if the conducting wire of the galvanometer circuit is considered as a part of the current generating system, the phenomenon is understood. However, there are other situations where an explanation is difficult

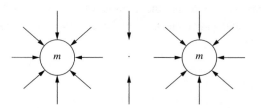

Fig. 2.33 Gravitational field of two mass particles

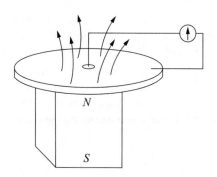

Fig. 2.34 Experiment with Faraday's disc

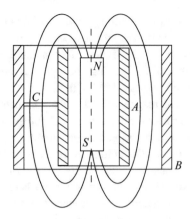

Fig. 2.35 Charged cylinder experiment

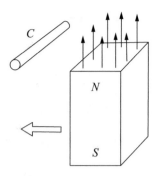

Fig. 2.36 Conductor and a magnet with translation

considering the lines of forces as physically real entities. Another experiment that decidedly proves that the lines of forces are not entities rigidly attached to the source magnet is briefly discussed below.

A cylindrical magnet is surrounded by two coaxial metallic cylinders A and B as shown in Fig. 2.35.

A and B are connected by a conducting rod C. Whenever the conductor C cuts the lines of forces, electrons flow making one of the cylinders to have an extra charge. The conductor can be cut off during the process, and the extra charge in one of the cylinders will be maintained and can be detected. When the cylinders are rotated and the magnet is kept stationary, a charge difference between A and B is developed as expected. When the cylinders are stationary and the magnet is rotated about its axis, no charge difference develops, implying that the lines of forces are not rigidly connected to the magnet. But when the cylinders and the magnet are rotated together, a charge difference is developed, again implying that the lines of forces are not rigidly connected with the magnet.

2.8.4 Field and Absolute Space

The experimental observations discussed in the previous section again bring back the question of the existence of 'absolute space'. The concept of fields, represented by the lines of forces, is one of the two possible ways to understand the phenomenon of 'action-at-a-distance' without violating spatiotemporal locality condition. The idea is that the effect, represented by a field, propagates through the universal medium 'aether'. Though there has been no direct evidence of the existence of aether, propagation of waves in the form of a disturbance in the aether can be conceived as a means of

transferring effects through a vacuum. Even the light is considered to propagate as waves in the aether medium. The other possible way to understand transfer of effects is through the flow of particles. Corpuscular theory of light gained acceptance because of this reason.

Now, whatever may be the medium filling all space, it is expected that it can also represent an entity like absolute space. Thus when any disturbance in this absolute space propagates, its motion is independent of the motion of other bodies including the source of the disturbance. That is why the speed of propagation of electromagnetic waves through a vacuum is a constant. The fields, described in the previous section, are also nothing but disturbances in the empty medium (may be called 'aether') . Thus, the existence of any field is characterized by the disturbance in the absolute space and, once created, has nothing to do with its source. Hence, the lines of forces described in the experiments in Sect. 1.8.3 have no connection with the source magnet and, therefore, do not share any motion. But an experiment with a conductor and a magnet with uniform translation, shown in Fig. 2.36, yields an unexpected result. A current develops! One possible explanation for this difference in results could be that rotational motion involves acceleration, whereas translation with constant speed does not.

However, there could be some link between the non-attachments of the field lines with the source and the existence of absolute space. Further discussion may tend to become more philosophical, and so the matter is closed at this point.

Doing experiments in the similar lines with gravitation is not possible. The similarity between gravitation and electric fields is derived from the similarity between Newton's law of gravitation and Coulomb's law of electrostatic force. The best testing laboratory for Newtonian gravitation is the solar system, and studying the motion of the celestial bodies deeper understanding of Newtonian gravitation can be achieved.

Post 'Principia' Developments

3.1 Early Concepts and Aristotelian Physics

'Principia' represents the culmination of conceptual evolution of science of motion based upon the developments during the previous two millennia. Of course, it has been shown in Chap. 1 that the final grand synthesis by Newton depended very much on the works of others before him and Newton did not develop his version of 'science of motion' *exnihilo*. Though it is very natural to expect that Newton's mechanics could have easy acceptance in England because of obvious reasons, further progress in mechanics did not take place in England! Why this decline in science took place in England is still not well understood. Further, developments in mechanics took place in continental Europe during the eighteenth century. These developments took place through an interplay of Newton's mechanics with the ideas originated by Huygens and Leibniz. This became possible by the emergence of great mathematicians like Euler and Lagrange.

3.1.1 Diffusion of Newton's Mechanics in Europe

Before taking up the issue of interplay between Newton's concepts with those of the continental scientists, it may be not out of place to have a quick look into the way Newton's mechanics did diffuse into the minds of continental scientists. This is, of course, of purely historical interest. There were mainly two factors that created some initial hurdle for ready acceptance of Newtonianism in the continental Europe. The first one was of technical nature; since Newton's mechanics dealt with force and acceleration, the treatment was purely geometrical. In fact that made Principia very difficult to follow and grasp. On the other hand, in the continent, the mechanics was under the influence of Leibniz in which the principal quantity was energy (living force)—a scalar quantity. The other difficulty arose because of a bitter

dispute between Newton and Leibniz regarding calculus. Leibniz's claim was that he discovered calculus independently but Newton with the Royal Society backing him charged Leibniz with plagiarism. Thus, one of the most important scientists in the continent in those periods, Johan Bernoulli, resisted the propagation of Newtonianism in the continent. He was an ardent follower of Leibniz with the deepest sense of respect for him. The first important work for introducing Newton's ideas in continental Europe was by Jacob Herrmann who published a book containing the ideas behind Newton's mechanics as presented in Principia. This volume, Phoronomia, published in 1716, was the first book on mechanics and for the first time presented the laws of motion in a different form. It was Daniel Bernoulli, son of Johann Bernoulli, who was the first to reformulate Newton's ideas in 1738 using the mathematics of Leibniz and his father Johann. The other channel through which the propagation of Newton's mechanics in Europe took place was through a very popular textbook by Francesco Algarotti published in 1737.

Diffusion of Newtonianism in continental Europe was also helped through the popularization of experiments in mechanics. The experimental set-ups were mostly manufactured in England and Holland. These experiments were conducted to test the different aspects of the laws of motion using impact and fall of bodies. One popular set-up was the Atwood's machine.

Figure 3.1 shows the basic scheme. If the moment of inertia of the wheel of radius r be I, then, the downward acceleration of the heavier mass m_2 can be expressed as

$$a = \frac{m_2 - m_1}{m_2 + m_1 + \frac{I}{r^2}} \cdot g$$

By suitably adjusting the values of m_1 and m_2, it was possible to make the coefficient of g as small as 1/64. Thus, easy experiments were possible even at school level.

In France, the strong influence of Cartenianism, and in Germany, the stiff resistance from Leibniz's students created

© Springer Nature Singapore Pte Ltd. 2018
A. Ghosh, *Conceptual Evolution of Newtonian and Relativistic Mechanics*,
Undergraduate Lecture Notes in Physics, https://doi.org/10.1007/978-981-10-6253-7_3

Fig. 3.1 Atwood's machine

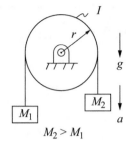

$M_2 > M_1$

the initial hurdle for Newtonianism to spread. In Italy, there was some degree of acceptance, and Holland was the first country to accept Newtonianism.

3.1.2 Multiplicity in the Concept of Force

A serious difficulty in the acceptance of Newton's mechanics in the continental Europe was a major difference in the cause of motion (or change of motion). During that period a considerable amount of confusion existed because of the multiplicity in the concept of force. There was serious doubt in the minds of the early eighteenth century scientists regarding the true character of the quantity associated with motion. Thus, often the concept of force, i.e. an entity that can act to either cause motion or, stop motion, used to refer to a number of quantities. Using the modern terms and symbols, these were

(i)   force (F)
(ii)   work ($W = F \cdot s$)
(iii) momentum ($p = m|v|$)
(iv)   energy ($E = mv^2$)
(v)   impulse ($I = F \cdot t$)
(vi)   power ($P = F \cdot v$)

Newton used the term 'force' almost in the modern sense, and it was a vector quantity. Thus, the whole treatment of mechanics by him was geometrical in nature. In continental Europe, the major fight was between the Cartesians (followers of Descartes) and the followers of Leibniz.

In France, the Cartesians used the term $m|v|$, i.e. the product of the mass of a body with speed (not velocity) to represent force. Descartes himself also used to represent force by the weight and the height raised implying mechanical work. However, a vague sense of conservation emerged and $m|v|$ became the appropriate quantity since in a very rough sense this quantity was considered to be conserved as indicated by the early experiments on collision of bodies.

Leibniz considered the matter and classified force into two types - 'living force' and 'dead force'. According to

him, 'The living force is that resides in a body when it is in uniform motion; the dead force that which a body at rest receives when it is urged either to move or to move more or less quickly, if these bodies are already in motion'. Huygens was a friend and disciple of Descartes and conducted successfully the collision experiments. He noted that the entity that is conserved in the collision process was mv^2, i.e. energy when the bodies were perfectly elastic. However, he did not propose the conservation of this quantity (termed as 'living force') as a principle initially. He stated later 'In any motion of bodies no force is lost or destroyed if some effect is generated and persists, for whose production the same force is required than that lost. I call force the power to raise weight. Thus a double force is that can raise the same weight to a double height'.

Johan Bernoulli spent considerable time in studying living and dead forces as defined by Leibniz. He considered massless springs and opined the capability of a spring in a compressed state to act as the dead force. But the same 'dead' force generates 'living' force when the spring expands.

Descartes' followers considered the quantity, mass X speed, as the force following their master. It should be noted that in one-dimensional collision problems this is same as the quantity 'momentum' and is conserved in collision process without requiring the bodies to be perfectly elastic. Therefore, according to them, $m|v|$ is a right quantity to represent 'force'. Newton's followers found this definition to be more acceptable as it matched with Newtonian mechanics where change in momentum is associated with the action of a force. On the other hand, the Newtonian mechanics had no place for a scalar quantity mv^2. The dispute continued for a few decades till it was realized by the scientists like D'Alembert and Euler that both quantities are important parameters in the description of motion.

An accurate mathematical representation of the science of motion was necessary to stop the controversy. This was possible primarily because of the work done by Euler and Daniel Bernoulli who synthesized Newton and Huygens' physics with the mathematics of Leibniz and Johan Bernoulli. Frankly speaking 'mechanics' became a discipline for study of motion primarily because of Euler's book 'Mechanica sive motus scientia analytice exposita'.

3.1.3 Degeometrization of Newtonian Mechanics

The gradual emergence of two primary parameters for representing motion—the momentum and kinetic energy led the further development in two distinct directions. The Newtonian mechanics based upon the vector quantities like force, momentum, acceleration and displacement became a branch suitable for handling vector quantities. On the other hand, the approach using a scalar quantity like energy developed following analytical approach. However, even Newton's geometric approach became couched in mathematical expressions and mathematical treatment of mechanics problems became not only popular but it gradually pushed the geometrical treatment out of scene.

Representation of physical quantities like velocity, acceleration and force is very natural using geometry as geometric lines represent both direction and magnitude. The application of law of parallelogram for adding vector quantities, developed by Galileo in studying composite motions, was also directly related to geometric representation. It has been shown that earlier application of infinitesimal geometry was abundant in studying mechanics problems. In this technique, certain properties or relationships should be revealed geometrically.

A very simple example is shown in Fig. 3.2. DAC is a circle with O as its centre. AB is a tangent to the circle. OB cuts the circle at point C. Now,

$$BC = OB - OC = OB - OA$$

Again

$$OB = \left(OA^2 + AB^2\right)^{\frac{1}{2}}$$
$$= OA\left[1 + \left(\frac{AB}{OA}\right)^2\right]^{\frac{1}{2}}$$

When AB tends to be very small (as in infinitesimal calculus) then

$$OB \approx OA + \frac{1}{2}\left(\frac{AB}{OA}\right)^2$$

since $AB/OA \ll 1$
Hence,

$$BC = OA + \frac{1}{2}\left(\frac{AB}{OA}\right)^2 - OA = \frac{1}{2}\left(\frac{AB}{OA}\right)^2$$

Finally,

$$BC \propto AB^2$$

This relation has ample use in mechanics as the displacement (in a straight line motion) is proportional to (time)^2 with a constant acceleration.

A major disadvantage of a geometric approach is the difficulty (and impossibility in many cases) in developing generalized solution of mechanics problems. Developing general algorithms is also not possible with a purely geometric solution and each problem needs to be solved individually. This serious shortcoming was noticed by the scientists of the continental Europe and desired to develop mathematical description of motion quantities. Again taking the example of a one-dimensional motion, the time–displacement diagram is shown in Fig. 3.3.

In a geometrical approach, the instantaneous speed is given by the average speed during a very short time interval at that instant. Thus, taking a small step Δt, the displacement is Δs. So the average speed at this infinitesimally small period is

$$v = \frac{\Delta s}{\Delta t} \text{ with } \Delta t \text{ very small}$$

This was formalized by the continental scientists as

$$v = \frac{ds}{dt} = \lim_{\Delta t \to 0} \frac{\Delta s}{\Delta t}$$

The above relation eliminates the need of using a geometric figure any more. If Δs can be expressed in term of Δt, v can be estimated. To handle vector quantities, the technique of splitting it in terms of components was developed.

As shown in Fig. 3.4, a vector A can be represented by an analytical (algebraic) expression as shown below

$$A = \hat{i}A_x + \hat{j}A_y + \hat{k}A_z$$

where A_x, A_y and A_z are the x, y and z components of A and $\hat{i}$, $\hat{j}$ and $\hat{k}$ are the unit vectors along the fixed directions O_x, O_y

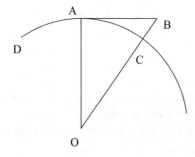

Fig. 3.2 Use of infinitesimal geometry

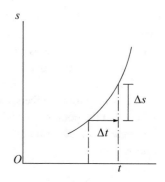

Fig. 3.3 Representing velocity by calculus

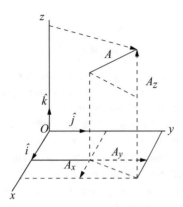

Fig. 3.4 Algebraic representation of a vector

Fig. 3.5 Motion of a point mass

and O_z. In this way, the vectors can be handled by employing algebra without referring to geometry. The use of a space-fixed Cartesian coordinate system x–y–z, for the purpose described in Fig. 3.4, was first employed by Euler in 1752. Thus, it was only after more than 60 years after the publication of Principia the second law of motion was cast in a format used subsequently till the present times. The fertility of the mechanics of Newton started being increasingly demonstrated once the powerful mathematical tools were employed. This finally led to the rejection of the geometrical approach to solve mechanics problem adopted by Newton, and the subject 'mechanics' became amenable to mathematical solution using infinitesimal calculus.

During this period, a number of branches of mathematics got enriched. One example is the concept of a function. It was again Euler in 1755 who gave a precise definition of a function as follows: 'If, x denotes a variable quantity, then all quantities which depend upon x in any way, or are determined by it, are called functions of x'. Apart from this, the continuous search for new principles in order to solve problems of mechanics led to the development of another important branch—variational calculus. This development took place in the latter half of the eighteenth century primarily in the hands of Euler and Lagrange.

For the representation of motion problems using mathematical tools, it was necessary to develop formalism by which an interrelation among various parameters is obtained. Euler led the way and developed the methodology that is described as 'the equation of motion'. Considering a point mass (i.e. a particle) subjected to a force p, the dynamical law of motion was recast by Euler in the following form

'A point can be moved along the direction AM (Fig. 3.5) and it is acted on, while it traverses the small distance Mm, by a force p pulling in the same direction; the increase in the speed, that the point meanwhile acquires, is as the product of the force by the short time, in which the element of distance Mm is traversed'.

Taking A as the mass of the particle, c as the speed, p the force and t as time, Euler gave the equation of motion as follows:

$$dc = xp\frac{dt}{A}$$

where x was a constant of proportionality.

3.2 Emergence of Analytical Mechanics

Since one of the primary motivation for developing the science of motion was to explain the motion of the planets, Newtonian mechanics was concerned with the mechanics of point mass. The aim was to determine the motion of an unconstrained point mass due to externally applied forces. Thus, the whole issue was of 'instantaneous' nature and the object was a point mass. However, the matter of characterizing a system based on its totality of all possible motions remained unaddressed in Newtonian mechanics. In Chap. 2, this was briefly mentioned.

In continental Europe, there were two schools—one formed by the followers of Descartes and Huygens and the other followed the thought process initiated by Leibniz. In both the schools, one consideration played a major role though in Newtonian mechanics the matter was absent. It was the concept of conservation as mentioned in the previous section. Cartesian scientists considered the quantity $m|v|$ to be conserved during dynamic interactions, whereas the Leibnizians took mv^2 as the quantity that remains conserved. Thus, Cartesian philosophy was closer to the Newtonian ideas, but Leibnizians' thinking was very different from that of Newtonian concept. Therefore, the branch of mechanics that started developing in Europe parallely to Newtonian mechanics took the scalar quantity mv^2 as the primary representative for a system in motion. This scheme also allows a system to be represented by an entity that is identified with a dynamical system irrespective of its any particular state of motion. In Chap. 2, this point was discussed and it was mentioned that such an entity was termed as the 'energy function'. To elaborate the matter the two particle system, shown in Fig. 3.6, may be considered.

Fig. 3.6 An isolated two particle system

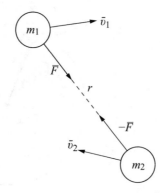

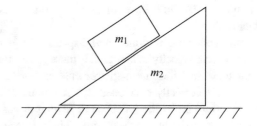

Fig. 3.7 Block on an moving wedge

The system is totally isolated from the rest of the universe. In the Newtonian approach, the accelerations of the two particles are determined taking the gravitational attraction into consideration. Knowing the initial motion of the two particles, the subsequent motions can be found out. For each prescribed initial conditions, separate solutions can be found out for the motions of the two particles. But an isolated system like the one described above there is something that uniquely identifies with the system irrespective of its individual motions. This was, perhaps, vaguely in the minds of Leibnizians and the scalar quantity mv^2 as a parameter of primary importance representing motion gained importance. Again considering the example discussed above, the quantity that remains unchanged for the system is the total energy for a given situation. Of course, Leibnizians were not anywhere near the understanding of the conservation of energy in the modern sense.

In the early part of the eighteenth century, dynamical problems used to be solved following the Newtonian force–acceleration-based approach. Of course, gradually the geometric representation got increasingly replaced by analytical calculus-based language mainly due to Varignon, Daniel Bernoulli, Hermann and Euler. The solutions were, thus, based upon a replacement of geometric nature of a problem by a mathematical equation of motion-based description. The idea that the condition of motion of a system can be considered independently of any specific motion status gradually emerged when new principles were discovered.

3.2.1 New Principles for Dynamical Problems

It was gradually becoming clear that the laws of motion, as encapsulated in Newton's Principia, were alone not enough for solving all dynamics problems. This forced the scientists of the continental Europe in the first half of the eighteenth century to look for new principles which govern motion of mechanical systems. Newtonian mechanics was directly applicable to the dynamics of unconstrained point mass only; many applied problems were not tractable. For

example, motion of a body moving on an inclined plane that is again free to move freely along another direction.

Figure 3.7 shows such a situation which is a commonly discussed popular problem. Johan Bernoulli analysed the problem by Newtonian approach introducing the concept of 'constraint forces'. However, this situation was very uncomfortable to the scientists; though in statics problems constraint forces were in common use, in dynamical problem introduction of constraint reaction was less obvious. In this scenario, the constant intellectual struggle led to the gradual emergence of two new principles—conservation of living forces (mv^2) and the least action principle.

As mentioned in Sect. 2.1.2, it was Huygens who first mentioned about the conservation of living force in a dynamical interaction. Later Johan Bernoulli studied the matter of conservation of living force and, analysing the case of a massless spring, also concluded that 'living force' can assume various forms. He surmised that in the compressed state a spring stores a living force which can release itself upon another body. This can be considered a very early and nebulous form of change of potential energy into kinetic energy (using modern language). Using this principle of conservation of 'living force', Bernoulli solved a number of dynamics problems which were difficult to solve following a direct Newtonian route.

The next important development for solving dynamics problems without using vectorial formulation was important towards the emergence of analytical mechanics. It started with Johan Bernoulli's 'principal of virtual velocities' perfected by D'Alembert. Subsequently, the emergence of the 'principle of last action' led to the final formulation by Lagrange.

3.2.2 Principle of Virtual Velocity and Virtual Work

Though the basic idea of virtual displacement and virtual work emerged much earlier, its effective application started primarily by Johan Bernoulli who generalized the principle. Of course, he developed this based on the ground work done by Varignon to treat the problem of static equilibrium of a system under the influence of several forces acting simultaneously.

Bernoulli conceptualized the small displacement given to a point as 'virtual velocity' although it is more appropriate to designate these infinitesimal displacements as 'virtual displacement'. He correctly concluded that if a point P be in static equilibrium under the influence of several forces then the total work done by the forces is zero. This is the principle of virtual work in modern technology. In all probability, the concept was arrived at by graphically finding the components of a virtual displacement along each acting force and summing the products of the forces and the corresponding components of a virtual displacement. Bernoulli considered the product to be positive when the direction of the force is same as that of the component of the virtual displacement and vice versa. He found the sum to be always zero when the forces had zero resultant. Such displacements were called 'virtual' as these were considered to be 'conceptual' and not 'real'. However, in modern mechanics 'virtual displacements' are also considered to take place with zero passage of time when applied to the problems of dynamic equilibrium.

Figure 3.8 shows a system of forces acting on a point that is given an infinitesimal displacement δs (the virtual displacement). The displacement is infinitesimal so that the geometric configuration does not undergo any noticeable alteration. So, the forces can be considered to remain unaltered. Using modern vector notation, the principle emerged in a straight forward manner as shown below.

Since the system of forces keep the point in static equilibrium $\sum_{i=1}^{n} F_i = 0$. Now, if δs be the 'virtual displacement' then the 'virtual work' is given by

$$\delta w = F_1 \delta s_1 + F_2 \delta s_2 + \cdots + F_n \delta s_n$$

where $\delta s_1, \delta s_2, \ldots, \delta s_n$ are the components (both +ve and −ve) of the virtual displacement along $F_1, F_2, \ldots, F_n$. The principle states that $\delta w = 0$. It is easy to find that

$$\delta w = \sum_{i=1}^{n} F_i \cdot \delta s = \delta s \cdot \sum_{i=1}^{n} F_i = \delta s \cdot 0 = 0$$

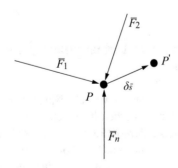

Fig. 3.8 Principle of virtual work

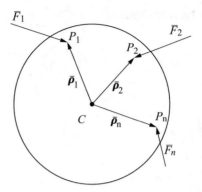

Fig. 3.9 Virtual work principle for an extended body

δs is not necessarily the same as a 'real displacement' since a real displacement causes the configuration to change (though small).

This principle was, of course, restricted to point objects. The development of the principle for application upon extended bodies or systems is a later development. The situation for a rigid body in static equilibrium under the action of n forces is shown in Fig. 3.9.

C is the centre of mass, and the body is in static equilibrium under the action of forces $F_1, F_2, \ldots, F_n$ applied at $P_1, P_2, \ldots, P_n$. $\rho_1, \rho_2, \ldots, \rho_n$ are the position vectors of these points with C as the origin. The condition for equilibrium of the body can be expressed as follows:

For force balance $\sum_{i=1}^{n} F_i = 0$
and for moment balance $\sum_{i=1}^{n} \rho_i \times F_i = M = 0$

Now, let small virtual linear and angular displacements be δs_c (of the centre of mass) and $\delta \theta$ (a small rotation of the body). Then the virtual displacement of P_i is given by

$$\delta s_i = \delta s_c + \delta \theta \times \rho_i$$

Hence, the virtual work done by the forces is as follows:

$$\delta w = \sum_{i=1}^{n} F_i \cdot \delta s_i$$

$$= \sum_{i=1}^{n} F_i \cdot (\delta s_c + \delta \theta \times \rho_i)$$

$$= \delta s_c \cdot \sum_{i=1}^{n} F_i + \sum_{i=1}^{n} F_i \cdot (\delta \theta \times \rho_i)$$

$$= \delta s_i \cdot \sum_{i=1}^{n} F_i + \delta \theta \cdot \sum_{i=1}^{n} F_i \times \rho_i$$

$= 0$ (in view of the conditions for static equilibrium

mentioned above.)

3.2.3 D'Alembert's Principle

The period between Newton and Lagrange was full of activities to develop mechanics as a branch of science. Though Euler followed the philosophy of Newton that proposed the rate of change of momentum to be proportional to the impressed force, D'Alembert was not in favour of such an approach. In his zeal to make some profound contribution to science, he divided sciences into two groups. One of these were based upon principles which D'Alembert considered to be necessarily true (and, perhaps, self evident). In the other group, according to him, the sciences were based upon experimental observations and hypotheses. D'Alembert considered 'mechanics' to belong to the first group and should be based upon absolute truth and independent of assumptions. To him, the principles in mechanics belong to the clan of 'necessary truths'.

D'Alembert summarily rejected Newton and Eulers' approach which, according to him, depended on the single and vague obscure axiom that the effect is proportional to its cause. His approach of solving dynamics problems was based upon composition of motion. He reasoned that when an object's motion is changed, the final motion is made up of the original motion and an acquired motion. Thus, the original motion could be considered to be composed of the final motion and the motion lost by the object.

Figure 3.10 indicates his idea with the help of vector representation of motion. Since the state of equilibrium was more understandable to D'Alembert, he wanted to frame all dynamical problems based upon the principle of equilibrium. The principle that he wanted to use for solving all dynamical problems can be described as follows. The problem statement is given first as mentioned below:

'Given a system of bodies arranged mutually in any manner whatever, let us suppose that a particular motion is impressed on each of the bodies that it cannot follow because of the action of the others, to find that motion that each body should take'.

The general procedure for solving the above general problem D'Alembert suggested the following approach (Ref. Fig. 3.11):

'Let A, B, C, etc. be the bodies that constitute the system and suppose that the motions a, b, c, etc. are impressed on them; let there be forces, arising from their mutual action, which change these into $\bar{a}, \bar{b}, \bar{c}$, etc. It is clear that the motion impressed on the body A can be compounded of the motion $\bar{a}$ which it acquires and another motion α. In the same way the motions b, c, etc. can be regarded as compounded of the motions $\bar{b}$ and β, $\bar{c}$ and χ, etc. From this it follows that the motions of the bodies A, B, C, etc. would be the same, among themselves, if instead of their having been given the impulse a, b, c, etc. They had been simultaneously given the twin impulsions $\bar{a}$ and α, $\bar{b}$ and β, $\bar{c}$ and χ etc. Now, by supposition, the bodies A, B, C, etc. have assumed, by their own action the motions $\bar{a}, \bar{b}, \bar{c}$, etc. Therefore the motions α, β, χ, etc. must be such that they do not disturb the motions $\bar{a}, \bar{b}, \bar{c}$, etc. in any way. That is to say, that if the bodies had only received the motions α, β, χ, etc. these motions would have been cancelled out among themselves, and the system could have remained in rest'.

The procedure for solving the problem is to decompose the impressed motions each into two motions such that the one set motions balance among themselves (i.e. when applied to the system it remains in rest). The remaining components of the motion represent the actual motion of the system under the action of the impressed motions. At this point, it is appropriate to mention that D'Alembert considered 'impulse' as the cause to generate motion. By 'motion', he implied the velocity. It is clear that though the principle suggested by D'Alembert is clear, its application is not easy. His approach for solving problems of motion has been described by many (including Lagrange) as a process of converting the dynamics problems into statics. Though D'Alembert himself did not subscribe to such an interpretation of his principle, but that has continued till the present day. Not only that this idea of converting a problem of dynamic equilibrium to a problem of static equilibrium by using D'Alembert's problem led to the opportunity of applying the principle of virtual work. This became the founding step towards the development of analytical mechanics. When these concepts were being developed a parallel development started that had the most extensive influence on classical mechanics.

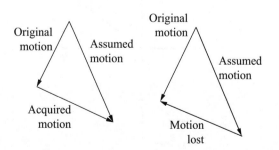

Fig. 3.10 Composition of motion in D'Alembert's scheme

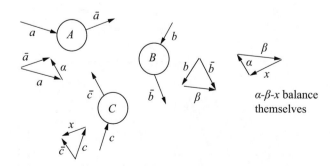

Fig. 3.11 General statement of D'Alembert's problem

3.2.4 Principle of Least Action

The laws of the refraction of light induced many to provide an explanation to the observed characteristics (Fig. 3.12). When a light beam passes from one medium to another, the direction of the beam changes. Descartes tried to explain the phenomenon by considering the speed of light to increase when it passes from a lighter medium to a denser medium (which was, of course wrong).

However, Descartes was opposed and Fermat brought the concept that Nature always follows the path of least resistance. Fermat extended it to the principle that a light beam should take minimum time for travelling from a point A in medium I to a point B in medium II.

He was successful in proving that the time taken by the light ray going along the straight paths ACB is more than another path AOB (Fig. 3.13). It is simple to demonstrate that the time taken along AOB is less than that along ACB when the speed of light $v_I > v_{II}$.

For example, considering $AC = BC$, $\theta = 45°$ and $OC = 0.1\,AC$, it is simple to show that the time along the path ACB is $1.67AC/v_{II}$ when v_I is assumed to be equal to $1.5v_{II}$.

However, it is necessary to prove that for the actual path AOB, the time taken is minimum. Fermat was able to do that geometrically in the year 1662. He got the relation between the incident angle θ_i and the refracted angle θ_r as a function of the speed of light in region I and II. However, the followers of Descartes did not refrain from criticizing Fermat. The history of science has recorded the bitter exchange between the two groups that followed for quite some time.

Subsequently, the Fermat's principle received serious objection from Leibniz also. According to him, why travel time has to be minimized. Philosophically minimizing the travel time cannot have any preference over minimizing the distance (when the path is straight). Philosophically what can be considered for minimization by the Nature is the total 'effort', given by the length of the path and the 'resistance' experienced. Though at this stage the whole idea was a bit

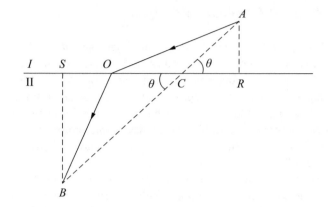

Fig. 3.13 Path of refracted light beam

nebulous, this laid the basic foundation for the concept of 'action' that got developed into a matured concept in the hands of Maupertuis more than half a century later.

Maupertuis proposed that while travelling from one medium into another one, the light chooses a path along which the 'quantity of action', is minimum. This could resolve the 'distance-time' debate raging the scientific world at that time. However, it is not very easy to physically conceptualize a concept like 'action', particularly in the absence of a clear concept of mechanical work. It should be noted that though the whole problem was related to the propagation of light, it soon got extended to the field of 'dynamics'.

Maupertuis considered the action to be represented by a product of the distance travelled and the velocity of propagation. Figure 3.14 shows a light ray travelling from point A to point B following the path ACB.

The angles of incidence and refraction are θ_i and θ_r, respectively. The speed of light propagation are v_1 and v_2 in the two regions as shown in figure. Maupertuis defined action as follows:

$$\mathcal{A} \propto AC \cdot v_1 + CB \cdot v_2$$

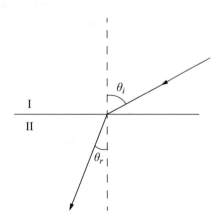

Fig. 3.12 Refraction of a light beam

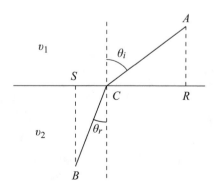

Fig. 3.14 Law of refraction from principle of least action

or,

$$\mathcal{A} \propto v_1 \sqrt{AR^2 + RC^2} + v_2 \sqrt{BS^2 + SC^2}$$

which is minimum when C is the correct position. Using the consideration that $\mathcal{A}$ is minimum then a small change in the position of C (all other points being fixed) will yield zero change in the value of $\mathcal{A}$. Or,

$$d(\mathcal{A}) = 0$$

Now,

$$d(\mathcal{A}) = v_1 \cdot \frac{2RC d(RC)}{\sqrt{AR^2 + RC^2}} + v_2 \frac{2SC d(SC)}{\sqrt{BS^2 + SC^2}} = 0$$

since AR and BS are constants. As $(SC + CR)$, i.e. SR is also constant

$$d(RC) = -d(CS)$$

Hence, $d(\mathcal{A}) = 0$ yields the following relation

$$v_1 \frac{RC}{\sqrt{AR^2 + RC^2}} = v_2 \frac{SC}{\sqrt{BS^2 + SC^2}}$$

or, $\frac{v_1}{v_2} = \frac{\sin \theta_r}{\sin \theta_i}$; the law of refraction.

This 'principle of least action' defined by Maupertuis was used by him in a problem of dynamics—impact of two bodies. This step can be considered to be a giant leap as the action principle became a major foundation stone for the subsequent developments in dynamics.

It was Euler in whose hands the principle of least action attained maturity for application to dynamical problems. Like most other philosophers, he also believed that all 'effects of Nature obey some law of maximum or minimum', i.e. extremum. He considered the product 'mass × displacement × velocity' as the appropriate parameter for defining action as follows:

$$\mathcal{A} = \int mv \, ds$$

From this point, the further development of analytical mechanics using this 'principle of least action' was done by Lagrange. In essence, he combined D'Alembert's principle to state a dynamical problem as a statics problem and then applied Bernoulli's principle of virtual velocity or virtual work (which is also a variational form). Along with this, he developed a very ingenuous method called 'the method of undermined multiplier' to get the complete solution for a generalized dynamical problem. The concept is presented in the next section.

3.2.5 Lagrangian Mechanics

The primary aim of analytical mechanics is to solve dynamical problems using only scalar quantities like energy. This greatly simplifies the procedure as handling vectors in a complex problem can be difficult. The three main points can be summarized here which show the distinction between the Newtonian approach and analytical approach.

1. Unlike Newtonian approach, in analytical mechanics, only scalar quantities are involved in the formulation of the problem.
2. In Newtonian approach, the equation of motion for each particle (of the system) are parallely handled with the equations of constraints. In analytical mechanics, the configuration of a system is described using 'generalized coordinates' which are consistent with the constraints. Thus, the number of generalized coordinates is equal to the number of degrees of freedom.

Figure 3.15 shows a pendulum. The configuration of the pendulum can be described by the position coordinates of the mass x and y and the constraint equation $x^2 + y^2 = l^2$. On the other hand, the angle θ (as a generalized coordinate) can completely specify the system configuration. Thus, there is no need for formulating constraint equations when generalized coordinates are used.

3. In Newtonian approach, each individual body's motion is decided by the external forces applied on the body and the constraint forces acting on the body during motion. But these constraint forces are not known 'a priori'. In analytical mechanics, the formulation is done in such a way that the constraint forces do not appear in the

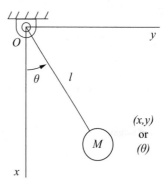

Fig. 3.15 Generalized coordinate

formulation at all. This is achieved by using the principle of virtual work. The constraints do not do any work when the system is given a virtual displacement satisfying the constraints. Figure 3.16 shows two cases, in one, the constraint is time independent, whereas in the other case the constraint surface is time dependent.

If a particle P is given a small displacement with the time independent constraint, the virtual displacement δs is same as the actual displacement ds. However, when the constraint is dependent on time, the actual displacement ds is not the same as a virtual displacement δs that takes place without the passage of any time (i.e. the constraint remains the same). It is seen from the figures that the product $\delta s \cdot R = 0$ since R (the constraint force) acts in a direction normal to δs in all cases. So, the constraint forces do not appear formulation at all.

The above points make the analytical formulation superior to the Newtonian formulation in case of complex systems. If there are particles with r_i ($i = 1, 2,..., N$) as the position vectors subjected to p constraint equations then there will be $N - p = n$ generalized coordinates q_i ($i = 1, 2, ..., n$). The total force acting on the ith particle is

$$F_i = F_{i_a} + R_i$$

where F_i is the force applied and R_i is the constraint force. Using D'Alembert's principle, the system will be in static equilibrium if each particle is imposed with force $-m_i \ddot{r}_i$ where m_i is the mass of the ith particle and $\ddot{r}_i$ is its acceleration. Then applying the principle of virtual work for a system in static equilibrium, the total virtual work must be zero. Thus,

$$\sum_{i=1}^{n} (F_i - m_i \ddot{r}_i) \cdot \delta r_i = \delta w = 0$$

$$\text{or,} \quad \sum_{i=1}^{n} (F_{i_a} + R_i - m_i \ddot{r}_i) \cdot \delta r_i = 0 \qquad (3.2.1)$$

$$\text{or,} \quad \sum_{i=1}^{n} (F_{i_a} - m_i \ddot{r}_i) \cdot \delta r_v = 0$$

since $R_i \cdot \delta r_i = 0$ for $i = 1, 2,..., N$. Hence, the constraint forces are eliminated from the formulation. Next, the above

relation is expressed in terms of generalized coordinates using the relations

$$q_i = q_i(r_1, r_2, \ldots, r_N); \quad i = 1, 2, \ldots, n$$

and

$$\delta q_i = \delta q_i(\delta r_1, \delta r_2, \ldots, \delta r_N); \quad i = 1, 2, \ldots, n$$

But it should be noted that all the δq_i's are mutually independent. Now, Eq. (3.2.1) is expressed in the following form using generalized coordinates:

$$\sum_{i=1}^{n} ()_i \delta q_i = 0 \qquad (3.2.2)$$

where $()_i$ is an expression in terms of the generalized coordinates and their time derivates for the ith generalized coordinate. As all δq_i's are mutually independent, (3.2.2) can be satisfied only if

$$()_i = 0; \quad i = 1, 2, \ldots, n \qquad (3.2.3)$$

Solving these n equations, the solutions for the n-generalized coordinates are obtained. This constitutes the basic scheme of Lagrange's dynamics. The equations of motion are called Lagrange's equation.

3.3 Dynamics of Rigid Bodies

Since one of the major impetus for the development of mechanics was to understand the motion of the heavenly bodies, the mechanics developed by Newton and others was relevant to the motion of particles. Many scholars refer to this as 'point mechanics' implying point like matter. The point mechanics also yielded correct results for the motion of bodies (extended objects) in translation, both rectilinear and curvilinear. The first need to consider the motion of extended bodies arose with the work on pendulum clocks by Huygens before Newton's Principia. Huygens' major concern was the isochronism of pendulum clocks. He was concerned about maintaining the constancy of period of the oscillation of his pendulum clock when such clocks were used in ships. The

Fig. 3.16 Virtual and actual displacement

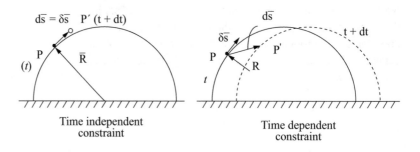

Time independent constraint   Time dependent constraint

continuous rocking motion of the ships leads to variations in the oscillation amplitude resulting in fluctuating oscillation period.

Of course, Huygens' study of compound pendulum was a very restricted form of investigating rigid body motion and the real development of rigid body dynamics was still a century in the future. Development of rigid dynamics started in the hands of Huygens and progressed in the hands of quite a few scientists of the continental Europe culminating with the final form developed by Euler. Huygens worked to find out the basic parameters of a simple pendulum that has the same oscillation period as that of a compound pendulum (i.e. a rigid body hinged at a point) of same mass. Of course, the whole analysis was of geometric nature (as the only mathematical tool to study motion of bodies). Proposition V of the fourth part of his famous book 'Horologium Oscillatorium' states

'Being given a pendulum composed of any number of weights, if each of these is multiplied by the square of the distance from the axis of oscillation, and the sum of these products is divided by the product of the sum of the weights with the distance of their centre of gravity from the same axis of oscillation, there will be obtained the length of the simple pendulum which is isochronous with the compound pendulum—that is, the distance between the axis and the centre of oscillation of the compound pendulum'.

Figure 3.17 shows a compound pendulum consisting of three masses A, B and C connected by a massless rigid rod hinged at O. The centre of gravity of A, B and C is point D. The weights (masses) at A, B and C are a, b and c. Next, a simple pendulum OG is considered with the mass at G whose magnitude g is equal to $(a + b + c)$. Left at the same position simultaneously the simple pendulum OG and the compound pendulum $OABC$, their oscillations will be isochronous if they reach identical position OX and OY simultaneously with equal velocities. Huygens showed that

such isochronism is achieved if the following relation is satisfied

$$OG = \frac{a \cdot OA^2 + b \cdot OB^2 + c \cdot OC^2}{(a + b + c) \cdot OD}$$

Though it took another century for the concept of moment of inertia to mature in the hands of Euler, it is clear that the numerator of the above expression in the RHS is nothing but the moment of inertia of the above compound pendulum. In modern language, the above equation can be written as follows:

$$l = \frac{I}{m\rho}$$

where l is the length of the equivalent simple pendulum, I is the moment of inertia about the axis through the point of suspension, m is the mass and ρ is the distance to the centre of mass from the point of suspension. This work by Huygens can be considered to be the first work relating to a body whose angular motion is relevant. But the concept of a rigid body as an agglomeration of particles of infinitesimally small size was still not there.

Jacob Bernoulli was the next scientist who investigated the motion of rigid bodies (i.e. an object with rotational motion). He realized that two points on a rigid body possess different accelerations and, so, there exists internal forces which ensure the required motions of the two particles (parts) due to the rotational motion of an extended body. This is the first hint of constraint forces developed during motion. Jacob Bernoulli found the correct principle of rigid body dynamics in plane motion. He started from the principle of a lever and realized that the body is in equilibrium when the sum of all moments of the forces about the hinge is zero. If the moments of the externally applied forces and the inertia forces are equated to zero, the principle of rigid body dynamics is obtained. In a sense, it was a mix of D'Alembert's principle and the law of lever.

Real breakthrough in rigid body dynamics was made by none other than Euler. He was one of the very few in the continental Europe who accepted Newtonian mechanics. The point mechanics developed by Newton led Euler to be attached to the basic principle of mechanics. In modern symbols, that is,

$$\dot{p} = F$$

Euler added to this a logical extension of the above principle

$$\dot{h} = r \times F$$

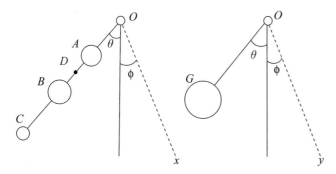

Fig. 3.17 Oscillation centre of a compound pendulum

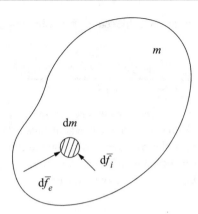

Fig. 3.18 Element of a rigid body

So the rate of change of angular momentum is equal to the moment of the applied forces. The above equation can be also written as

$$I\dot{\omega} = M$$

and moment of inertia concept attains its rightful place. Euler also reached the proper definition of a rigid body in which the distance between any two points on the body remains constant. The concept of a rigid body as an agglomeration of infinite number of infinitesimal elements was another major input by Euler. He conceptualized a rigid body as shown in

Fig. 3.18. The element of the body with mass dm is subjected to externally acting force df_e and a resultant force df_i exerted by its surrounding elements.

Thus, Euler applied the basic principle of mechanics

$$dm \cdot a = df_e + df_i = df$$

Euler also realized that when summed over the whole body, the internal forces cancel among themselves leaving only the externally applied forces on the body.

In a major intellectual leap, Euler conceptualized the general motion of a rigid body as a combination of a motion of the centre of mass and rotation of the body about the centre of mass. He introduced the three Euler angles to quantify the motion of a body around the centre of mass. He, thus, greatly simplified the complex problem of rigid body motion. Euler also distinguished between centre of mass and centre of gravity. He defined rotational inertia and also demonstrated the existence of three principal axes about which the uniform rotation is possible for a free body. He also noticed that the products of inertia terms also disappear greatly simplifying the equations of motion. Using the Euler angles, he could write the angular velocities with respect to the body-fixed principal axes and the final equations of motion of a rigid body were greatly simplified.

Special Theory of Relativity

4.1 Introduction

Like in all fields of science, the development of relativistic mechanics is also indebted to many scientists though its final formulation was in the hands of Albert Einstein in 1905. Before proceeding further, it is desirable to specify the subtle difference between the 'Principle of Relativity' and 'theory of relativity'. The Principle of Relativity is a fundamental philosophical statement whereas the theory of relativity represents the mathematical formulation. In Sect. 2.6, the Principle of Relativity was presented. The concept of relativity is pretty old and goes back to Galileo who first proposed the indistinguishability of physical phenomena taking place in two different frames of reference moving at uniform speed relative to each other. In modern language, the 'Principle of Relativity' can be stated as follows:

> It is not possible to distinguish between two inertial frames through any experiment conducted internally.

Sometimes, an alternate way of defining the principle is to state that it is impossible to determine the motion (i.e. velocity) of an inertial frame by conducting internal experiments. This very deep-rooted fundamental principle can give rise to many insights which can yield logical situations to manage problems in Newtonian dynamics. In Sect. 2.6, it was demonstrated how it was utilized by Huygens for solving the collision problem. In the next section, it will be shown how Euler used this principle for solving the complex problem of generalized rigid body motion.

4.1.1 Space–Time in Newtonian Mechanics

Newton believed in absoluteness of space and universality in time. Let two events $E_1(r_1, t_1)$, $E_2(r_2, t_2)$ be considered in Newtonian mechanics. r_1 and r_2 are the locations where the two events are taking place at times t_1 and t_2. Graphically they can be represented as shown in Fig. 4.1.

So, a series of successive events separated by infinitesimal time can be represented by a continuous line as illustrated in Fig. 4.2.

Such diagrams are called world-line diagrams. In Newtonian scheme, time is universal so all frames have similar clocks keeping identical rates of progress of time. Since 'dt', a time interval, is same in all frames, acceleration also acquires an absolute character in Newtonian mechanics. Since the velocities of a particle observed from two different inertial frames can be related by

$$v = v' + V$$

where V is the relative velocity of the primed frame with respect to the unprimed frame, and

$$\mathrm{d}v = \mathrm{d}v'$$

as V is a constant. Again in Newtonian scheme,

$$\mathrm{d}t = \mathrm{d}t'$$

where these are the time intervals in the unprimed and the primed frames. Hence,

$$a = \frac{\mathrm{d}v}{\mathrm{d}t} = \frac{\mathrm{d}v'}{\mathrm{d}t'} = a'$$

Thus in Newtonian mechanics, the dynamics of a phenomenon which depends on acceleration (because $F = ma$) is described by equations those transform following the Galilean transformation (from unprimed to primed frame of reference)

$$a' = a$$

$$v' = v + V$$

$$t' = t$$

Leibniz (and a few others in the continental Europe) did not accept the absolute character of space, and according to

© Springer Nature Singapore Pte Ltd. 2018
A. Ghosh, *Conceptual Evolution of Newtonian and Relativistic Mechanics*,
Undergraduate Lecture Notes in Physics, https://doi.org/10.1007/978-981-10-6253-7_4

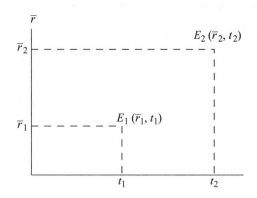

Fig. 4.1 Two events described in Newtonian scheme

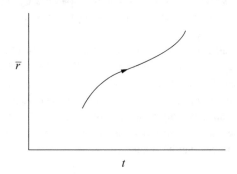

Fig. 4.2 World line of a series of events

Leibniz, space and time are nothing but relations between objects and events. Hence, all geometrical and kinematic motions are relative. However, since no one suspected against the universality of time, the Galilean transformation appeared to be valid.

So, the different inertial frames have different motions but had the identical clocks. As a result, the time played the role of a 'parameter' in Newtonian mechanics. However, it is very interesting to note that in 1788, Lagrange proposed dynamics to be phenomenon in a four-dimensional space–time structure! Thus, one can see the early concept of a four-dimensional world amalgamating space with time as early as the second half of the eighteenth century!

4.2 Euler's Work on Relativity: Confrontation of Dynamics with Optics

As noted earlier in this book, it is Euler who cast the science of motion in a modern framework. His involvements with the Principle of Relativity were on problems he investigated. In one, he solved the rigid body motion using the Principle of Relativity, and in other, he was the first to bring optics face-to-face with mechanics.

4.2.1 Principle of Relativity in Solving Rigid Body Dynamics Problem

The general motion of a rigid body in three dimensions was an extremely complicated problem before Euler applied the Principle of Relativity for simplifying the description. Euler could split the general motion of a rigid body into a translation of the centre of mass and a rotational motion of the body about the centre of mass. This is explained in Fig. 4.3. A rigid body's initial and final positions are shown in Fig. 4.3a marked by (1) and (2).

Euler split this general motion first into a translation of the centre of mass C with a uniform speed from C to C' and a rotation of the rigid body from the intermediate position $(2)'$ to (2) about point C, the centre of mass. He reasoned that as a pure translation with constant velocity implies fixing the body onto an inertial frame of reference, Principle of Relativity ensures dynamic equivalence. Using the Principle of Relativity, he could convert the untractable complex motion of a rigid body into a pure translation and a pure rotation rendering the problem easier to handle.

For a freely moving rigid body the general motion can be very complex to comprehend; but from Newton's laws its centre of mass has to move with a constant velocity since the linear momentum cannot change. Figure 4.4 shows the rigid body freely moving with respect to the inertial frame x-y-z.

The centre of mass O moves with a constant velocity u. If another frame x'-y'-z' is considered with its CM as the origin and is given a translational motion, then x'-y'-z' also constitutes an inertial frame. So the motion of the rigid body in the x'-y'-z' frame also satisfies the same equation as in the frame x-y-z. Euler could completely solve the dynamics of freely moving rigid bodies by employing the Principle of Relativity. A similar feat was achieved by Huygens 75 years ago for solving the collision problem.

4.2.2 Euler's Work on the Problem of Stellar Aberration

Once telescopic astronomy started, the observation of the stars became a real occupation of the astronomers. The primary objective was to detect any parallax of the starts due to the orbital motion of the earth as proposed by Copernicus, Kepler and Galileo but still not accepted by the scientific community (and the Church, of course) in general. From the mid-seventeenth century, astronomers led by the French astronomer Jean Picard, noticed a peculiar yearly motion of the stars (describing loops) against the backdrop of the fixed celestial sphere. In the early eighteenth century, the English astronomer James Bradley studied the phenomenon for years. All stars were seen to describe elliptic loops in unison

Fig. 4.3 Splitting the motion of a rigid body into a pure translation and a pure rotation

(a)   (b)

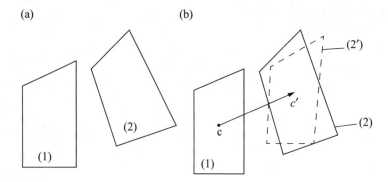

The telescope on a stationary base needs to be tilted by an angle θ with the horizontal. If the telescope is mounted on a moving earth, it needs to be tilted a bit more towards the direction of motion of the telescope. This effect is explained further in Fig. 4.7. The velocity diagram shows the finite speed of light c and the instantaneous speed of the telescope v (Fig. 4.6).

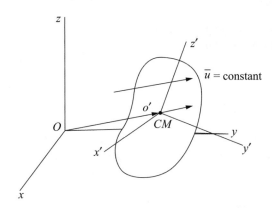

Fig. 4.4 Motion of the centre of mass of a freely moving rigid body

This phenomenon was also another evidence that light moves with a finite speed. As Bradley measured the magnitude of the aberration angle as $20''$ of arc, knowing the orbital speed of the earth as 30 km s^{-1}, he estimated the speed of light as $3.09 \times 10^8 \text{ m s}^{-1}$, quite close to the modern value.

Euler investigated the phenomenon and came to the problem of confrontation of optics with mechanics. He believed light to be a wave phenomenon and also in the existence of ether as a medium to propagate light. As a result, Euler considered light to travel with constant speed irrespective of the motion of the source. Euler analysed the following three situations:

keeping in tune with the earth's orbital motion. Initially, it was suspected to be the expected parallax. But the directions of the displacements were not in the right direction for those to represent parallax.

As shown in Fig. 4.5, the displacement due to parallax at the position should be towards left. But the earth in this position moves into the plane of the paper. Bradley found the observed displacement was always in the direction of the earth's instantaneous velocity due to orbital motion. Thus, parallax was ruled out. In 1725, Bradley came up with the explanation that was referred to as the 'aberration of star light'. The explanation is given in Fig. 4.6.

(i)   The object (i.e. the star) moves
(ii)   The observer (i.e. our earth) moves
(iii) Both move

Figure 4.8a shows the case when the object moves with a velocity s. The observer S is fixed to the frame as shown in Fig. 4.8a. In the frame of S, the light travels from the star with a constant speed c to reach S, covering a distance d as shown. Now,

$$h \sin \theta = l \sin \alpha \qquad (4.2.1a)$$

$$l \cos \alpha + h \cos \theta = d \qquad (4.2.1b)$$

where

$$d = h \cdot \frac{c}{s} \qquad (4.2.1c)$$

and c is the speed of light.

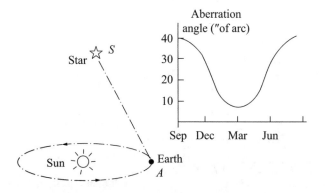

Fig. 4.5 Star and the orbiting earth

Fig. 4.6 Explanation of the aberration of stars

(a)   (b)

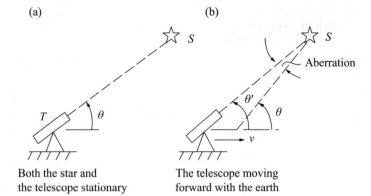

Both the star and   The telescope moving
the telescope stationary   forward with the earth

From the above equations, one easily obtains the following result for the aberration angle α

Fig. 4.7 Velocity diagram to explain the aberration

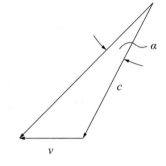

$$\tan \alpha = \frac{s \sin \theta}{c - s \cos \theta} \qquad (4.2.2)$$

The result for case (ii) was obtained by Euler through the application of the Principle of Relativity. Euler gave a velocity to the whole system S, a velocity $-r$, where r is the velocity of the observer and used (4.2.2) with the object speed as r. According to the Principle of Relativity, this reduces the system to the first one, and according to the principle, the result should be the same as that in the first case given by (4.2.2). Thus, for a case when the observer moves with a velocity r

$$\tan \alpha = \frac{r \sin \theta}{c - r \cos \theta} \qquad (4.2.3)$$

Euler did not stop here. He wanted to get the result by adding the velocities as shown in Fig. 4.8b. From the figure, one gets

$$\frac{c}{\sin \theta} = \frac{r}{\sin \alpha}$$

$$\sin \alpha = \frac{r}{c} \sin \theta$$

Hence,

$$\cos \alpha = \sqrt{1 - \sin^2 \alpha} = \frac{1}{c} \sqrt{c^2 - r^2 \sin^2 \theta}$$

Thus,

$$\tan \alpha = \frac{r \sin \theta}{\sqrt{c^2 - r^2 \sin^2 \theta}} \qquad (4.2.4)$$

The result is different from the one obtained by applying the Principle of Relativity along with a wave model for light, whereas (4.2.4) represents the result according to Newtonian mechanics applied to light moving as particles. This was the first instance where light appeared to contradict Newtonian mechanics. The concept of a stationary ether was also another outcome of this study.

Fig. 4.8 Euler's investigation of the stellar aberration phenomenon

(a)   (b)

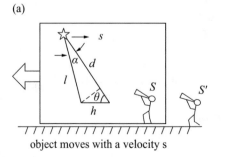

object moves with a velocity s

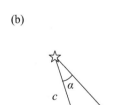

Result from velocity addition

4.3 Efforts to Detect Ether Speed: The Null Result of Michelson–Morley Experiment

4.3.1 Early Attempts

As the wave theory of light gained popularity and acceptability, the existence of luminiferous ether became essential to all models of optics and experiments with light. Since the ether is supposed to be in absolute rest and it flows through all objects freely, the moving earth should face an ether wind (flowing through the earth and all objects on it with a velocity opposite to that of the earth with respect to the stationary ether). Soon scientists started to attempt to detect the absolute velocity of the earth. Earth's orbital speed around the sun was known, but the velocity with which the solar system moves was not known.

The first such attempt at determining the velocity of the earth racing through the ether wind was made by the French scientist Francois Arago in 1818. His idea was quite ingenuous and based upon the theory of transmission of light in glass. It was known that a prism bends a light ray passing through it and the amount of bending depends on the refractive index of glass. From the wave theory of light, it is known that the refractive index is the ratio of the speed of light in vacuum to the speed of light in the glass.

Thus, in the case (a) of Fig. 4.9, the refractive index is

$$\mu = c/c'$$

where c and c' are the speeds of light in vacuum and in glass, respectively. When the prism moves with a velocity v (as shown in Fig. 4.9b), the speed of the light with which it approaches the glass is $c - v$. Inside the prism, the speed of light becomes $c'' \propto c' - v$. Thus, the refractive index should become

$$\mu \approx \frac{(c - v)}{(c' - v)}$$

and the angle θ' of bending of light should be different. Hence, Arago concluded that a prism on the earth should deflect light with different angles indicating the motion of the earth. The orbital velocity of the earth changes with time, and a signature of that was expected to be shown in the deflection of light by a prism on the earth. Arago designed a suitable experiment to detect the earth's motion, but he was surprised to notice no effect of the motion of the earth through the ether.

Arago's friend, Fresnel, suggested a peculiar scheme to explain the observed result. He suggested an additional amount of ether to be present inside the moving prism glass over and above the uniform ether present all over. Thus, there was an excess of ether inside the prism. The scheme suggested by Fresnel was very complex and, to some extent, weird! But the suggestion could explain the null result obtained by Arago. According to the conventional wisdom at that time, the speed of light inside a medium of refractive index μ, moving with a velocity v, was given by

$$c' = \frac{c}{\mu} + v$$

that was used by Arago. Fresnel suggested a modified form of the above equation. He suggested a formula for the speed of light inside a moving medium as follows:

$$c' = \frac{c}{\mu} + fv$$

where f was a correction factor called Fresnel's drag coefficient given by $f = \left[1 - \left(\frac{1}{\mu^2} \right) \right]$. However, wired could have been the assumptions made by Fresnel this formula yielded results which agreed with experimental observations! Soon after, in 1851, Fizeau, another expert in the field of experiments on light propagation, conducted an experiment to verify Fresnel's formula. His scheme is illustrated in Fig. 4.10.

A light beam was split up by a partial mirror into two branches A and B. A was guided to move against the flow velocity of water in the two tubes, whereas beam B moved along the flow velocity. When the two beams A and B got combined, interference fringes were formed. The flow of the water was stopped for initial calibration and the shift in the fringe pattern was noted when water flow was resumed. From the experimental result, a drag coefficient of 0.48 was obtained compared to the predicted theoretical value of 0.43! This was considered to support Fresnel's hypothesis of partial dragging of ether (inside a moving medium). But this

(a)   (b)

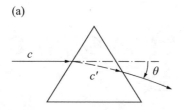   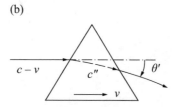

Fig. 4.9 Bending of light by a stationary prism and a moving prism

Fig. 4.10 Fizen's ether-drag
experimental set-up

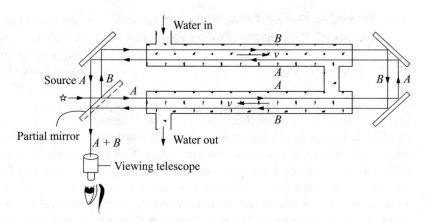

was against the observations on starlight aberration studied
by Bradley and Euler. That indicated a stationary ether.

Much later, the Fizeau experiment was repeated by
Michelson at the suggestion of Lord Kelvin. In 1885,
Michelson and Morley repeated the experiment and con-
firmed the result obtained by Fizeau more than 30 years ago!

4.3.2 Michelson–Morley Experiment

The attempt to measure the velocity of the solar system
through a stationary ether started as early as 1879 when
James Clark Maxwell suggested to conduct experiments
with eclipses of Jupiter's moon Io at different configurations
of the earth and Jupiter in the solar system as indicated in
Fig. 4.11.

In the configuration (a), eclipse of Io by Jupiter will be
observed earlier as the earth is approaching Jupiter. On the
other hand in configuration (b), the viewing of the eclipse
will be delayed as the earth is moving away from the light
signals from Jupiter's location. Maxwell received the
detailed tables of the eclipse timings in 1879, and he thanked
American astronomer Todd for those. But a few months later
Maxwell died without analysing the tables. Maxwell's letter
was presented in the Royal Society of London and later
published in Nature. Thus, it came to the notice of Michelson, an American physicist.

Maxwell's suggestion was of purely academic interest as
the detection of earth's motion through the ether needed
extremely small time intervals (approximately 10^{-15} s)
beyond the scope of contemporary experimental technology.
But Michelson developed a new instrument, an interferometer, capable of detecting extremely small length elements.
Since the speed of light is about 3×10^8 m s^{-1} in 10^{-15} s, it
travels a distance 0.3×10^{-6} m. This is comparable to the
wavelength of visible light which is about 0.5×10^{-6} m.
Using interferometry, extremely small effects could be
detected by Michelson's apparatus.

Michelson decided to conduct a terrestrial experiment to
detect any ether wind because of the earth's motion through
it. The basic idea is illustrated in Fig. 4.12. The arrangement
consists of two mirrors M_1 and M_2 fixed on a platform
equidistant from a light source at O.

When the set-up is stationary (with respect to the
luminiferous ether), light beams originated at O by a flash
travel the distance d to M_1 and M_2 in equal times d/c and
after being reflected by the mirrors reach back the source
point after a period d/c. Thus, both the beams arrive back at
O after a period $2d/c$. When the set-up has a velocity v (towards right, say), Michelson's expected chain of events is
depicted in Fig. 4.13.

A flash (by a source at O) takes place at $t = 0$. The two
beams, A and B, travel towards the mirrors M_1 and M_2 as
shown. After a time Δt, B reaches M_2 and is reflected;

Fig. 4.11 Earth–Jupiter
configuration for the suggested
eclipse experiment

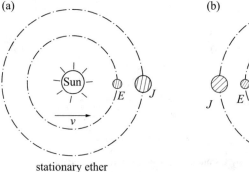

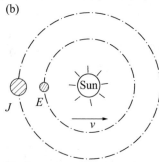

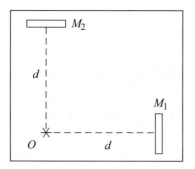

Fig. 4.12 Basic idea behind Michelson's experiment

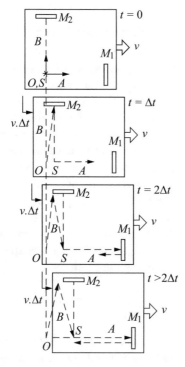

Fig. 4.13 Michelson's experiment with two mirrors moving with respect to stationary ether

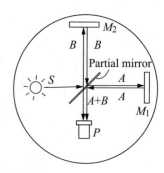

Fig. 4.14 Essential features of Michelson's experiment

A and B is expected to show up in an interference fringe pattern that can detect a shift of the waves relative to each other even by a small fraction of the wave length. Michelson expected to observe a shift in the fringe by rotating the whole set-up causing the beam paths to change their orientations relative to the direction of the earth's velocity relative to the stationary ether. The same experiment was conducted again in 1887 in Cleveland, Ohio, in collaboration with Morley; the accuracy attempted was much higher than the 1881 experiment. But, as the whole scientific community knows now, the result did not show any effect and no motion of the earth was detected. The obvious conclusion was that the ether is being dragged by the earth contradicting the observed stellar aberration. So the science community faced three possibilities—stationary ether, partially moving ether and completely dragged ether.

To explain the null result, the Irish scientist Fitzgerald gave a suggestion in his lectures in 1889. According to the suggestion, all objects contract in length for its motion through the ether, contraction being in the direction of motion. The lengths are reduced by a factor $\sqrt{\left(1 - \frac{v^2}{c^2}\right)}$. In 1892, H.A. Lorentz came up with a similar suggestion, and this effect became known as Lorentz–Fitzgerald contraction. Lorentz being the greatest expert on Maxwell's theory of electromagnetism also continued his research on the applicability of the Principle of Relativity; he found that electromagnetism does not satisfy Galilean relativity.

4.4 Electromagnetism: Challenge to the Principle of Relativity

Though the phenomenon of static electricity and magnetism were known to the scientists for a long time, the real progress in the study of the subject started after Michael Faraday investigated these two phenomena in 1831. Before him, the Danish scientist Hans C. Oersted first discovered in 1820 that a compass needle is deflected by electric current flowing through a conductor signifying that a magnetic field is created by an electric current flowing through a conductor as indicated in Fig. 4.15.

A continues its journey towards M_1. After another interval of time Δt, the reflected beam B reaches the source S. A may have been reflected by M_1 but still has not reached S. After another interval of time, the reflected beam A reaches back at S. Thus, the two beams reach back their original source at different times, the difference depending on the velocity v.

Michelson conducted his first experiment in a set-up placed deep inside the foundation of the observatory at Potsdam, to make the sensitive equipment free from disturbance. The essential feature of Michelson's experimental set-up is shown in Fig. 4.14.

A beam of light from the source is split into two beams A and B by the partial mirror S. After getting reflected by M_1 and M_2, the beams recombine and are examined for interference by the pickup P. The difference in travel distance of

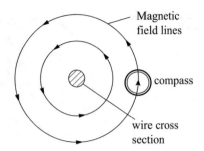

Magnetic
field lines

compass

wire cross
section

Fig. 4.15 Current in a wire developing a magnetic field

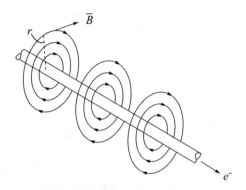

Fig. 4.16 Magnetic field produced by a straight conductor

This was the first instance when the two phenomena were found to be related. The conclusion in the modern sense is that moving charge causes a magnetic field to develop (electric current as a flow of negatively charged electrons is a much later discovery).

The reverse phenomenon that magnetism gives rise to electric current was discovered by Faraday. Of course, the discovery was also made by American scientist Henry and Russian scientist Lenz independently soon after Faraday. The field concept, introduced by Faraday, helped to conceptualize the action-at-a-distance phenomena involved with electricity and magnetism. The Scottish scientists James Clark Maxwell, 40 years younger to Faraday, developed the mathematical formalism for the fields produced by electricity and magnetism. His work demonstrated the symmetry between the two tightly interwound phenomena and a theory of electromagnetism slowly emerged. With his theory, Maxwell could demonstrate that an electromagnetic wave travels at a fixed speed through empty space. The basic essence of his theory can be described as shown below.

Considering the symmetry, Maxwell came to the opinion that varying magnetic field develops electric field and a varying electric field also generates a magnetic field. If E be the electric field at a point is space, the force developed on a charge q is given by the relation

$$F = qE \qquad (4.4.1)$$

The electric field due to a charge Q at position r (with Q as the origin) is known to be

$$E = \frac{1}{4\pi\varepsilon_0}\frac{Q}{r^3}r \qquad (4.4.2)$$

where $4\pi\varepsilon_0$ is as universal constant whose magnitude in SI units is given by 1.111×10^{-9}. A current-carrying straight conductor produces a magnetic field as indicated in Fig. 4.16 (a three-dimensional view of Fig. 4.15).

If the current magnitude be I, then the magnitude of the magnetic field B is given by

$$B = 2\left(\frac{\mu_0}{4\pi}\right)\frac{I}{r} \qquad (4.4.3)$$

where r is the distance of the point from the conducting wire. The magnitude of the universal constant $\left(\frac{\mu_0}{4\pi}\right)$ in SI unit is 10^{-7}. Maxwell's theory incorporated a symmetric relationship between electricity and magnetism. A varying electric field developed a magnetic field, and a varying magnetic field generates an electric field. So a travelling electric field creates a varying magnetic field at a certain point causing a corresponding magnetic field to develop (that also travels with the electric field wave as shown in Fig. 4.17).

Now depending on the values of ε_0 and μ_0, the magnitude of B developed by E has to be adequate again to generate the electric field wave. Thus, the travelling speed, on which the changing rates of E and B at a point depend, has to be unique to establish the required exact correspondence. This unique speed is found to be given by

$$c = \frac{1}{\sqrt{\varepsilon_0\mu_0}} = 2{,}997{,}924 \text{ km s}^{-1} \qquad (4.4.4)$$

which was exactly the speed of light. Thus, Maxwell declared that light is nothing but electromagnetic wave. Though there was enough indication already, Einstein's Special Theory of Relativity firmly unified the two apparently distinct phenomena of electricity and magnetism. This will be discussed in the next session.

Fig. 4.17 Electric field wave accompanied by a corresponding magnetic field wave

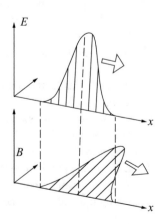

Analysing interaction between moving charges, English physicist Oliver Heaviside showed in 1888 that the force exerted by a charge q_1, moving past a charge q_2 at a distance r, with a speed v (Fig. 4.18) is given by

$$\frac{\varepsilon_0 q_1 q_2}{r^2} \frac{1}{\sqrt{1 - \frac{v^2}{c^2}}} \qquad (4.4.5)$$

Strangely, the factor $\left(1 - \frac{v^2}{c^2}\right)^{\frac{1}{2}}$ resembles the Lorentz boost in Einstein's Special Theory of Relativity!

But it is worth noting that Maxwell's electromagnetism is relativistic in character though it was formulated much before 1905, the year Einstein published his first paper proposing the Special Theory of Relativity. What Einstein's theory could do was to bring the rest of classical mechanics on the same platform as that of electromagnetism.

Electromagnetic phenomenon put forward a number of challenges before the scientific community so far as the Principle of Relativity was concerned. In Newtonian mechanics, force was linked to the acceleration of a body. The acceleration of a body remains unchanged by Galilean transformation. This and the constancy of mass and universality of time result in the expected invariance of dynamical phenomena referred to any inertial frame. This satisfied the Principle of Relativity, considered so important from philosophical point of view. However, in electromagnetic interaction between two bodies, the interactive force depends on the velocity also. When both electric and magnetic forces are present, then the total force on a particle of charge q moving with a velocity v is given by the Lorentz's force law

$$\boldsymbol{F} = q\boldsymbol{E} + q\boldsymbol{v} \times \boldsymbol{B} \qquad (4.4.6)$$

where $\boldsymbol{E}$ is the electric field and $\boldsymbol{B}$ is the magnetic field. The above law was first proposed by Heaviside in 1889 and was independently derived by Lorentz in 1895. Since the force law contains a term with velocity, its invariance is not satisfied under Galilean transformation. So, according to Newtonian mechanics, the 'Principle of Relativity' is violated.

Thus, towards the end of the nineteenth century, the scientists were confronted with two major problems—(i) inability to determine earth's motion through ether and

(ii) violation of the Principle of Relativity by the electromagnetic phenomena according to Newtonian mechanics. Lorentz attempted a modification of the Galilean transformation rule to explain the null result of Michelson–Morley experiment. In the rule of Galilean transformation

$$\boldsymbol{V'} = \boldsymbol{V} + \boldsymbol{v}$$
$$\boldsymbol{a'} = \boldsymbol{a}$$
$$t' = t$$
$$m' = m$$

To bring in a new rule for the transformation of t' he used

$$t' = t - \frac{vx}{c^2}$$

where the moving frame has a velocity $\boldsymbol{v}$ along the x (and the coincident x')-axis. Thus, this new time, called by Lorentz as 'local time', depended on the location x. When this modified law of Galilean transformation was applied to Maxwell's equations, the primed equations had the same form as the unprimed ones except for the second-order terms in $\left(\frac{v}{c}\right)$. With this, he demonstrated the failure of experiments to detect earth's motion up to the first-order terms in $\left(\frac{v}{c}\right)$.

This kind of patchwork solution to the problems of science was objected to by the French mathematician Henri Poincare in 1895. Lorentz's rule was developed to take care of Fresnel's idea to explain null results in experiments to detect an ether wind (up to first-order terms), and the apparent length contraction proposed by Fitzgerald and Lorentz. He wondered what would happen if future experiments yielded null result for second-order terms also. (In fact, by 1904, more sophisticated experiments yielded null result for earth's motion in stationary ether.) This had a major impact on Lorentz's thinking, and in 1904, he proposed a transformation law avoiding the patchwork approach. He incorporated both the concept of local time and also the Fitzgerald contraction. So, the transformation rules became (when the primed moving frame and the stationary unprimed frame are as shown in Fig. 4.19):

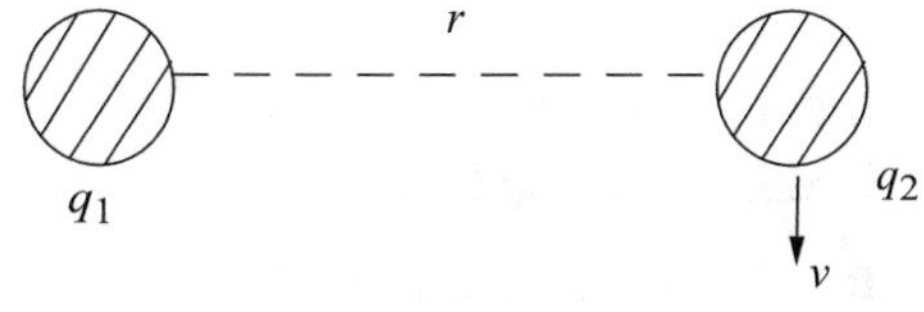

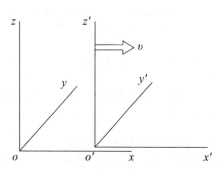

Fig. 4.18 Force due to a moving charge

Fig. 4.19 Frames of reference for Lorentz's transformation rule

$$x' = \gamma(x - vt)$$
$$y' = y$$
$$z' = z$$
$$t' = \gamma\left(t - \frac{vx}{c^2}\right) \qquad (4.4.7)$$
$$\gamma = \frac{1}{\left(1 - \frac{v^2}{c^2}\right)^{\frac{1}{2}}}$$

With these transformation rules, Maxwell's equations in the primed and unprimed frames take the same form satisfying the Principle of Relativity. It is interesting to note that these transformation rules were used by the English scientist Joseph Larmor in 1898. Something similar was used by the German scientist W. Voigt in 1887! However, keeping in view of the enormous contributions by Lorentz, Poincare named these equations as Lorentz transformation in 1905. These transformation equations satisfied the Principle of Relativity completely including all higher-order terms.

So, the big question was to accept the Principle of Relativity along with Lorentz transformation or accept Newtonian mechanics. In 1905, Poincare submitted a detailed paper on the transformation rules with arguments based primarily on electromagnetism. In the same year, Albert Einstein published his famous paper 'On the Electrodynamics of Moving Bodies' that is considered as the foundation for Special Theory of Relativity. The name 'Theory of Relativity' has been coined primarily because Einstein accepted the 'Principle of Relativity' rejecting Newtonian mechanics.

4.5 Einstein's Special Theory of Relativity

Unlike Lorentz and Poincare, Einstein attempted to bring all physical phenomena under the umbrella of Principle of Relativity. Thus, though the rules of transformation derived by him were the same as found by Lorentz (and others like Poincare), his approach was very fundamental. He showed that not only electromagnetism but also Newtonian mechanics (as a matter of fact any physical phenomenon) satisfies the same transformation rules and the Principle of Relativity (that was supreme in the eyes of Einstein). He proposed two basic principles on which the whole structure of Special Theory of Relativity stands. These are as follows:

Principle 1: The laws of physics take the same form in all inertial frames
Principle 2: The speed of light in vacuum has the same value in all inertial frames

Both these principles look apparently, somewhat, obvious, but the combination yields revolutionary results. Earlier the Principle of Relativity was considered primarily for mechanical phenomena, but Einstein not only unified electromagnetism with mechanical phenomena but also brought all physical phenomena under the same umbrella. The principle of constancy of the speed of light may also appear somewhat orthodox. Because light as a wave needed a medium (the luminiferous ether) and it was considered natural that its transmission speed depends on ε_0 and μ_0 as mentioned before. But Einstein's principle was 'fundamentally' different though it looked somewhat similar. The constancy of transmission speed in ether led to the conclusion that light's speed does not depend on the motion of the source. But Einstein completely eliminated the concept a medium like ether. In his scheme of things, 'constancy' of the speed of light was a 'fundamental principle', not a derived conclusion like before. It was considered a fundamental property of light that can be transmitted through 'vacuum' without any medium. To illustrate this, most important and fundamental point further the following example can be considered.

Figure 4.20 shows a situation where two identical space vehicles are A and B where B is moving with respect to A with a speed v. A and a fixed object (say a star) are in relative rest. The two observers conduct experiments inside their vehicles to measure the speed of light by flashing lamps at the light torches kept in the front end of the ships as indicated in Fig. 4.20. They also measure the speed with which light from S passes their space ships. Since the speed of light does not depend on the motion of the source, light waves move with the same speed coming out of S and the lamps in A and B. If light from S travels in ether with speed c, B will measure it as $c + v$. In the same way, the speed of light from the source in B should be also c. B should measure the speed as $c + v$. On the other hand, A finds the speed of light from S and the lamp inside A as c. These violate the first principle according to which A and B both should measure the light speed as c. Thus, in the absence of any luminiferous ether medium, the speed of light is constant as a principle, not because of the fixed properties of medium on

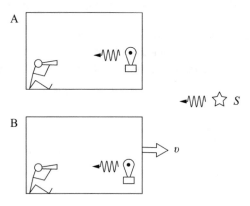

Fig. 4.20 Experiment with the principle of constancy of light

which the speed of wave propagation depends. The two principles look very simple (and also, perhaps, somewhat obvious) to begin with, but their combination yields the solution to the perplexing situation existing towards the end of the nineteenth century.

As the objective of this volume is not to discuss the theory in detail, only a few points are going to be taken up for discussion.

4.5.1 Lorentz's Transformation from the Two Principles of Relativity

It is not out of place to briefly show how the equations of Lorentz transformation emerge from the two basic principles proposed by Einstein.

Without losing any generality, the Cartesian coordinate attached to two inertial frames be so oriented that the relative velocity v is along the x and x' axes which are coincident as shown in Fig. 4.21.

The y and y' axes and the z and z' axes are always parallel to each other. Since the concept of a universal time (as in Newtonian mechanics) is no longer tenable, any momentary event (say, a flash of light) ε is described by its location and the instant of time recorded in a clock attached to the location in the frames. It is assumed that every point in each of the two frames has an observer and a clock. Furthermore, the clocks of a frame are suitably synchronized. A detailed description of the process of clock synchronization can be found in any standard textbook and is skipped here. Besides, taking Einstein's two basic postulates as true, the space is assumed to be homogenous and isotopic. It is further assumed that when $t = t' = 0$, the origin O' of frame S' is coincident with the origin O.

As the space is assumed to be homogenous and isotopic, any transformation rule between the coordinates of an event ε described in S' and those for describing the same event must be through equations which are linear. Similarly, any dependence on t has to be also linear. The most general form

of linear transformation equation describing coordinates (x', y', z', t) in terms of the coordinates (x, y, z, t) is as follows:

$$
\begin{aligned}
x' &= a_{11}x + a_{12}y + a_{13}z + a_{14}t \\
x' &= a_{11}x + a_{12}y + a_{13}z + a_{14}t \\
z' &= a_{31}x + a_{32}y + a_{33}z + a_{34}t \\
t' &= a_{41}x + a_{42}y + a_{43}z + a_{44}t
\end{aligned}
\tag{4.5.1}
$$

In abridged form, using matrix notation (4.5.1) can be written in the following manner:

$$
\begin{Bmatrix} x' \\ y' \\ z' \\ t' \end{Bmatrix} = [a] \begin{Bmatrix} x \\ y \\ z \\ t \end{Bmatrix} \quad \text{where} \quad [a] = \begin{bmatrix} a_{11} & a_{12} & a_{13} & a_{14} \\ a_{21} & a_{22} & a_{23} & a_{24} \\ a_{31} & a_{32} & a_{33} & a_{34} \\ a_{41} & a_{42} & a_{43} & a_{44} \end{bmatrix}
$$

It is obvious that for v approaching zero, the coefficients must tend to satisfy the following conditions:

$$
a_{ii} = 1 \text{ and } a_{ij} = 0; \quad i = 1, 2, 3, 4 \text{ and } j = 1, 2, 3, 4.
\tag{4.5.2}
$$

Or in other words, the transformation tends to become Galilean.

Since x' and x coincide at all times, the points on these axes must satisfy the conditions as given below:

$$
\begin{aligned}
y = 0, &\quad z = 0 \text{ for all times, and,} \\
y' = 0, &\quad z' = 0 \text{ for all times}
\end{aligned}
$$

These conditions can be satisfied only with the following transformation equations for y' and z':

$$
\begin{aligned}
y' &= a_{22}y + a_{23}z \\
z' &= a_{32}y + a_{33}z
\end{aligned}
$$

Or, in other words, $a_{21} = a_{24} = a_{31} = a_{34} = 0$. In the same manner, the $x - y$ plane and the $x' - y'$ plane must always coincide and so should be the case with $x - z$ plane and $x' - z'$ plane. These conditions imply that for $z = 0$, z' should be also zero and for $y = 0$, y' should be also zero. These can be satisfied only if the transformation rules for y and z are as given below:

$$
\begin{aligned}
y' &= a_{22}y \\
z' &= a_{33}z
\end{aligned}
$$

implying that $a_{23} = a_{32} = 0$. Now to determine a_{22} and a_{33}, the first postulate can be applied which ensures reciprocity (i.e. what the observer in S' notices must be the same as what the observer in S notices). Hence, the principle can be satisfied if

$$
a_{22} = a_{33} = 1
$$

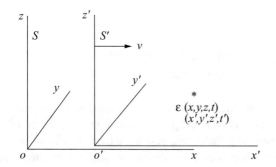

Fig. 4.21 Two inertial frames

or, $y' = y$

and $z' = z$ (4.5.3)

The remaining two transformation equations for x' and t' can be found out as described below.

First, it is understood that t' cannot depend on y and z. Otherwise, clocks placed symmetrically at $+y$, $-y$ or $+z$, $-z$, will show different times since the equations have to be linear in the coordinates. Hence,

$$a_{42} = a_{43} = 0$$

i.e. $t' = a_{41}x + a_{44}t$ (4.5.4)

Next, it is obvious that the description of point O' in S' given by $x' = 0$ must correspond to the condition $x = vt$ in S frame (since originally O' coincided with O and moves with a constant velocity v as mentioned). This can be represented mathematically by the following equation:

$$x' = a_{11}(x - vt)$$ (4.5.5)

What remains to be done next is to determine the coefficients a_{11}, a_{41} and a_{44}. This exercise can be done by considering the second postulate proposed by Einstein—the constancy of the speed of light in all inertial frames considers a flash of light at $t = t' = 0$ when O' coincided with O momentarily. The light wave front propagates in the form of the surface of a sphere whose radius increases at the rate c, the speed of light. Now, both observers must have the same description of the spherical wave front (expanding at the same rate c as per the second postulate). The equations for the wave front spherical surfaces described in the two frames are

$$x^2 + y^2 + z^2 = c^2 t^2$$
and $x'^2 + y'^2 + z'^2 = c^2 t'^2$ (4.5.6)

Now, substituting x', y', z' and t' in the second of the above two equation using (4.5.3), (4.5.4) and (4.5.5) and rearranging the terms, one gets

$$(a_{11}^2 - c^2 a_{41}^2)x^2 + y^2 + z^2 - 2(va_{11}^2 + c^2 a_{41}a_{44})xt$$
$$= (c^2 a_{44}^2 - v^2 a_{11}^2)t^2$$

The above equation can yield the first of the two equations (4.5.6) if and only if the following conditions are satisfied.

$$a_{11}^2 - c^2 a_{41}^2 = 1$$
$$va_{11}^2 + c^2 a_{41}a_{44} = 0$$
and $c^2 a_{44}^2 - v^2 a_{11}^2 = c^2$

Solving the three equations, the coefficients a_{11}, a_{41} and a_{44} come out as follows:

$$a_{11} = \gamma$$
$$a_{41} = -\frac{v}{c^2}\gamma$$

and

$$a_{44} = \gamma$$

where

$$\gamma = \frac{1}{\sqrt{1 - \frac{v^2}{c^2}}}$$

Thus, substituting all the coefficients a_{ij} ($i = 1, 2, 3, 4$; $j = 1, 2, 3, 4$), the final form of the transformation equations is as follows:

$$\begin{aligned} x' &= \gamma(x - vt) \\ y' &= y \\ z' &= z \\ t' &= \gamma\left(1 - \frac{vx}{c^2}\right) \end{aligned}$$ (4.5.7)

The equations above are nothing but the Lorentz transformation. One can check the condition of relative reciprocity by finding the reverse transformation equations as follows:

$$\begin{aligned} x &= \gamma(x' + vt') \\ y' &= y \\ z' &= z \\ t &= \gamma\left(t' + \frac{vx'}{c^2}\right) \end{aligned}$$ (4.5.8)

It is also verified that for small values of v, i.e. when $\frac{v}{c} \ll 1$, (4.5.7) leads to the equations of Galilean transformations

$$\begin{aligned} x' &= x - vt \\ y' &= y \\ z' &= z \\ t' &= t \end{aligned}$$

It is very important to note that the above equations were derived in a completely general way from the two postulates by Einstein and using the property of Euclidean space and time.

4.5.2 Special Relativity in Electromagnetic Phenomenon

Earlier electricity and magnetism used to be considered to be two different physical phenomena. Einstein's theory of special relativity shows that both are the different manifestations of the same phenomenon—electromagnetism. Two simple

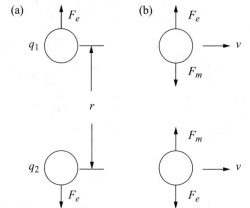

Fig. 4.22 Two moving charges

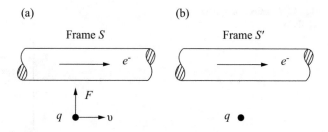

Fig. 4.23 Conductor and an electric charge

cases are being presented below to illustrate the matter. The two charges, shown in Fig. 4.22a, rest in an inertial frame separated by some distance r.

According to Coulomb's law, they are subjected to a repelling force (assuming both charges to be of equal polarity)

$$F_c = \frac{\varepsilon_0 q_1 q_2}{r^2}$$

Next, let the situation be observed by another inertial observer who is moving with velocity v towards left with respect to the stationary charges. Applying Maxwell's equation shows that the two moving charges produce magnetic fields along with electric fields at each other charge's location. Thus, they will be acted upon by attractive magnetic force F_m and repulsive electrical force F_e. But strangely, the net repulsive force $(F_e - F_m)$ is different from F_e. So application of Maxwell's equation to phenomenon described in different inertial frames yields different results. The question arises in which description of the electric and magnetic fields is real. Earlier to Einstein, it was often presumed that there exists a special frame of reference in which the electric and magnetic fields appear in their correct form. Assumed existence of ether gave rise to such conclusion in favour of a privileged reference frame. But special relativity shows that the two observers in the above experiment will be unable to detect any difference between the changes in the

separation of the two charges. The timescales in the two frames are different like the masses. (It has not been discussed so far). Thus, the Principle of Relativity remains satisfied.

Another example can show how Special Theory of Relativity helps to preserve the Principle of Relativity. Figure 4.23 shows a straight conductor (of infinite length, only a short portion of which is being considered) carrying a steady electric current.

Now when an electric charge moving with a speed v is kept near the conductor, it will experience a force F according to Lorentz's law. Now if another observer S', moving with a uniform speed v (along with charge), finds the charge to be stationary, will he detect any attraction of the charge towards the conductor? If not, the Principle of Relativity is violated. Special Theory of Relativity resolves the problem as explained below.

The conductor can be represented by fixed protons and a series of drifting electrons as shown in Fig. 4.24a. Let the drift velocity of the electrons be v.

This generates a magnetic field B, and the moving charge q experiences a force $\overline{F} = -qv \times B$ as shown. Since the conductor with moving electrons is electrically neutral (as the numbers of +protons and −electrons are equal), the moving charge does not experience any electrical force qE. For the sake of simplicity (It can be shown for a general case also but that makes the mathematics complicated. The basic idea behind this simple case is good enough.), let the moving frame S' move with a uniform velocity v as indicated in Fig. 4.24b. So, the charge q (and also the electrons) appears to be motionless in the frame. On the other hand, the

Fig. 4.24 Current-carrying conductor and moving charges

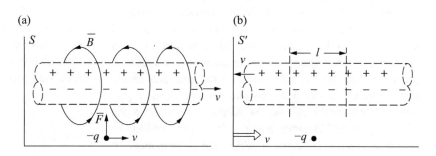

Fig. 4.25 Effect of special
relativity

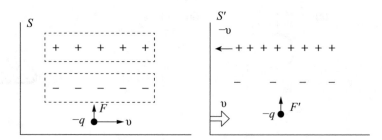

positively charged protons appear to move in the opposite
direction with a speed $-v$. Considering the Lorentz transformation as per the requirement in Special Theory of Relativity, the length l of a relevant section of the conductor
protons appears to be shorter but the length of the electron's
part becomes longer compared to the length when they were
moving is S as shown in Fig. 4.25.

Thus, the number of protons become more than that of
the electrons in the relevant section of the conductor. This
makes the conductor to acquire a resultant positive charge,
and the negative charge experiences an attractive force F'.
Detailed analysis shows F and F' to be equal. So, the
observers in S frame and S' frame find no difference in the
results of this experiment satisfying the Principle of Relativity. It also shows how Special Theory of Relativity unifies
electricity and magnetism. Depending on the point of view,
an effect can be either electrical or magnetic. Unifying
electricity with magnetism is a major achievement of Special
Theory of Relativity.

4.5.3 Need for a Relativistic Mechanics

It has been shown how Special Theory of Relativity restored
the Principle of Relativity for all electromagnetic phenomena. However, it should not be forgotten that Special Theory
of Relativity must bring mechanics (science of motion of
objects) also under the umbrella of the Principle of Relativity. If one attempts to apply Lorentz transformation to the
laws of Newtonian mechanics, the Principle of Relativity is
not satisfied as velocity is ingrained in the Lorentz transformation equations. This led Einstein to make some alterations of Newtonian mechanics.

For mechanics to also satisfy the Principle of Relativity, it
is necessary that laws of mechanics remain same under
Lorentz transformation. A very fundamental law of
mechanics is the invariance of total linear momentum of an
isolated dynamical system. The simplest case of a dynamics
problem of an isolated (where no external force acts on the
system) system is the elastic collision of two identical particles. A symmetric collision of identical particles in the
laboratory frame S is shown in Fig. 4.26. Let the collision be
almost a grazing collision in which the x-component of

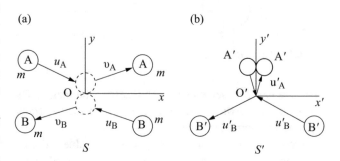

Fig. 4.26 Elastic collision seen in two inertial frames

velocities in frame S is much larger than the y-component
velocities of the particles.

If the total momentum of the colliding particles before
and after the collision is zero, then in frame S $u_{A_y} = -v_{A_y}$
and $u_{B_y} = -v_{B_y}$ (Fig. 4.26a). When seen from the S' (which
is moving towards right with speed u_{A_x}), the situation is
shown in Fig. 4.26b. Since the momentum conservation rule
remains valid in both the frames, it is easy to show that

$$u'_{A_y} = u'_{B_y}$$

according to Newtonian mechanics. If Lorentz transformation rules are employed, one gets a different result. Avoiding
the algebra, one gets

$$u'_{A_y} = u'_{B_y} \left(\frac{1}{1 - \frac{u'_{x_B} v}{c^2}} \right)$$

where $v = u_{A_x}$. This is different from the Newtonian result.
So, if one computes linear momentum using the definition

$$\boldsymbol{p} = m\boldsymbol{u}$$

the momentum is conserved in one frame but not in the
other. This problem can be removed and linear momentum
conservation law remains valid in both frames after applying
Lorentz transformation if one uses the following equation

$$m_B = \frac{m_A}{\sqrt{1 - \left(\frac{u'_{x_B}}{c}\right)^2}} \qquad (4.5.9)$$

Thus, mass becomes dependent on velocity in relativistic mechanics according to the following equation

$$m = \frac{m_0}{\sqrt{1 - \left(\frac{v}{c}\right)^2}} \qquad (4.5.10)$$

where m_0 is called the rest mass, the mass measured in a frame in which the particle is at rest, and m is the mass of the same particle as estimated when it moves with a speed v. Further analysis using relative mechanics using the new form of relativistic force law

$$F = \frac{m_o u}{\sqrt{1 - \left(\frac{v}{c}\right)^2}}$$

finally yields the famous mass–energy equivalence relation $E = mc^2$. It is well known how this simple equation establishing the equivalence of mass and energy revolutionised science and also the human civilization. However, it is also interesting to note a vague reference to a similar concipt found in Wilhelm Weber's work in the 19th century! Refer to "Wilhelm Weber's Werke" Volume IV, Page 571, published in 1894 by Verlag von Julius Springer, Berlin.

General Theory of Relativity and Extension of Mach's Principle

5.1 Introduction

Before starting the presentation on the conceptual evolution of the extension of the Principle of Relativity to all frames of reference, and mechanics (along with all other physical phenomena) is freed from the shackles of inertial frames, a brief note on the titles of the two theories propounded by Einstein can be very useful. Commonly, the titles given to these theories are 'Special Theory of Relativity (STR)' and 'General Theory of Relativity (GTR)'. But as Mendel Sachs commented in his introductory article, 'The Mach's Principle and the Origin of Inertia from General Relativity'[1] a critical analysis of these two titles can help in grasping the basic aims Einstein had for developing these theories. According to him, the meanings of the two names are somewhat confusing to those who are given an introductory exposure to the subject. Sachs comments that a better way to name these theories could be 'Theory of Special Relativity' and 'Theory of General Relativity'. This is so as the adjective should be ascribed to the word 'relativity' instead of 'theory'. Because the natures of the 'relativity' are 'special' and 'general' in the two theories. Thus, a fresh reader gets a clearer idea about the basic aim of the transition from the restricted special relativistic characteristic of physical laws to totally unrestricted (i.e. general) relativistic property of physical laws.

Any relationship between gravitation and acceleration was not thought of by physicists till Mach touched upon the outer boundaries of the subject in the second half of the nineteenth century as described in Sect. 2.1.4. But it was Huygens who first demonstrated that gravitational attraction can be countered by inertia force of an accelerating body (Sect. 1.12.4) suggesting a vague notion of the 'principle of equivalence'. Further progress in the concept had to wait for the development of a more focussed concept of 'mass' by Newton.

Till the phenomenon of electromagnetism came to play a centre stage role in science, there was no conflict of philosophical nature among the mechanistic phenomena following the laws of motion and science was considered to be complete. Occasional difficulties created by optical effects were not important enough to search for alternative theories. The laws of motion, as proposed in Newtonian mechanics and Galilean transformation, satisfied the 'Principle of Relativity' to great delight of the physicists and science philosophers.

As shown in Chap. 4, it was Einstein's firm belief in the beauty and unity among the natural phenomena described by the laws of physics that forced him to abandon Newtonian mechanics (based upon the concept of absolute space and time) and adopt the 'Principle of Relativity' satisfied by all inertial frames. His theory was restricted to the description of all physical phenomena in inertial frames only satisfying covariance of physical laws. Since his theory of relativity was applicable to only a special class of frames of reference—inertial frames—it is called 'special relativity'. As expected, Einstein's absolute faith in the 'Principle of Relativity' in its most generalized sense paved the path for developing a theory that led to the application of 'Principle of Relativity' to all types of frames of reference. This generalization of the applicability of the theory led to the name 'general relativity'.

The later part of the chapter presents a recent development in which Mach's principle is extended. In Mach's principle, an accelerating mass is expected to interact with the field created by the matter–energy distribution in the whole universe. In the extended version of the principle, such an interaction depends also on the velocity of the particle. The concept depends on the possibility of the existence of a mean-rest-frame of the universe that is considered to be infinite and quasistatic. A number of interesting results are obtained, but the model is phenomenological in character. A number of eminent relativists feel that there is new physics involved, and serious research may reveal its connection with the relativistic mechanics. As the concept of

---

[1]Sachs, M. and Roy, A.R. (ed) 'Mach's Principle and the Origin of Inertia', Apeiron, Moutreal, 2003.

© Springer Nature Singapore Pte Ltd. 2018
A. Ghosh, *Conceptual Evolution of Newtonian and Relativistic Mechanics*,
Undergraduate Lecture Notes in Physics, https://doi.org/10.1007/978-981-10-6253-7_5

inertial frames and one's inability to detect the velocity of such a frame is abandoned, some small deviations from the existing formulations of STR and GTR are expected to be present.

Since the evolution of the General Theory of Relativity is a very well researched field, this chapter is included in the book more for the sake of completeness. Furthermore, the presentation cannot contain the mathematical formulation in details as it is extremely involved and mathematically complex. A good base in tensor calculus is essential for the purpose. However, the basic physical principles involved will be taken up and discussed.

5.2   Transition to General Relativity

The subject 'General Relativity' is far more complex than the 'Special Theory of Relativity'. In comparison with the special relativity, the development of general theory was far more tangled and complicated. One important point that needs to be noted is that the conceptual development of special theory was the work of many hands Einstein playing a vital role, of course. Many scholars believe that the discovery of Special Theory of Relativity was inevitable even without Einstein. On the contrary, the General Theory of Relativity was the outcome of Einstein's own concept and work. Thus, had there been no Einstein, perhaps, the world would have still no such theory.

The topic, evolution of the General Theory of Relativity, is an extensively researched field. A very large number of historians of science, philosophers and physicists have worked on the subject, and, a huge amount of material is available. Unfortunately, this has resulted in certain amount of confusion among the students as many experts present somewhat contradictory views. As mentioned earlier, the theory's primary creator was, undoubtedly, Einstein although some contemporary work existed. Particularly, it should be remembered that Hilbert arrived at the field equation following a different route and published it five days before (20th November 1915) Einstein published his result on 25th November 1915. But the situation is made complicated to a student of science by contradictory comments in texts. At the beginning of the discussion on General Relativity, an author writes[2] 'There are a few common myths about General Relativity that we need to dispel before presenting the theory. It is often said that General Relativity extends Special Relativity as follows: in Special Relativity all inertial reference frames (i.e. all Lorentz frames) are equivalent, and in General Relativity all frames of reference

are equivalent; or in Special Relativity there is a physical distinction between accelerated and unaccelerated motion, but in General Relativity there is none; or in Special Relativity space–time has an inertial structure that is not a function of the distribution of masses, but in General Relativity it is a function of the distribution of masses (thus vindicating Mach). These claims are all false'. But another text[3] writes 'General Relativity follows on rather naturally from special relativity and results in even more profound changes to our concepts of space and time than those follow from special relativity'. Again in another publication[4] presenting the original papers by Einstein in the introduction, it is mentioned 'Einstein's first theory is restricted in the sense that it only refers to uniform rectilinear motion and has no application to any kind of accelerated movements. Einstein in his second theory extends the Relativity Principle to cases of accelerated motion'. Furthermore, in a more recent text,[5] the chapter on General Relativity starts as follows: 'General Theory of Relativity stemmed from Einstein's attempt to incorporate the Newtonian theory of gravity to the frame work of Special Theory of Relativity". So, one is made unnecessarily perplexed about how the General Relativity started or what was the driving force behind Einstein's grand creation.

A recent review paper by Norton[6] discusses the matter in great details. At the very beginning, he writes '—, the question of precisely what Einstein discovered remains unanswered, for we have no consensus over the exact nature of the theory's foundations'. The situation is further aggravated as Einstein himself changed his stand on various principles adopted by him for developing the theory. Apart from this, the basic founding principles, used to develop the General Theory of Relativity, are presented in different formats and with varying descriptions.

5.2.1   Minkowski's Four-Dimensional Space–Time Continuum

The Special Theory of Relativity finally developed by Einstein was in a purely algebraic form as depicted in Chap. 4. The space and time coordinates are variables, associated with an event, in one inertial frame of reference and which

[2]Mandlin, T.—'Philosophy of Physics: Space and Time', Princeton University Press, 2012.

[3]Longair, M.—'Theoretical Concepts in Physics', Cambridge University Press, 2003.
[4]Saha, M.N. and Bose, S.N.—'The Principle of Relativity: Original Papers by A. Einstein and H. Minkowski', (Translated into English), University of Calcutta Publication, 1920.
[5]Kopeikin, S., Efroimsky, M. and Kaplan, G.—'Relativistic Celestial Mechanics of Solar System', Wiley-VCH, 2011.
[6]Norton, John D.—'General covariance and the foundations of general relativity: eight decades of dispute', Rep. Prog. Phys. 56(1993), 791–858.

transform according to Lorentz's transformation rules when one wants to describe the same event in another inertial frame of reference. Though popularly Einstein is given the credit for conceptualizing a four-dimesnional world, in reality, the idea of a four-dimensional structure of space–time was first proposed by H. Minkowski in the year 1908. He suggested that it is advantageous to cast the relativistic mechanics using the three spatial and one temporal quantities (describing an event) in a combined form as the four coordinates of an event in a four-dimensional space–time continuum. Thus, the algebraic form of special theory received a geometric point of view. For the development of the General Theory of Relativity, the concept of four-dimensional space–time continuum was essential and was adopted by Einstein.

For every inertial frame, there exists synchronized clocks registering time t and every location is ascribed with three spatial coordinates x, y and z. Thus, an inertial frame may be conceived as a four-dimensional continuum with t, x, y, z as the four coordinates of an event. To provide the same length dimension to all the coordinates, the temporal coordinate is represented by ct with c as the speed of light which is a constant. So, a two-dimensional slice of the four-dimensional continuum can be shown as in Fig. 5.1.

The y and z dimensions have been suppressed and only the x-ct coordinates are relevant. A line in this graph represents the motion history of a particle (if the occupation of a point by the particle at any instant is taken as an event). Such a line is called the world line. Usually, the scales of the x and the ct axes are kept same. Thus, the slope of the world line at any instant A (Fig. 5.1) given by $\tan \theta$ represents the instantaneous speed of the particle. For the world line (1), the speed is varying. A straight world line making $45°$ with the ct (or, x) axis represents the world line of a light beam as it indicates $x = ct$. Thus, it is clear that no world line can have a slope, θ, more than $45°$ as that will amount to something moving at a speed faster than that of light.

If two events A and B are separated by Δx, Δy, Δz and Δt in an inertial frame then the quantity

$$\Delta s^2 = \Delta x^2 + \Delta y^2 + \Delta z^2 - (c\Delta t)^2 \quad (5.2.1)$$

defines the interval Δs. The same two events when observed from another frame of reference with coordinates (x', y', z', ct'), the interval is given by

$$\Delta s'^2 = \Delta x'^2 + \Delta y'^2 + \Delta z'^2 - (c\Delta t')^2 \quad (5.2.2)$$

It can be shown that the interval between two events, defined in this way, is invariant under Lorentz transformation. The transformation rule is

$$x' = \beta(x - vt), \quad y' = y, \quad z' = z t' = \beta\left(t - \frac{vx}{c^2}\right)$$

with $\beta = \frac{1}{\sqrt{1 - \frac{v^2}{c^2}}}$, v being the velocity of the primed frame moving with respect to the unprimed frame along the x axis, the axes being all aligned. Now,

$$\Delta s'^2 = \Delta x'^2 + \Delta y'^2 + \Delta z'^2 - c^2 \Delta t'^2$$

where

$$\Delta x' = \beta(\Delta x - v\Delta t), \quad \Delta y' = \Delta y, \quad \Delta z' = \Delta z, \quad \Delta t'$$
$$= \beta\left(\Delta t - \frac{v\Delta x}{c^2}\right)$$

A straight forward algebra shows that

$$\Delta s'^2 = \Delta s^2 \quad (5.2.3)$$

Minkowski's four-dimensional space–time is flat and isotropic as evident from (5.2.1). It is very interesting to note that from this geometric perspective a theory based upon the Principle of Relativity becomes trivial. The invariance of the quantity Δs ensures that a theory based on this space–time structure automatically satisfies the Principle of Relativity. Thus, the subtle difference between Einstein's and Minkowski's approaches is that Einstein based his theory on the principle of covariance of physical laws, whereas Minkowski's approach was based upon the invariance of the space–time interval between two events.

5.2.2 Principle of Equivalence

The basic idea of this principle germinated from Huygens' time. But Galileo's observation of the simultanuity of free fall gave the empirical evidence. However, any further progress in the matter had to wait for the emergence of the concept of 'mass' in Newton's 'Principia' as discussed in Sect. 1.15.1.

Newton recognized the equivalence between two different properties of matter—inertial mass and gravitational mass. All experiments so far have demonstrated the equivalence to

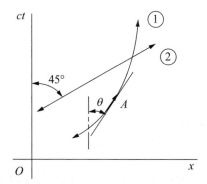

Fig. 5.1 World lines

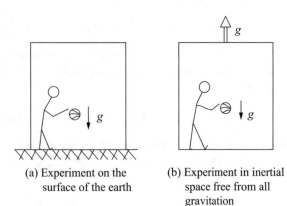

(a) Experiment on the surface of the earth

(b) Experiment in inertial space free from all gravitation

Fig. 5.2 Accelerating frame generates gravity **a** experiment on the surface of the earth, **b** experiment in the inertial space free from all gravitation

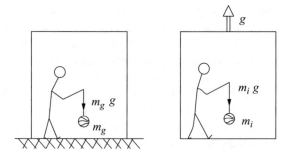

Fig. 5.3 Force felt in the gravitational field and in the accelerated frame

an extremely high degree of accuracy. As Einstein's basic aim was to include gravity in his theory of relativity, he developed a preliminary connection using this principle suggesting exact equivalence between inertial mass and gravitational mass. He suggested a thought experiment as described below.

Figure 5.2 shows two carriages containing persons each carrying a heavy ball. In Fig. 5.2a, the carriage is resting on the surface of a large massive body that produces a strong gravitational field (uniform and parallel) causing the ball (when released) to drop with an acceleration g in the downward direction. On the other hand, the carriage shown in Fig. 5.2b is in inertial space far removed from all gravitating bodies. The man and the carriage are being pulled in the upward direction with an acceleration g as shown. When the ball is released (with zero speed) by the person, it stays stationary as it is in inertial space and initial velocity is zero. But as the person moves up with acceleration, g, relative to him, the ball appears to move downwards with an acceleration g. Thus, to the person, the carriage appears to be in a gravitational field. There is no internally conducted experiment that can distinguish between situations shown in the two cases of Fig. 5.2. This hypothesis was termed by Einstein as the 'principle of equivalence'.

When the person hangs the ball from his hand as shown in Fig. 5.3, the force felt in the gravity field is $m_g g$ where m_g is the gravitational mass of the ball. In the other case, the force on the hand is $m_i g$, the force required to pull the ball with acceleration g along with the carriage. Because of the equivalence of inertial and gravitational mass $m_i = m_g$. Hence, the force felt by the experimenter is same in both the situation. This showed Einstein that gravity field can be created by acceleration. So, according to the extended version of the Principle of Relativity, it is impossible to speak of acceleration of a frame of reference in an absolute sense. This is similar to the situation in Special Theory of Relativity

that does not permit to attach any absolute velocity to a frame of reference. However, it is still not clear if this can lead to a generalization of the concept of covariance of physical rules. At this stage, Einstein adopted the philosophy of Ernst Mach that suggests the inertia of an object to be the manifestation of its gravitational interaction with the matter present in the rest of the universe. Einstein, thus, termed this to be the 'relativity of inertia'. Gradually, this hypothesis became known as 'Mach's principle'.

5.2.3 Freely Falling Frames

Freely falling frames play a very important role in developing the General Theory of Relativity and need some more detailed discussion. But before that it is essential to point out the spatial and temporal restrictions to be observed. The equivalence between an accelerating frame of reference and that in a gravity field, suggested in Sect. 5.2.2, is not truly exact. Figure 5.4 shows two frames of reference—frame A is in the gravitational field of a body and frame B is accelerated in a region free from any gravitational field.

If two particles 1 and 2 are released from rest, the observer can observe a subtle difference between the subsequent movements of the test particles in the two frames of reference. In frame A, the particles not only fall down but also slowly come closer as indicated in the figure. On the other hand, in frame B, the particles' paths are perfectly

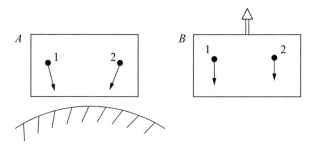

Fig. 5.4 Difference between a frame in a gravitational field and an accelerating frame

parallel. But if the region of space allowed for the experiment be very small and the duration of the observation be also very small, this distinction between the observed motions is not perceptible. Hence, for the equivalence between a gravitational field and uniform acceleration to hold good, the observations must be limited to an extremely small region of the space–time manifold.

Now if a frame of reference falls freely in a gravitational field, any particle released in such a frame an observer finds it to move uniformly with the speed it was started initially. This is so as the observer and the particle both fall with the same acceleration causing their relative acceleration to be zero. Thus, such a freely falling frame satisfies all the characteristics of an inertial frame, if the observations are kept limited to a small region of the space–time manifold. Thus, a freely falling frame is a locally inertial frame.

It can be also shown that freely falling frames are locally inertial frames for not only laws of mechanics but for all laws of physics. To show the equivalence for electrical phenomena, let two laboratories be considered. One of those being at rest on the earth's surface (or in any gravitational field as well) and the other on an accelerating rocket. The acceleration due to gravity in the first laboratory is g, and the uniform acceleration of the second laboratory is also equal to g (Fig. 5.5).

Let electrically charged particles A and B interact between themselves electromagnetically and also fall freely. The uncharged particle C also falls freely. Three similar particles A', B' and C' are considered in the experiment conducted in the accelerating frame shown in Fig. 5.5b. Now, the principle of equivalence suggests that the trajectories described by the particles in the two experiments are same (if the size of the space and the duration of the experiment are both very limited as mentioned before). Hence,

$$\begin{aligned} \boldsymbol{R}_A(\mu) &= \boldsymbol{R}'_{A'}(\mu) \\ \boldsymbol{R}_B(\mu) &= \boldsymbol{R}'_{B'}(\mu) \\ \boldsymbol{R}_C(\mu) &= \boldsymbol{R}'_{C'}(\mu) \end{aligned} \qquad (5.2.4)$$

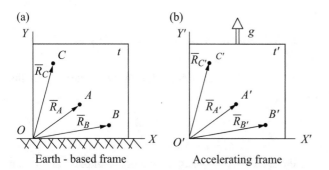

Fig. 5.5 Experiments in earth-based and accelerating frames

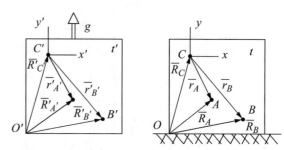

Fig. 5.6 Relating observations in accelerated frame with those in earth-based frame

for all values of μ where $\mu = t$ and $\mu = t'$. Next, change the origin of the frames of reference to the uncharged particle C' in the accelerating frame as indicated in Fig. 5.6.

Now, the frame $x' - y'$ attached to C' is an inertial frame as C' is a particle in free space subjected to no force, i.e. no acceleration. The position vectors $\boldsymbol{r}'_{A'}$ and $\boldsymbol{r}'_{B'}$ as observed in the $x' - y'$ frame can be expressed as follows:

$$\begin{aligned} \boldsymbol{r}'_{A'}(t') &= \boldsymbol{R}'_{A'}(t') - \boldsymbol{R}'_{C'}(t') \\ \text{and} \quad \boldsymbol{r}'_{B'}(t') &= \boldsymbol{R}'_{B'}(t') - \boldsymbol{R}'_{C'}(t') \end{aligned} \qquad (5.2.5)$$

Since $x' - y'$ frame is inertial, the trajectories of A' and B', observed from this frame, must conform to the predictions of Special Theory of Relativity. Putting (5.2.4) and (5.2.5) together and examining Fig. 5.6 one gets.

$$\boldsymbol{r}'_{A'}(\mu) = \boldsymbol{R}'_{A'}(\mu) - \boldsymbol{R}'_{C'}(\mu) = \boldsymbol{R}_A(\mu) - \boldsymbol{R}_C(\mu) = \boldsymbol{r}_A(\mu)$$

$$\begin{aligned} \text{or,} \quad \boldsymbol{r}'_{A'}(\mu) &= \boldsymbol{r}_A(\mu) \\ \text{Similarly } \boldsymbol{r}'_{B'}(\mu) &= \boldsymbol{r}_B(\mu) \end{aligned} \qquad (5.2.6)$$

for all values of μ ($=t$ and t'). Since $\boldsymbol{r}'_{A'}(\mu)$ and $\boldsymbol{r}'_{B'}(\mu)$ are results of experiments in an inertial frame $x' - y'$, and the results of experiments in a freely falling frame in a gravitational field conform to each other, the freely falling frame in a gravitational field being an inertial frame for all physical phenomena.

Considering a uniformly accelerated frame in the scheme had been the first attempt by Einstein to bring gravitation within the preview of relativity. Taking the 'principle of equivalence' to be true Einstein was able to discover a few features about gravitation. The first one is about the influence of gravity on the running rate of clocks.

Figure 5.7 shows two experiments—one conducted on the earth and the other in space inside an accelerating frame. Let there be two experimenters A and B. A is on the bottom of the laboratory, and B is above A at a height of h. The gravity field's strength is g. If B sends light signals at the every tick of his clock and the interval between two consecutive ticks be Δt_B what should be the interval between the consecutive light signals received by A? The answer can be

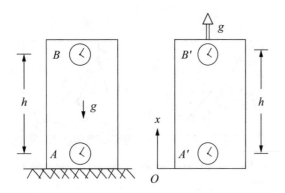

Fig. 5.7 Effect of gravity on the running of clocks

found out by using the 'principle of equivalence'. Thus, let a similar experiment conducted in an accelerating laboratory be considered. The acceleration is equal to g and the distance between the experimenters A' and B' be equal to h. A little simplified (and, so, a little approximate) analysis is presented below.

Let the accelerating laboratory be moving in the x-direction as indicated taking O as the position ($x = 0$) of A' at $t = 0$. Without losing any generality, the starting velocity can be taken as zero.[7] So, the position of A' and B' as functions of t are given by

$$x_{A'}(t) = \frac{1}{2}gt^2$$
$$\text{and } x_{B'}(t) = h + \frac{1}{2}gt^2 \tag{5.2.7}$$

Now, let B' send the first light signal at $t = 0$, and it is received by A' at time t_1. Then considering the distance travelled by the light signal

$$x_{B'}(O) - x_{A'}(t_1) = ct_1 \tag{5.2.8}$$

The second signal emitted after a time $\Delta t_{B'}$ ($= \Delta t_B$ in the experiment conducted in the earth laboratory) is received by the observer A' after a time $\Delta t_{A'}$ after it receives the first signal, i.e. at time $t_1 + \Delta t_{A'}$. Hence,

$$x_{B'}(\Delta t_{B'}) - x_{A'}(t_1 + \Delta t_{A'}) = c(t_1 + \Delta t_{A'} - \Delta t_{B'}) \tag{5.2.9}$$

since the signal was emitted at $\Delta t_{B'}$ and was received at $t_1 + \Delta t_{A'}$. Considering the time intervals between consecutive ticks to be small in an approximate analysis, the higher order terms in $\Delta t_{A'}$ and $\Delta t_{B'}$ can be neglected. Using (5.2.7) in (5.2.8), one gets

$$h - \frac{1}{2}gt_1^2 = ct_1 \tag{5.2.10}$$

and

$$h - \frac{1}{2}gt_1^2 - gt_1\Delta t_{A'} = c(t_1 + \Delta t_{A'} - \Delta t_{B'}) \tag{5.2.11}$$

Subtracting (5.2.11) from (5.2.10)

$$gt_1\Delta t_{A'} = c(\Delta t_{B'} - \Delta t_{A'})$$

or,

$$\Delta t_{A'} = \frac{c \cdot \Delta t_{B'}}{gt_1 + c}.$$

From (5.2.8) and (5.2.7), one can obtain $t_1 = \frac{h}{c}$. Using this in the above equation, the following relation between $\Delta t_{B'}$ and $\Delta t_{A'}$ is obtained:

$$\Delta t_{A'} \approx \Delta t_{B'} \cdot \left(1 - \frac{gh}{c^2}\right) \tag{5.2.12}$$

Thus, the ticking interval is shorter for A' meaning that the clock B' will appear to run at faster rate. According to the principle of equivalence, the clock at a higher altitude (of B in Fig. 5.7) appears to run faster when observed by A in the presence of a gravitational field. Again as $\Delta t_{A'} < \Delta t_{B'}$, it is obvious that observer B' will find A''s clock to go slower.

This phenomenon is reflected on the frequency of light travelling against gravity. The frequency of light emitted by A reaches B (at a higher gravitational potential) with lower frequency. If the signal emitted at A posses a frequency v, it appears to be $(v - \Delta v)$ to the observer B. Since $v \propto \frac{1}{\Delta t}$, (5.2.12) can be used to show that when light travels against gravity its frequency is reduced. Thus, the frequency v is reduced to Δv where

$$\frac{\Delta v}{v} = \frac{gh}{c^2}$$

[$\Delta t_A = \Delta t_B\left(1 - \frac{gh}{c^2}\right)$ or, $\Delta t_B = \Delta t_A\left(1 + \frac{gh}{c^2}\right)$. So, $\frac{1}{v_B} = \frac{1}{v_A}\left(1 + \frac{gh}{c^2}\right)$ or, $v_B = v_A\left(1 - \frac{gh}{c^2}\right)$.

The reduction in frequency $\Delta v = v_A - v_B = v_A\frac{gh}{c^2}$. Hence, $\frac{\Delta v}{v} = \frac{gh}{c^2}$, taking $\Delta v \ll v_A, v_B$.]

As $v\lambda = c$ for electromagnetic waves where λ is the wavelength

$$\frac{\Delta v}{v} = -\frac{\Delta \lambda}{\lambda}$$

Hence, Δv representing a reduction implies $\Delta \lambda$ as an increase. So, the wavelength of light increases according to the rule

---

[7]If not, an inertial frame moving with the starting speed of the laboratory can be imagined and in that inertial frame the starting speed of the laboratory is zero.

$$\frac{\Delta\lambda}{\lambda} = \frac{gh}{c^2} \qquad (5.2.13)$$

when moving against gravity.[8] This phenomenon is called gravitational redshift. Einstein's prediction of redshifting of light has been observationally confirmed. In (5.2.13), gh is nothing but the difference of gravitational potential $-\Delta\varphi$ between the two positions. Hence, the equation can be written in the following form also:

$$\frac{\Delta\lambda}{\lambda} = -\frac{\Delta\phi}{c^2} \qquad (5.2.13a)$$

This also leads to a similar rule for time dilation due to higher gravitational potential. If the difference between two events measured at a higher potential be dt', then its relation to the interval dt (at a lower potential) is given by

$$dt' = dt\left(1 + \frac{\Delta\phi}{c^2}\right) \qquad (5.2.13b)$$

Another effect of gravity on light was predicted by Einstein employing the 'principle of equivalence'. Figure 5.8 shows again the two laboratories, one in the gravitational field the other being in free space but accelerated.

If a light beam leaves its source and traverses across the laboratory that is accelerating upwards, the apparent trajectory of light with respect to the laboratory bends downwards. This is a kinematic effect as illustrated in Fig. 5.9.

So, though the light travels in a straight path, with respect to the observer in the laboratory that is accelerating upwards, it appears to bend downwards. According to the 'principle of equivalence', an observer must notice the same bending of the light beam when an experiment is conducted in a stationary laboratory in the presence of a gravitational field. However, there is an interesting consequence of bending of light wave in the presence of gravitation. As is established the speed of light above has to be more so that the wave veers downwards. This effect of gravitation on the speed of light caused a puzzle. The resolution of this puzzle could not be satisfactorily achieved by Einstein during the early period of the development of General Theory of Relativity. Einstein suspected that the geometry of the space–time manifold is not flat (or, Euclidean) in an accelerated frame, or, in the presence of gravity. That must have been the reason of bending of the light beam, thought Einstein.

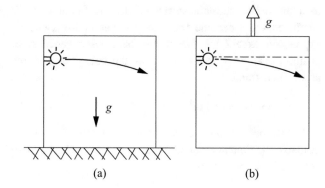

Fig. 5.8 Bending of light by gravity

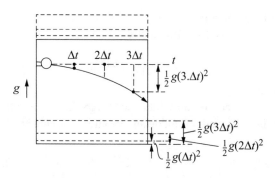

Fig. 5.9 Apparent bending of light in an accelerating laboratory

5.2.4 Uniformly Accelerating Frames and the 'Entwurf' Theory

Although Minkowski proposed the concept of four-dimensional space–time manifold in 1908, Einstein did not take any notice of it till 1912. Facing the difficulties encountered in handling the problems related to frames with gravity field, he finally took to the geometric representation of space–time. He, along with his friend and collaborator Grossman, prepared a paper that was entitled 'Entwurf einer verallgemeinerten Relativitästheorie und einer Theorie der Gravitation'. It is popularly referred to as the 'Entwurf' theory.

It was known to them that the generalized expression for the Minkowski's invariant interval after transforming it into an arbitrary coordinate system x_μ ($\mu = 1, 2, 3, 4$) can be written in the following differential form

$$ds^2 = \sum_{\mu,\nu=1}^{4} g_{\mu\nu}x^\mu x^\nu \qquad (5.2.14)$$

As a first case, Einstein and Grossman developed the expression for the invariant transforming Minkowski's invariant interval

$$ds^2 = c^2 dt^2 - dx^2 - dy^2 - dz^2 \qquad (5.2.15a)$$

---

[8]It is not any trickery of clocks, but it is due to the effect of gravity when light wave traverses space–time in the presence of gravity.

to a uniformly accelerated frame.[9] It was found that the coefficients $g_{\mu\nu}$ become the same as that in Minkwoski's expression except the fact that c becomes a function of position. Taking x', y', z' and t' as the coordinates in the accelerated frame,

$$\mathrm{d}s^2 = c^2(x', y', z')\,\mathrm{d}t^2 - \mathrm{d}x'^2 - \mathrm{d}y'^2 - \mathrm{d}z'^2 \qquad (5.2.15b)$$

It has been shown earlier that using the equivalence principle the time interval $\mathrm{d}t'$ (in the presence of gravitational potential) can be expressed as

$$\mathrm{d}t' = \mathrm{d}t\left(1 + \frac{\Delta\phi}{c^2}\right)$$

Taking the gravitational potential $\varphi(r) = 0$ when $r \to \infty$, r being the distance from a point source $\Delta\varphi = \varphi(r)$. So,

$$\mathrm{d}t' = \mathrm{d}t\left[1 + \frac{\phi(r)}{c^2}\right] \qquad (5.2.16)$$

This $\mathrm{d}t'$ can be now introduced into the expression for invariant $\mathrm{d}s'$, and the following equation is obtained

$$\mathrm{d}s'^2 = \mathrm{d}t^2\left[1 + \frac{\phi(r)}{c^2}\right]^2 c^2 - \mathrm{d}x'^2 - \mathrm{d}y'^2 - \mathrm{d}z'^2 \qquad (5.2.17)$$

where $\phi(r) \equiv \phi(x', y', z')$

So, the dependence of c on position is the effect of the presence of gravity. The obvious conclusion drawn by Einstein and Grossman was that $g_{\mu\nu}$ can represent the presence of gravity. But the procedure did not take care of a general situation as the space–time manifolds represented by (5.2.15a) and (5.2.15b) are both flat. Einstein's ultimate objective to develop a General Theory of Relativity based upon the principle of general covariance (according to which the laws of physics are expressed in forms which are same for all space–time coordinate systems) faced a formidable mathematical challenge. Thus, in the General Theory of Relativity, the presence of gravitation had to be taken care of by the terms $g_{\mu\nu}$ which were called metric tensor components.

Consideration of a rotating frame shown in Fig. 5.10 convinced Einstein that it was essential to consider the non-Eucledean character of space–time manifold when gravity is present. Point A, located at the periphery of the rotating circular disc, has a finite velocity in the tangential direction with respect to the centre O. Now, if one places a small rod 1 of length Δ at O along the radial direction, its length is unaffected by the rotation of the disc as γ at the location of the rod is equal to unity. The situation remains unaltered as the rod 1 is placed at different locations on the

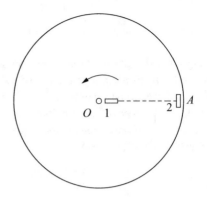

Fig. 5.10 A rotating frame

radial line OA (because the velocity of the rod is perpendicular to the length of the rod). Let the total length OA takes 'n' such rods and the radius of the disc is then considered as n units. Now, when an identical rod 2 is placed at the circumference, its length will be contracted because γ is non-zero. If 'm' such rods are required to cover the whole circumference, the ratio of m and n will be less than 2π. Or, $m/n < 2\pi$. On the other hand, if another stationary disc is placed just below the rotating one and the same experiment is conducted $m/n = 2\pi$. Thus, an accelerating frame that is non-inertial does not follow the Eucledean flat geometry. This convinced Eintein to look for non-Eucledean geometry and he concluded that gravity curves the space–time continuum.

Einstein started working along with his former friend Grossman who introduced Einstein to tensor calculus that became essential to deal with non-Eucledean geometry of curved space–time. In 1913, they published the 'Entwurf' paper. Unfortunately, the resulting equations out of this theory failed to demonstrate general covariance. As is known Einstein developed his 'hole argument' to demonstrate that a generally covariant equation is physically uninteresting and abandoned his principle of covariance. Perhaps he wanted to justify the results obtained from his 'Entwurf' theory.

5.2.5 The Field Equation and Final Formulation

It is a well-known fact that Einstein struggled very hard to finally arrive at the final solution to his objective. Localized form of the principle of equivalence could be useful only for the immediate vicinity of a point in space–time. Such localized frames satisfied Lorentz transformation maintaining covariance. But no extended frame of reference can represent a non-uniform gravitational field. Einstein had to develop a way to stitch all the local frames through a general formulation. It was realized that matter–energy introduces curvature to the space–time; the curved space-time then

---

[9]According to the 'principle of equivalence' such a frame could represent a frame in a uniform gravitational field.

controls the motion of bodies in it. So, using an analogy with electromagnetism where the charge distribution decides the field which, then, governs the motion of the charged particles.

The concept of a straight line (or shortest distance) is replaced by geodesics in curved space–time. So, the requirement is to develop an equation that determines the curvature of space–time due to the presence of matter. The motion of the test particles in this field is along the geodesics. Thus, the concept of gravitational force is eliminated.

Einstein had to struggle a lot for three intense years. His aim was to determine the metric tensor components $g_{\mu v}$ with given mass energy distribution. An equivalent situation in Newtonian gravity is the Poisson's equation which is given below:

$$\nabla^2 \phi = 4\pi G \rho \qquad (5.2.18)$$

where ρ is the matter density. If (5.2.14) is compared with (5.2.17) in the case of a uniform gravity

$$g_{00} = \left(1 + \frac{\phi}{c^2}\right)^2 \approx \left(1 + \frac{2\phi}{c^2}\right)$$

as $\frac{\phi}{c^2} \ll 1$.

In the expanded form of (5.2.14) using $ct = x^0$, $x = x^1$, $y = x^2$, $z = x^3$, g_{00} represents the coefficient of the first term. Using the above expression,

$$\nabla^2 g_{00} = \nabla^2 \frac{2\phi}{c^2} = \frac{2}{c^2} \nabla^2 \phi = \frac{8\pi G \rho}{c^2} \qquad (5.2.19)$$

The above approximate approach hints that the derivatives of $g_{\mu v}$ can be related to the energy momentum tensor $T_{\mu v}$ whose first term T_{00} is equal to ρ.

After considerable struggle, Einstein realized that he has to use Ricci tensor $R_{\mu v}$ to represent the space–time curvature and the curvature scalar[10]

$$R = g^{\mu v} R_{\mu v}$$

to develop his field equation. There was no definite way to arrive at the field equation. Einstein being the genius he was, guessed the right equation in the following form:

$$R_{\mu v} - \frac{1}{2} g_{\mu v} R = \frac{8\pi G}{c^2} T_{\mu v} \qquad (5.2.20)$$

This equation shows the relationship between the components of the metric tensor $g_{\mu v}$ and the mass–energy distribution.

---

[10]$g^{\mu v}$ should be considered the same as $g_{\mu v}$.

Once $g_{\mu v}$ are known, the geodesics can be determined from the extremum principle

$$\delta \int_A^B ds = 0 \qquad (5.2.21)$$

for the shortest path (geodesic) between points A and B in the space–time diagram. The field Eq. (5.2.20) represents 10 partial differential equations in 10 unknown metric tensor components. In four dimensions, a second rank tensor has 16 components. But $g_{\mu v}$ being a symmetric tensor, the number of independent components is 10. Using (5.2.17)

$$ds^2 = c^2 \left[1 + \frac{\phi(r)}{c^2}\right]^2 dt^2 - dx^2 - dy^2 - dz^2$$

$$\approx c^2 \left(1 + \frac{2\phi}{c^2}\right) dt^2 - dx^2 - dy^2 - dz^2$$

$$\text{or,} \quad \frac{ds}{dt} \approx c \left[\left(1 + \frac{2\phi}{c^2}\right) - \frac{v^2}{c^2}\right]^{\frac{1}{2}}$$

since $\frac{\sqrt{dx^2 + dy^2 + dz^2}}{dt} = v$. As c is a constant, (5.2.21) results in (for weak fields and small speeds)

$$\delta \int_A^B ds = \delta \int_A^B (U - T) \, dt = 0$$

which is Hamilton's principle in Newtonian mechanics.

The author finds it a little strange to note the similarity between the outcome of Einstein's General Theory of Relativity and the philosophy of Aristotle proposed around three and half millennia ago. In Aristotle's scheme, objects moved because of their tendency to go to their respective 'natural' places. Such motions were termed as 'natural motion' by him. The final outcome of General Theory of Relativity also eliminates the concept of any force due to gravitational attraction; instead motion of objects in the presence of gravitating matter is a natural phenomenon which causes objects to follow their respective geodesies.

5.3 Extension of Mach's Principle

Mach's principle is one of the most tantalizing (and controversial at the same time) topics in science of motion. It even influenced a scientist of Einstein's stature. However, since it remained as a principle only, its utilization remained confined to philosophical discussions primarily. Some definite progress started only after a quantitative phenomenological model was proposed by Sciama as discussed in Sect. 2.1.5.

But the primary aim, i.e. to demonstrate that inertia is the manifestation of dynamic gravitational interaction of a body with the local gravito-inertial field created by the mass–energy distribution in the whole universe, has not been fulfilled. It was shown in Sect. 2.1.5 that Sciama's acceleration dependent inertial induction model of Mach's Principle can show the inertial mass to be approximately equal to the gravitational mass. Considering the wide range of quantities involved in the analysis, such an approximate result gives credibility to the principle's validity. However, an exact equivalence between gravitational and inertial mass needs an extra domain to the principle. An exact equivalence cannot be achieved through an extreme fine-tuning of the parameters of our universe. Thus, one needs to look forward to having some feedback mechanism in the quantitative model of inertial induction. Keeping in mind that gravitation is a phenomenon that acts on gravitation it is worthwhile to search for such a feedback mechanism. It is demonstrated in the rest of this section that such a mechanism can be incorporated in the model if the concept of 'velocity-dependent inertial induction' is added to 'acceleration dependent inertial induction' model proposed by Sciama. Furthermore, one has to also pay attention to the unique status given to acceleration. Why acceleration is resisted but not uniform velocity has remained an unanswered question since the advent of the subject 'dynamics' from Galileo's time.

The subject 'dynamics' has evolved based upon the concepts like displacement, velocity, acceleration and jerk. But maybe it is possible to represent motion in a totally different manner where individual time derivatives of different orders need not be considered individually and it may be possible to do away with the concepts of velocity, acceleration, etc. for the description of motion. However, till such a formulation is developed in the future, one is constrained to follow the established conventional route using the concepts of velocity, acceleration, jerk, etc. to represent the instantaneous state of motion of a body.

5.3.1 Velocity-Dependent Inertial Induction

In Sect. 2.1.5, Sciama's acceleration-dependent inertial effect was quantified by the term $\frac{Gm_1m_2a}{c^2r}$ where m_1 and m_2 are the gravitational masses of two interacting particles at a distance r, a representing the relative acceleration between the two particles. If it is considered that the relative velocity between the two particles also generates a force depending on the relative velocity then another term $\frac{Gm_1m_2v^2}{c^2r^2}$ can be considered where v is the relative velocity between the two particles. At this stage, the consideration of higher order derivatives (beyond acceleration) is kept in abeyance.

Figure 5.11 shows two particles A and B at a distance r where $\boldsymbol{r}$ is the position vector of B with respect to A. The

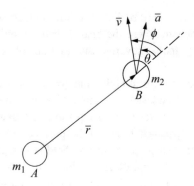

Fig. 5.11 Velocity- and acceleration-dependent forces between two particles

velocity and acceleration of B with respect to A are $\boldsymbol{v}$ and $\boldsymbol{a}$, respectively, making angles ϕ and θ with $\boldsymbol{r}$. The total force $\boldsymbol{F}$ on m_2 at the instant can be written as follows:

$$\boldsymbol{F} = -G\frac{m_1m_2}{r^2}\hat{u}_r - G\frac{m_1m_2}{c^2r^2}v^2f(\phi)\hat{u}_r - G\frac{m_1m_2}{c^2r}af(\theta)\hat{u}_r$$

$$(5.3.1)$$

where $\boldsymbol{r} = \hat{u}_r r$. $f(\phi)$ and $f(\theta)$ represent the effect of the inclination of the velocity and acceleration vectors with $\boldsymbol{r}$. G is the coefficient of gravitational interaction and can depend on r. The exact form of the inclination effects is not known *a priori* but these must satisfy the following conditions:

$$\begin{aligned} f(\phi) &= f(\theta) = 1 & \text{for } \theta = \phi = 0 \\ f(\phi) &= f(\theta) = 0 & \text{for } \theta = \phi = \pi/2 \\ f(\phi) &= f(\theta) = -1 & \text{for } \theta = \phi = \pi \end{aligned} \quad (5.3.2)$$

Thus, as a first attempt the following functions are used:

$$\begin{aligned} f(\phi) &= \cos\phi \cdot |\cos\phi| \\ f(\theta) &= \cos\theta \cdot |\cos\theta| \end{aligned} \quad (5.3.3)$$

The first term on the R.H.S of (5.3.1) is the Newtonian static term, the third term represents the acceleration-dependent inertial induction proposed by Sciama and the second term represents the effect of velocity-dependent inertial induction as proposed by Ghosh.[11] The proposed model is phenomenological and approximate in nature. The inclination effects depicted through (5.3.3) are also adhoc assumptions keeping the basic requirement given by (5.3.2) satisfied. However, as the magnitude of the function $f(\theta)$ remains confined between $+1$ and -1, it cannot introduce any major effect.

It is expected that the second term can give rise to a servomechanism effect and the need of any fine-tuning of

---

[11]Ghosh, A.—'Velocity Depended Inertial Induction: An Extension of Mach's Principle', Pramana (Journal of Physics), v. 23, 1984, p. L671.

universal parameters is eliminated. To determine the total force on a particle of gravitational mass m due to its interaction with the whole universe (or, more precisely the field created by the mass–energy of the rest of the universe at the location of the test particle), it is essential to adopt an appropriate model of the universe. As discussed in Sect. 2.2.1, the universe can be considered to be infinite and homogenous, quasistatic and non-evolving as a whole. In such a situation, a mean rest frame of the universe can be considered and, therefore, the velocity v of the test particle (for the purpose of determining the universal interaction) is with respect to the mean rest frame of the universe. Furthermore, every element (of course elements of the infinite universe could be large in size and with no motion with respect to the mean rest frame of the universe) can be considered to be stationary. There is no conceptual problem with the acceleration a of the test particle with respect to the mean rest frame of the universe.

At this point, it is desirable to discuss the point regarding the concept of an absolute velocity (relative to the stars). Sciama[12] discussed the matter in length and showed the presence of any sign of velocity with respect to the stars to be absent based upon the result of K-meson decay experiment that did not indicate any effect due to the velocity. However, it is shown later that the effect is extremely small and below the sensitivity of the experiment. It should be noted here that the presence of a velocity-dependent effect does not violate the Principle of Relativity according to Sciama as the force is due to interaction with the stars that can be considered to be an effect due to outside experiment and not a truly internal one. In fact, according to this model, the mean rest—frame of the universe can be treated as an absolute frame (giving some similarity with Newton's absolute space) .

To find out the total force acting on the test particle moving with velocity $v(=\hat{u}_v v)$ and $a(=\hat{u}_a a)$, the interactive force with an element of the universe dM at a distance r has to be integrated over the whole infinite universe. Thus, this force F, felt by the test particle, is given by

$$F = 0 - 2 \int_0^\infty \int_0^{\frac{\pi}{2}} \hat{u}_v \frac{G \cdot 2\pi r^2 \rho \, \sin\phi \cdot v^2 \cdot m f(\phi) \cos\phi \, \mathrm{d}\phi \, \mathrm{d}r}{c^2 r^2}$$

$$- 2 \int_0^\infty \int_0^{\frac{\pi}{2}} \hat{u}_a \frac{G \cdot 2\pi r_1^2 \rho \, \sin\theta \cdot a \cdot m f(\theta) \cos\theta \, \mathrm{d}\theta \, \mathrm{d}r_1}{c^2 r_1^2}$$

$$(5.3.4)$$

where ρ is the average mass–energy density of the universe. The integration of the first term (the Newtonian static term) yields zero because of symmetry. The above equation can be rewritten in the following form:

$$F = -\hat{u}_v \frac{mv^2}{c} \int_0^\infty \frac{\chi G \rho}{c} \, \mathrm{d}r - \hat{u}_a \frac{ma}{c^2} \int_0^\infty \chi G r_1 \rho \, \mathrm{d}r_1 \quad (5.3.5)$$

where

$$\chi = 4\pi \int_0^{\frac{\pi}{2}} \sin\phi \, \cos\phi \, f(\phi) \, \mathrm{d}\phi = 4\pi \int_0^{\frac{\pi}{2}} \sin\theta \, \cos\theta \, f(\theta) \, \mathrm{d}\theta$$

$$(5.3.6)$$

Once the form of the function $f(\theta)$ is known, χ can be found out. To complete the integration in (5.3.5), it is essential to know how G varies as a function of the distance between two interacting bodies. Unlike, in Newtonian mechanics, it is not a constant. Equation (5.3.5) can be written as

$$\mathbf{F} = -\hat{u}_v \frac{mv^2}{c} \kappa - \hat{u}_a \frac{ma}{c^2} \int_0^\infty \chi G r_1 \rho \mathrm{d}r_1 \quad (5.3.7)$$

where $\kappa = \int_0^\infty \chi \frac{G\rho}{c} \, \mathrm{d}r$, represents a drag coefficient and the first term on the R.H.S of (5.3.7) is nothing but a drag force on any object moving through space in the universe. The magnitude of the drag force is given by

$$\kappa \frac{mv^2}{c},$$

and it acts in a direction opposite to v. It is a major departure from conventional mechanics. Due to this drag, even the agents of transporting gravitation (may be called gravitons) are subjected to this drag. The intensity of gravitational interaction can be assumed to be proportional to the energy of gravitons arriving at the location of the passive gravitating mass. So

$$G \propto E \left(= \text{gravitons mass} \times c^2\right)$$

It can be easily shown that due to a cosmic drag $\kappa \frac{mv^2}{c}$, the energy E takes the following form

$$E = E_0 \exp(-\kappa r/c),$$

and consequently

$$G = G_0 \exp(-\kappa r/c) \quad (5.3.8)$$

[12]Sciama, D.W.—'Physical Foundations of General Relativity', Heinemann Educational Books Ltd., London, 1972, pp. 20–21.

where E_0 and G_0 are the values when $r = 0$. G_0 is equal to 6.67×10^{-11} m³ kg⁻¹ s⁻². Substituting G from (5.3.8) into the expression for κ, one gets

$$\frac{\chi G_0 \rho}{c} \int_0^\infty \exp\left(-\frac{\kappa r}{c}\right) dr = \kappa$$

$$\text{or,} \quad \frac{\chi G_0 \rho}{c} \cdot \frac{c}{\kappa} = \kappa$$

$$\text{or,} \quad \kappa = (\chi G_0 \rho)^{\frac{1}{2}} \qquad (5.3.9)$$

Using this value of κ and substituting G from (5.3.8) in (5.3.7) yields the total force acting on the test particle as follows:

$$\boldsymbol{F} = -\hat{\boldsymbol{u}}_v \kappa \frac{mv^2}{c} - \hat{\boldsymbol{u}}_a \frac{ma}{c} \int_0^\infty \chi \rho G_0 \exp\left(-\frac{\kappa r_1}{c}\right) r_1 dr_1$$

$$= -\hat{\boldsymbol{u}}_v \kappa \frac{mv^2}{c} - \hat{\boldsymbol{u}}_a \frac{ma}{c} \cdot \frac{\chi G_0 \rho}{\kappa^2}$$

$$= -\hat{\boldsymbol{u}}_v \kappa \frac{mv^2}{c} - ma \qquad (5.3.10)$$

It is very interesting to note that the predominant force is given by exactly —ma (as per Newton's second law), and this inertial mass is nothing but the originally assumed gravitational mass of the test particle establishing exact equivalence of the gravitational and inertial masses of a body. Furthermore, it is also worth noticing that this equivalence is independent of the universe's basic parameters like G_0 and ρ, eliminating the necessity of any fine tuning.

However, one notices the modified force law (5.3.10) has an extra velocity-dependent drag force that acts on any object moving even with uniform velocity with respect to the mean rest frame of the infinite quasistatic universe. To estimate the magnitude of this force, one needs to know χ and ρ. Development of proper field equations in the case of a moving body can yield the correct result. But for an approximate analysis,[13] the form of $f(\theta)$ and $f(\phi)$ given in (5.3.2) can be adopted. With this

$$\chi = 4\pi \int_0^{\frac{\pi}{2}} \sin\theta \cos^3\theta \, d\theta = \pi$$

and

$$\kappa = (\pi G_0 \rho)^{\frac{1}{2}} \qquad (5.3.11)$$

Substituting the magnitudes of G_0 and ρ as 6.67×10^{-11} m³ kg⁻¹ s⁻² and 7×10^{-27} kg m⁻³, respectively, the value of κ becomes

$$\kappa = 1.21 \times 10^{-18} \text{ s}^{-1} \qquad (5.3.12)$$

The gravitational coefficient G drops according to the equation $G = G_0 \exp(-\lambda r)$ where $\lambda = \kappa/c = 0.4 \times 10^{-26}$ m⁻¹. Since λ is far less compared to the limit set by Laplace ($\lambda \leq 10^{-17}$ m⁻¹), the effect is not detectable from the motions of the solar systems objects. The value of G drops by only 0.0004% while traversing across the Milky Way galaxy with a diameter of 10^5 l.y. The 'cosmic drag' on an object of 1 kg mass at a velocity 1 m s⁻¹ with respect to the mean rest frame of the universe is equal to 0.4×10^{-26} N. It is beyond the capability of contemporary science and technology to detect this. The static gravitation force among the components of an experimental set-up will completely overwhelm the 'cosmic drag' force.

5.3.2   Some Features of Velocity-Dependent Inertial Induction

Three important outcomes of the proposed model based upon the interaction of a test particle with the matter–energy present in the rest of the universe are (i) variation of the gravitational coefficient G with distance, (ii) existence of a drag force on a particle moving with constant velocity, called the cosmic drag and (iii) the exact equivalence of inertial and gravitational mass vindicating the philosophy that inertia is nothing but the manifestation of dynamic gravitational interaction.

As mentioned in the previous section, the variation of G with distance is extremely small for detection. But the existence of the cosmic drag can be investigated by considering the fastest moving objects, i.e. photons travelling longest possible distances, i.e. light from the distant galaxies. Next, the effect of cosmic drag on photons from distant galaxies is investigated.

Cosmic drag on photons travelling through space: The mass and energy of a photon satisfy the following equations

$$E = mc^2 \text{ and } E = h\nu = hc/\lambda$$

where h is the Planck's constant, ν is the frequency and λ is the wavelength. When a photon starts its journey from a distant galaxy, it is subjected to a drag κmc according to (5.3.10) as $a = 0$ and $v = c$. This drag causes the energy E of the photon to decrease leading to decrease in frequency ν and increase in λ. Thus, the photon is redshifted. A simple equation is obtained as follows:

---

[13]It should be noted that the final result given by (5.3.10) is independent of the form of the inclination effect.

$$\frac{\mathrm{d}E}{E} = -\frac{\kappa}{c}\,\mathrm{d}x$$

where $\mathrm{d}x$ is the distance travelled in a short interval and $\mathrm{d}E$ is the decrease in photon energy during this period. The solution to the above equation is

$$\frac{\lambda}{\lambda_0} = \exp\left(\frac{\kappa}{c}\chi\right) \qquad (5.3.13)$$

where λ_0 is the wavelength at the source and λ is the wavelength of the photon after it travels through a distance x. So the light from a galaxy at a distance x from the earth will be redshifted by an amount according to the following rule:

$$z = \frac{\lambda - \lambda_0}{\lambda_0} = \exp\left(\frac{\kappa}{c}x\right) - 1 \qquad (5.3.14)$$

When $\kappa x/c \ll 1$, the above relation can be simplified to the following linear rule:

$$z \approx \frac{\kappa}{c}x \qquad (5.3.15)$$

If this redshift is assumed to be due to a Doppler effect due to an assumed recessional speed v, the above equation yields

$$v = zc \approx \kappa x \qquad (5.3.16)$$

Hence, κ is nothing but the Hubble constant whose current estimate is not much different from the theoretically obtained value of κ, $1.21 \times 10^{-18}\ \mathrm{s}^{-1}$! This can be considered to be a strong support for the observed cosmological redshift without any universal expansion.

Relative magnitude of acceleration and velocity-dependent induction effects: The model has three terms in the interaction between two particles. An examination of (5.3.1) reveals that the velocity- and acceleration-dependent terms are much smaller in magnitude compared to that of the static Newtonian term because of the c^2 term in the denominators. Again between the two inertial inductions terms, the acceleration dependent term is a long range force because its denominator contains a term r in its first power. On the contrary, the velocity-dependent term has r^2 in its denominator. As a result, for localized interactions, the velocity-dependent term is more predominant in its effect. Whereas, the contribution of the acceleration dependent term is more in the effects due to interaction with the far off bodies. Therefore, the contributions of the distant matter are less significant compared to the velocity-dependent interaction with the nearby matter. Since the magnitude of the velocity-dependent term predominates over the effects of the distant matter, it may be possible to detect the effects of velocity-dependent inertial induction of local nature. In fact, quite a few such results for photon-matter and matter-matter

Table 5.1 Comparison of inertial induction effect

Interacting system	Velocity-dependent induction effect (mv^2/c^2)	Acceleration-dependent induction effect
Earth (near surface)	~ 10	$\sim 10^{-8}\ ma$
Sun (near surface)	~ 275	$\sim 10^{-7}\ ma$
Milky way galaxy (near sun)	~ 200	$\sim 10^{-6}\ ma$
Universe	3.63×10^{-10}	$= ma$

inertial induction of local nature have already been found to explain a number of phenomena. Table 5.1 shows the comparison of the effects due to velocity and acceleration-dependent interaction.[14]

Transfer of angular momentum: It should be carefully noted that the velocity-dependent term is not just dependent on velocity (like some velocity-dependent term in mechanics), but always acts 'against' the velocity. Another important result of this term is that this interaction can cause transfer of angular momentum between a spinning central body and an orbiting body (like a planet and its satellite, or, the sun and the planets). Such a mechanism of momentum transfer is not available in conventional mechanics. This has been able to explain a number of phenomena in the solar system dynamics.

5.3.3 Concluding Remarks

The extended concept of inertial induction shows a number of interesting results explaining quite a few unexplained or ill explained phenomena. However, the magnitude of the velocity- and acceleration-dependent inertial induction is extremely small for detection through terrestrial laboratory experiments. But indirect support for the proposed effect is not unsubstantial, and it hints the possibility of new physics. It has been suspected by a number of physicists that there is deep-routed physics but unearthing it may be a difficult task. Whether the phenomenon can come out as an extension of the General Theory of Relativity is not known. But an attempt can be made by developing a field-based interaction in which the space–time fields are produced by moving bodies.

At the end, it must be emphasized that though the magnitude of the velocity-dependent inertial induction effect is extremely small to influence the motion of bodies studied by

---

[14]Ghosh, Amitabha, 'Origin Inertia: Extended Mach's Principle and Cosmological Consequences', Apeiron, Montreal, 2000.

the scientists, its impact on our understanding of the nature of the universe is paramount. In the literature, it has been suggested that a careful study of the planet mars can help to either throw away the hypothesis or accept it. If mars is found to have a secular retardation (of the order of 1.25×10^{-22} rad s^{-2}) as predicted by the proposed theory, it will be difficult to explain it using the conventional 'tidal friction' mechanism as mars does not have any sizeable satellite (like the moon in case of earth) to pickup mars' angular momentum. But it is a big project and, perhaps, an organization like NASA can take up the endeavour.

Bibliography

1. Barbour, J.B.: Absolute or Relative Motion. Cambridge University Press, Cambridge (1989)
2. Cohen, J.B.: The Birth of a New Physics. Penguin, Harmondsworth (1992)
3. Ghosh, A.: Origin of Inertia: Extended Mach's Principle and Cosmological Consequences. Aperion, Montreal (2000)
4. Dugas, R.: A History of Mechanics. Dover Publication, New York (1988)
5. Speiser, D (Essays): Discovering the Principles of Mechanics 1600–1800. Birkhäuser, Basel (2008)
6. Capecchi, D.: The Problem of the Motion of Bodies. Springer, Berlin (2014)
7. Lange, M.: An Introduction to the Philosophy of Physics. Blackwell Publishers, London (2002)
8. Crowe, J.M.: Mechanics from Aristotle to Einstein. Green Lion Press, SantaFe, New Mexico (2007)
9. Assis, A.K.T.: Relational Mechnics. Apeiron, Montreal (1999)
10. Maudlin, T.: Philosophy of Physics. Princeton University Press, Princeton (2012)
11. Sciama, D.W.: The Physical Foundation of General Relativity. Heinemann Educational Books Ltd., London (1972)
12. Cohen, R. S., Seeger, R. J. (ed.): Earnst Mach—Physicist and Philosopher. D. Reidel Publishing Co., Holland, Dordrecht
13. Longair, M.: Theoretical Concepts in Physics, 2nd edn. Cambridge University Press, Cambridge (2003)
14. Hesse, M.B.: Forces and Fields. Philosophical Library, New York (1962)
15. Westfall, R.S.: The Construction of Modern Science. Cambridge University Press, Cambridge (1977)
16. Arnold, V.I.: Huygens & Barrow, Newton and Hooke. Birkhäuser Verlag, Basel (1990)
17. Weinberg, S.: To Explain the World—The Discovery of Modern Science. Allen Lane (an imprint of Penguin Books) (2015)
18. Understanding Space and Time—Block1. The Open University Press (1979)
19. Understanding Space and Time—Block2. The Open University Press (1979)
20. Understanding Space and Time—Block4. The Open University Press (1979)
21. Sachs, M., Roy, A.R. (eds.): Mach's Principle and the Origin of Inertia. Aperion, Montreal (2003)
22. Mach, E.: The Science of Mechanics: A Critical and Historical Account of its Development (Translation by McCormack, Th J.). The Open Court Publishing Co. (1960)
23. Damerow, P., Freudenthal, G., McLughlin, P. and Renn, J.: Exploring the Limits of Preclassical Mechanics. Springer, Berlin
24. Reichenbach, H.: The Philosophy of Space & Time. Dover Publication Inc. New York (1958)
25. Eddington, A.: Space, time and gravitation. Harper & Row Publishers, New York (1959)
26. Einstein, A.: Relativity—The Special and General Theory. Pi Press, New York (2005)
27. Einstein, A.: The Meaning of Relativity. Princeton University Press, Princeton (1988)
28. Einstein, A., Minkowski, H.: The Principle of Relativity (Translated by Saha, M.N., Bose, S.N.). Calcutta University Press (1920)
29. Hoffman, B.: Relativity and Its Roots. Dover Publication Inc. New York (1999)
30. Berry, M.V.: Principles of Cosmology and Gravitation. Overseas Press (2005)
31. Ray, C.: The Evolution of Relativity. Adam Hilger, Boston (1987)
32. Hartle, J. B.: Gravity—An Introduction to Einstein's General Relativity. Pearson Education (2003)
33. Schutz, B.: A First Course in General Relativity. Cambridge University Press, Cambridge (2009)
34. Corwell, B.: Relativity, Light and Matter. Fullerton, California (2009)
35. Resnick, R.: Introduction to Special Relativity. Wiley-India (2011)
36. Kopeikin, S., Efroimsky, M., Kaplan, G.: Relativistic Celestial Mechanics of the Solar System. Wiley-VCH (2011)
37. Norton, J.D.: General covariance and the foundation of general relativity: eight decades of disputes. Rep. Prog. Phys. **56**, 791–858 (1993)
38. Levrini, O.: Reconstructing the basic concepts of general relativity from an educational & cultural point of view. Sci. Educ. **11**, 263–278 (2002)
39. Munera, H.A. (ed.): Should the Laws of Gravitation be Reconsidered?—The Scientific Legacy of Mauric Allais. Apeiron, Montreal (2011)
40. Feynman, R.P.: The Character of Physical Law. Penguin Books (1992)
41. Kuhn, T. S.: The Copernican Revolution, Harvard University Press, Baltimore (1999)
42. Galilei, G.: Two New Sciences (Translation by Drake, S.). Wall & Emerson Inc. (1989)

© Springer Nature Singapore Pte Ltd. 2018
A. Ghosh, *Conceptual Evolution of Newtonian and Relativistic Mechanics*,
Undergraduate Lecture Notes in Physics, https://doi.org/10.1007/978-981-10-6253-7

Index

A

Aberration, 89, 90, 92, 93
Absolute
    motion, 37, 47, 49, 53–55
    space, 37, 47–50, 53, 59, 74, 103, 113
    time, 37, 47, 49, 103
Abu Mashar, 71
Acceleration, 5, 19, 23, 24, 28, 33, 34, 36, 38, 42, 45, 48–52, 56, 58,
        60–64, 71, 75, 77, 79, 84, 85, 87, 95, 106, 107, 112, 115
Acronychal, 17
Action-at-a-distance, 19, 59, 70–72, 74, 94
Aether (ether), 72, 74
Agriculture, xiii
Alexandria, 2
Algarotti, Francesco, 75
Amplitude, 24, 85
Anaxagoras, 1
Anaximander, 1
Anaximenes, 1
Anima motrix, 19
Aphelion, 16
Apollonius, 6
Apside, 15, 17, 18
Arago, Francois, 91
Area law, 15–19, 43
Aristotle, 1, 2, 4, 12, 19, 111
Astrology
    judicial, 4
    natural, 1, 5, 6, 12, 13, 19, 23, 27, 65, 75, 96, 111
Astronomy, 2
Atwood's machine, 75

B

Babylon, 1, 5
Background radiation, 55
Bernoulli
    Daniel, 46, 75, 79
    Jacob, 46, 76, 79
    Johann, 46, 75
Big-Bang theory, 53
Borelli, Giovanni Alphonso, 42
Boscovitch, Roger, 72

Bradley, 88, 89, 92
Brahe, Tycho, 4, 14, 15, 22
Bucket experiment, 48–50
Burdian, Jean, 11

C

Calculus
    variational, 78
Cancer, 4
Capricorn, 4
Cassiopeia, 14
Celestial
    equator, 4, 6
    sphere, 4, 88
Center of gravity, 31, 48, 85
Centrifugal force, 32–36, 50, 58
Centripetal acceleration, 32–35, 39
Charge
    negative, 51, 100
    positive, 51
Chord, 24, 50
Civilization, 1, 11
Collinear, 40
Collision, 28–32, 37, 38, 44, 60, 62, 70, 76, 88, 100
Conservation
    energy, 32, 64, 66, 70, 79
    momentum, 29, 32, 70, 100
Conservative system, 65
Contact force, 70, 71
Continuum, 53, 105, 110
Copernicus, 11–13, 21, 22, 88
Corpuscular theory, 74
Cosmic microwave background radiation, 53, 55
Cosmology, 1, 14
Coulomb, Augustin de, 72
Coulomb's law of electrostatic force, 51, 74
Covariant, 110

D

D'Alembert, 46, 76, 81, 84, 85
Descartes, 11, 28–30, 32, 36, 37, 46, 59, 70, 76, 78, 82

© Springer Nature Singapore Pte Ltd. 2018
A. Ghosh, *Conceptual Evolution of Newtonian and Relativistic Mechanics*,
Undergraduate Lecture Notes in Physics, https://doi.org/10.1007/978-981-10-6253-7

Dielectric constant, 51
Dipole anisotropy, 55
Displacement, 4, 23, 34, 35, 40, 53–55, 65, 69, 77, 80, 84, 89, 112
Distance
    angular, 2
Distance law, 15, 16, 19
Diurnal, 4, 12
Doppler effect, 55, 115
Dynamics, 4, 5, 14, 20, 23, 28–30, 37, 38, 45, 47, 48, 59, 64, 79, 83,
        85, 87, 88, 112, 115

E

Eccentric, 7, 8, 15
Eclipse, 2, 50, 92
Ecliptic, 4, 6, 15
Egypt, 1
Einstein, Albert, 87, 96
Electrical
    induction, 72
Electromagnetic, 64, 72, 73, 94, 95, 100, 108
Electromagnetism, 93–96, 103, 111
Electrostatic, 5, 72, 73
Element
    air, 1
    fire, 1
    water, 1
Ellipse, 17, 18, 40, 42, 43
Energy
    gravitational, 25
    kinetic, 32, 45, 60, 66
    mechanical, 31, 60, 64, 65, 67, 77
    potential, 25, 64–67, 79
Energy function, 65, 66, 78
Entwurf theory, 109, 110
Epicycle-Deferent, 6, 7
Equant, 8, 10, 16
Equation of motion, 40, 65, 78, 83
Equinoctial, 2, 4, 6, 7
Equinox, 6, 39
Equivalence principle, 110
Eratosthenes, 2
Euclidean geometry, 48
Euler, Leonhard, 46, 72
Evolution, 4, 14, 53, 75, 104

F

Faraday, Michael, 72
Faraday's Disc, 73
Fermat, 82
Field
    electric, 74, 94, 95, 99
    equation, 104, 111
    gravitational, 29, 50, 73, 106–110
    magnetic, 73, 93, 94, 99
    velocity, 72
FitzGerald, 93, 95
Fizeau, 91, 92
Focus
    of ellipse, 43
Force
    dead, 60, 76
    impressed, 37, 38, 48, 56

innate, 37, 38, 45
inverse square, 40, 41, 43
living, 60, 76, 79
motive, 37, 38, 45
Form invariance of physical laws, 63
Frame of reference
    absolute, 37, 45, 53, 54
    inertial, 63, 66, 104, 106
    non-inertial, 66
    relative, 99
Free fall, 5, 11, 23, 24, 26, 28, 34, 36, 42, 57, 105
Frenicle, 70
Frequency, 108, 114
Fresnel, 91, 95
Friction, 1, 4, 11

G

Galilean, 23, 27, 61, 87, 93, 95, 98, 103
Galilei, Galileo, 20
Galvanometer, 73
Generalized coordinates, 83, 84
Geocentric, 4, 12, 21, 22
Geodesic, 111
Geometro-quantitative, 6
Geometry
    Eucledian, 48
G-field, 50
Gilbert, W., 5, 70
Gravitation, 5, 19, 36, 37, 39, 42, 45, 50–52, 55, 57, 64, 70–72, 74,
        103, 107, 109, 110, 112–114
Gravitons, 113
Gravity, 4, 19, 23, 24, 27, 31, 32, 34, 35, 37, 56, 64, 86, 106–110
Great circle, 4, 6
Grossman, 109, 110

H

Halley, 36, 42, 45
Hamilton's principle, 111
Heaviside, Oliver, 95
Heliacal, 12
Helio
    astral, 7
    centric, 4, 7, 12, 13, 21
    static, 15
Hellenistic, 1, 2, 5, 41
Henry, 94
Herman, Jacob, 75
Hesse, Mary, 72
Hilbert, 104
Hipparchus, 2, 5, 6, 8, 11
Hooke, Robert, 42
Hubble's constant, 52
Huygens, Christiaan, 30, 59
Hydrodynamics, 72
Hypotenuse, 2
Hypothesis, 4, 6, 15, 17, 21, 50, 106, 116

I

Impetus, 5, 11, 28, 38, 72, 84
Impulse, 30, 40, 60, 81
Incline, 23, 24, 26

Inertia, 4, 19, 23, 25, 26, 29, 30, 37, 48, 50, 52, 55, 57, 58, 85, 86, 103, 106, 112
Inertial induction, 50, 52, 112, 115
Interference, 91, 93
Interferometer, 92
Intrinsic, 53, 55
Invariant, 37, 60, 64, 66, 68, 105, 109, 110
Io, 92
Isolated system, 61, 64, 79

J
Jerk, 112
John of Holland, 11
Jupiter, 21, 92

K
Kant, Emanuel, 72
Keplar, Johannes, 14, 20
Kinematics, 23, 32
Kinetic energy, 60, 64, 76, 79
K-meson, 49

L
Leibniz, G.W., 45, 46, 49, 59
Lenz, 94
Lines of forces, 72–74
Longitude, 17, 18
Lord Kelvin, 92
Lorentz, 93, 95, 96, 100
Lorentz-FitzGerald contraction, 93
Luminosity, 1

M
Mach, Ernst, 50, 106
Mach's principle, 50–53, 71, 103, 106, 111, 112
Magnet, 19, 73, 74
Magnetism, 5, 19, 94–96, 100
Mass
    gravitational
        active, 56
        passive, 56, 113
    inertial, 37, 45, 56–58, 105, 106, 112, 114
Maupertuis, 82, 83
Maxwell, James Clerk, 72, 92, 94
Mechanics, 1, 11, 23, 27, 30, 36, 45, 46, 49, 50, 55, 57, 59–61, 72, 75–79, 81, 83, 85, 87, 88, 90, 95, 96, 100, 101, 103, 107, 113
Mercury, 11, 12, 53
Merton
    College, Oxford, 11, 23
    rule, 23, 24
Michelson, 92, 93
Michelson-Morley experiment, 91, 92, 95
Minkowski, H., 105
Moment of inertia, 75, 85, 86
Momentum
    angular, 69, 70, 86, 115, 116
Morley, 92, 93

Motion
    absolute, 37, 49, 53, 55
    compound, 25, 26
    natural, 1, 10, 21, 27
    relative, 12, 32, 47, 48, 50, 53, 59
    uniform circular, 1, 6, 10, 21
    uniform rectilinear, 5, 29, 104
    uniform straight line, 35
    violent, 1, 27

N
Natural philosophy, 59
Newtonian, 19, 44, 47, 48, 50–53, 55–57, 59–61, 74, 76, 78, 79, 83–85, 87, 88, 90, 95–97, 100, 103, 104, 111–113, 115
Newton, Issac, 37
Norton, John D., 104

O
Observation, 2, 5–8, 10, 12, 14–17, 19, 21, 22, 27, 53, 60, 73, 74, 81, 88, 91, 92, 105, 107
Oersted, Hans C., 93
Orbit, 4, 7, 8, 12, 13, 15–19, 22, 36, 40–44
Orbital, 4, 6, 7, 10, 12, 13, 15–20, 27, 36, 39, 42, 70, 88, 89, 91
Oresme, Nicole, 11, 12, 23
Orion, 21
Oscillation, 57, 72, 84, 85
Ovoid, 17

P
Parallax, 2, 4, 5, 14, 20, 22, 88, 89
Parallelogram of addition of motive force, 38
Paris Académie, 70
Pendulum
    conical, 36, 57
    spherical, 57
Perfect cosmological principle, 53
Perihelion, 16, 53
Philosopher, 1, 2, 4–6, 12, 22, 26, 28, 37, 39, 41, 42, 47–50, 54, 57, 60, 71, 83, 103, 104
Philosophy, 1, 5, 6, 32, 41, 46, 49, 53, 57, 71, 78, 81, 106, 111, 114
Phoronomia, 46, 75
Photon, 114, 115
Physical science, 5, 14
Physis, xiii
Picard, Jean, 88
Planes, 15, 19, 23, 24
Planets, 1, 4, 5, 7, 11–16, 18–21, 36, 39, 42, 70, 71, 78, 115
Plato, 1
Poincare, Henri, 95
Poles
    north, 73
    south, 73
Potential energy
    gravitational, 25, 65, 108
    strain, 65
Power, 1, 18, 21, 76, 115
Precision, 6, 20, 39
Predictions, 6, 8, 14–16, 19, 20, 67, 107, 109
Principia, 30, 44–46, 59, 64, 75, 79, 84, 105

Projectile, 1, 5, 23, 26, 27, 39
Ptolemy, 4, 6, 8, 11
Pythagoras, 1

Q
Quasi-static, 53, 103, 113, 114

R
Reciprocity, 97, 98
Redshift, 53, 109, 114, 115
Reflection, 59
Relative, 4, 10, 12, 21, 27, 30, 32, 37, 47–51, 53–55, 59, 60, 70, 73, 87, 93, 96, 97, 101, 106, 107, 112, 113, 115
Relativity
    Einstein, 27, 53, 58, 60, 61, 94–96, 98, 111
    Galilean, 23, 27, 30, 61, 93
    general theory, 30, 35, 50, 53, 58, 103–106, 109–111, 115
    special theory, 50, 60, 61, 94–96, 99, 100, 103, 104, 106, 107
Rest, 19, 23–25, 29, 37, 38, 49–55, 61, 62, 65, 67, 73, 76, 79, 81, 91, 95, 96, 101, 103, 106, 107, 112–114
Retrograde, 7, 12
Revolution, 11, 14, 21, 30
Robervel, 70
Rotation, 4, 7, 19, 27, 32, 50, 57–59, 70, 85, 86, 88, 110
Royal society, 38, 39, 42, 45, 75, 92

S
Sachs, Mendel, 103
Satellite, 115, 116
Scaliger, J.C., 70
Schlick, M., 50
Scholar, 11, 20, 84, 104
Sciama, D.W., 50
Science, 1, 4, 5, 10–14, 19–23, 25–30, 32, 34–37, 39, 41, 42, 44, 45, 47, 49, 53, 55, 59, 60, 64, 70, 72, 75, 76, 78, 81, 82, 87, 88, 93, 95, 103, 104, 111, 114
Scientific, 1, 4, 5, 11, 14, 20, 22, 27, 30, 42, 53, 71, 82, 88, 93, 95
Secular retardation, 116
Seyne, 2
Socrates, 1
Solar
    year, 6, 50
Solar system, 19, 21, 27, 74, 91, 92, 114, 115
Solenoid, 73
Solstice, 6
Space-time, 53, 87, 88, 104, 105, 107, 109–111, 115
    curvature, 53, 110, 111
Spatio-temporal, 70–72, 74
Speed
    angular, 7, 8, 57
Sphere
    stellar, 4
Spherical, 2, 13, 52, 57, 98
Spin, 4, 12, 32, 34
Spyglass, 20, 21
Stationary, 1, 6, 21, 48, 52, 53, 72–74, 89–93, 95, 99, 106, 109, 110, 113
Sun, 1, 2, 4–8, 11–19, 21, 27, 36, 39, 40, 42, 49, 50, 70, 115
Supernova, 14

Symmetric energy tensor, 50
Symmetry, 30, 59, 61–63, 67, 94, 113

T
Temporal, 27, 71, 105, 106
Tensor
    metric, 110, 111
    momentum, 111
    rank, 111
    Ricci, 111
Terrestrial
    equator, 6
Thales, 1
Tidal friction, 116
Time
    homogenity, 48, 59, 62, 66
Tisserand, 52
Todd, 92
Trajectory, 39–41, 109
Transformation
    Galilean, 60, 68, 87, 88, 95, 98, 103
    Lorentz, 96–98, 100, 105, 110
Translation, 47, 69, 70, 74, 84, 88
Triangulation, 15, 17
Trigonometry, 4

U
Universal gravitation, 37, 41, 45, 51, 72
Universe
    expanding, 53
    homogenous, 52–54
    isotropic, 52–54
    stationary, 1, 52, 53
University of
    Padua, 20
    Paris, 11
    Pisa, 20

V
Varignon, 79
Vector, 15, 55, 65, 68, 70, 76, 77, 80, 81, 112
    angular, 7, 58
Velocity-dependent inertial induction, 112, 115
Venus, 11, 12, 21
Vicarious hypothesis, 17
Virtual
    diaplacement, 79, 80, 84
    velocity, 79, 83
    work, 60, 79–81, 83, 84
Vis insita, 37
Voigt, W., 96

W
Wavelength, 92, 108, 115
Weber, 52
Weight, 11, 31, 37, 45, 55, 58, 70, 76
Wren, Christopher, 42

Printed in the United States
By Bookmasters